Groovy Science

Groovy Science : Knowledge, Innovation, and American Counterculture

David Kaiser and W. Patrick McCray, Editors

The University of Chicago Press :: Chicago and London

David **Kaiser** is the Germeshausen Professor of the History of Science and professor of physics at the Massachusetts Institute of Technology. He is the author of *Drawing Theories Apart*, also published by the University of Chicago Press, and *How the Hippies Saved Physics*. **W. Patrick McCray** is professor in the department of history at the University of California, Santa Barbara. He is the author of *The Visioneers* and *Keep Watching the Skies*.

The University of Chicago Press, Chicago 60637
The University of Chicago Press, Ltd., London
© 2016 by The University of Chicago
All rights reserved. Published 2016.
Printed in the United States of America

25 24 23 22 21 20 19 18 17 16   1 2 3 4 5

ISBN-13: 978-0-226-37288-4 (cloth)
ISBN-13: 978-0-226-37291-4 (paper)
ISBN-13: 978-0-226-37307-2 (e-book)
DOI: 10.7208/chicago/9780226373072.001.0001

Library of Congress Cataloging-in-Publication Data

Names: Kaiser, David, editor. | McCray, Patrick (W. Patrick), editor.
  Title: Groovy science : knowledge, innovation, and American counter-culture / David Kaiser and W. Patrick McCray, editors.
  Description: Chicago ; London : The University of Chicago Press, 2016. | Includes index.
  Identifiers: LCCN 2015040309 | ISBN 9780226372884 (cloth : alk. paper) | ISBN 9780226372914 (pbk. : alk. paper) | ISBN 9780226373072 (e-book)
  Subjects: LCSH: Science—Social aspects—United States. | Counter-culture—United States—History—20th century.
  Classification: LCC Q175.5 .G756 2016 | DDC 303.48/3097309046—dc23 LC record available at http://lccn.loc.gov/2015040309

♾ This paper meets the requirements of ANSI/NISO Z39.48–1992 (Permanence of Paper).

Contents

Introduction

David Kaiser and W. Patrick McCray

groovy \ˈgru-vi\ *adj* 1: marvelous, wonderful, excellent. 2: hip, trendy, fashionable.

Something was happening. Otherwise, why would Theodore Roszak be agreeing with Edward Teller? What could the left-leaning history professor and favorite of Berkeley-area coffeehouse radicals possibly have in common with a nuclear-weapons designer whose dark visions helped inspire Stanley Kubrick's sinister *Dr. Strangelove*? Something was happening. But what was it?

As the 1970s began, Teller and Roszak were about as far apart politically as two men could possibly be. Yet they found some quantum of concordance in their assessment that young Americans were relating to science and technology in a manner very different from that of just a decade earlier. When he surveyed the development of science in the United States since 1945, Teller concluded that "all is not well." The problem was especially acute among "our young people," for whom "thorough and patient understanding is no longer considered relevant." To Teller—the very apotheosis of the military-industrial complex—this endemic "lack of interest" was "the surest sign of decadence."[1]

Roszak concurred. As he wrote in his 1969 book, *The*

Making of a Counter Culture, the youth of his day were fleeing science "as if from a place inhabited by plague," seeking "subversion of the scientific world view" itself.[2] Roszak's counterculture—a monolithic term he deployed to capture what was actually an inchoate collection of individuals, interests, and ideologies—was antirationalist and opposed to both sterile technocracy and impersonal "Big Science." Instead of astrophysics, engineering, or molecular biology, youthful members of the counterculture, Roszak held, were more likely to embrace astrology, Eastern religions, and chemically enhanced spirituality.

Starting from opposite ends of the political spectrum, Teller and Roszak both arrived at a common diagnosis: by the late 1960s, American youth—and American society in general—had turned away from the science and technology that had underpinned the expansion, power, and prosperity of the United States since 1945. But Roszak and Teller were *wrong* in their assessment. America's youth had not turned away wholesale from science and technology. Everyone from vocal campus radicals to members of Nixon's "silent majority" could see that a new type of science, a new way of engaging with technology, was emerging. Something new was percolating from 1970s-era dorm rooms, 'zine publishers, hot tubs, and experimental communities. This book explains *what* and *why*.

Our argument in this edited collection is that members of many strands of the American counterculture, although wary of a larger society that seemed to prize conformity, consumerism, and planned obsolescence, were *not* antiscience. Rather, we argue, many young people who self-identified as part of the counterculture in the United States, stretching from the late 1960s through the early 1980s, dismissed examples of science and technology that struck them as hulking, depersonalized, or militarized—a rejection of Cold War–era missiles and mainframes rather than of science and technology per se. Instead of the traditional science and technology promulgated by giant government programs, defense contractors, and corporate research labs, the counterculture embraced something new and different. We call this alternative *groovy science*.

Indeed, this book's cover art reflects the blend of the conventional and the countercultural. Here is a stereotype of an Establishment scientist—white, male, hair closely cut—but his activities are blurred, distorted, and shifted by colorful, sometimes polarizing, and often groovy forces emanating from outside the lab.

Some of the most iconic, colorful, and controversial figures in the American countercultural movements were enamored of modern science and technology: people like psychedelics enthusiasts Timothy Leary and Andrew Weil, human-potential mavens Werner Erhard (of *"est"* fame) and Michael Murphy (cofounder of the Esalen Institute), and *Whole Earth Catalog* visionary Stewart Brand.[3] Each of them actively sought out the latest advances in quantum mechanics, chaos theory, cybernetics, ecology, and beyond, often paying handsomely to encourage the work and learn of it firsthand. Following their lead, many young people who self-identified as members of the counterculture tended toward small-scale "appropriate technology" rather than the massive projects of the Sputnik era. When Leary solicited essays from "tuned-in" physicists on quantum entanglement in 1976, for example—while sitting in a California jail on drug charges—he was seeking the comprehensive worldviews of modern science, not the room-sized mainframes of Cold War techno-science; but his tastes can hardly be dismissed as "antiscientific" for all that.[4] Likewise, the long-haired freethinkers who bought the makings for Buckminster Fuller–inspired geodesic domes from Brand's *Whole Earth Catalog* or who eagerly snatched up copies of popular books about cybernetics and the brain by Grey Walter, Ross Ashby, and Stafford Beer were hardly fleeing from science, no matter how their activities might have appeared to observers like Teller and Roszak.[5]

We therefore borrow the term "groovy" to label this cluster of activities and aspirations. Originally a slang term used by African American jazz musicians in the 1930s and 1940s, by the late 1960s the word had become more closely associated with "hippies" and counterculture. The songwriting duo Paul Simon and Art Garfunkel released their "59th Street Bridge Song" (also known as "Feelin' Groovy") in 1966; a few years later the Grateful Dead produced a cover. In 1971 a clinical psychologist at the University of Southern California explained in *The Underground Dictionary* that among the largely white, middle-class celebrants of American counterculture, "groovy" had come to mean "great, fantastic, joyful, happy." The term faded from common usage during the 1980s, and hence it provides a useful marker for the "long 1970s" on which we focus here.[6]

The essays in this volume do not aim to cover all aspects of the history of science in the United States during the 1970s. Rather, they lay a foundation for addressing the shifting practices and cultural valences of science and technology during this recent and underexplored period. Examples of groovy science can be found among mainstream researchers who sought ways to engage with different communities and nontraditional forms of science, as well as among fringe thinkers who sought a place in the mainstream's spotlight. Groovy science, in other words, reflects the social exploration, experimentation, and eclecticism that were emblematic of the counterculture(s) during one of the most colorful periods of recent American history. To further characterize groovy science, we have highlighted four themes, around which this volume is organized: *conversion*, *seeking*, *personae*, and *legacies*.

Conversion. Although the era was marked by widespread protests against the military-industrial-academic complex, Cold War influences ran deep. One inspiration for this volume is the opportunity to revisit Cold War historiography. For nearly three decades, historians have interrogated the massive transformations of American science and technology that unfolded between the early 1940s and the early 1960s. From the ballooning of Pentagon-backed research to the fevered suspicions of domestic anti-Communism, the early nuclear era witnessed unprecedented realignments of science, politics, culture, and identity.[7] The relationships between power, patronage, politics, and practice are far less understood for the late 1960s and 1970s. The essays in the opening section thus provide a bridge between some well-understood topics and new intellectual territory. While rejecting the "megamachine," some forms of groovy science adapted resources and forms of knowledge that had been characteristic of earlier Cold War science and turned them toward new ends. In "Adult Swim: How John C. Lilly Got Groovy (and Took the Dolphin with Him), 1958–1968," for example, D. Graham Burnett charts how Lilly derived his interests in consciousness and interspecies communication—culminating in his famous experiments with dolphins, LSD, and sensory-deprivation tanks—from military-sponsored research into "brainwashing" and "mind control." Peter Neushul and Peter Westwick, in "Blowing Foam and Blowing Minds: Better Surfing through Chemistry," document how formerly buttoned-down aerospace engineers in Southern California leaned on their expertise with space-age materials to launch the "shortboard" revolution in surfing. And in "Santa Barbara Physicists in the Vietnam Era," Cyrus C. M. Mody describes how a relatively new academic department, which had been founded in the era of seemingly limitless Defense Department largesse, adapted to the new realities of the 1970s by crafting a low-budget, hands-on, experiential curriculum centered on "socially relevant" topics like environmental and biomedical applications, as well as on a few devices meant to help elucidate mystical and parapsychological phenomena.

Seeking. Many participants pursued some version of groovy science as a kind of secularized, quasi-spiritual quest. The search for "authenticity," "self-actualization," and sustainable self-sufficiency drove experiments in everything from psychology to environmental management to medical care. In "Between the Counterculture and the Corporation: Abraham Maslow and Humanistic Psychology in the 1960s," Nadine Weidman describes Maslow's decade-long dance with the flowering counterculture. The Brandeis University professor was by turns fascinated, inspired, and ultimately deeply ambivalent about the youth movement, even as many young people, encountering his work at places like the Esalen Institute, appropriated his ideas about "humanistic psychology" and his critiques of the traditional scientific method. Henry Trim, in "A Quest for Permanence: The Ecological Visioneering of John Todd and the New Alchemy Institute," reveals some of the symbiotic ties between self-professed countercultural scientists, eager to carve out a kind of "technoecological sublime," and traditional institutions like regional governments and Establishment foundations that helped to foot the bill. In "The Little Manual That Started a Revolution: How Hippie Midwifery Became Mainstream," Wendy Kline follows an eclectic group of "tuned-in" seekers who were eager to appropriate indigenous and medical knowledge to launch a home birth movement.

Personae. Regardless of how one participated in it, the counterculture was a product of mass media as much as cultural and political forces. And, of course, these communication technologies and strategies were themselves products of the Cold War. With far-reaching media attention to the counterculture came the propensity for hero worship and fame seeking. Groovy science often orbited around charismatic individuals, self-proclaimed guru types who crafted hybrid identities: at once scientific experts and countercultural iconoclasts. As Michael Gordin discusses in "The Unseasonable Grooviness of Immanuel Velikovsky," bookish elders like Velikovsky, with their brazenly heterodox theories of ancient history and celestial mechanics, became heroes to restless college students across North America. Meanwhile, hedonist charmers like Timothy Leary championed space exploration—and even space colonization—as the ultimate "high," as W. Patrick McCray charts in "Timothy Leary's Transhumanist SMI²LE." At the same time, cultural entrepreneurs like Hugh Hefner looked to ethology—the study of animal behavior—to bolster their own notions of "human nature" and "natural" gender roles, as Erika Milam documents in "Science of the Sexy Beast: Biological Masculinities and the *Playboy* Lifestyle."

Legacies. Many once-radical ideas have since been absorbed into the mainstream, their countercultural roots obscured, ignored, or forgotten. In "Alloyed: Countercultural Bricoleurs and the Design Science Revival," Andrew Kirk argues that today's trends in sustainability and architectural design grew out of countercultural experiments inspired by maverick, omnidisciplinal Buckminster Fuller. In "How the Industrial Scientist Got His Groove: Entrepreneurial Journalism and the Fashioning of Technoscientific Innovators," Matthew Wisnioski traces glossy, heroic portraits of today's socially engaged "knowledge workers" to an ideological shift among American engineers, triggered by the disruptions of the late 1960s, when they scrambled to define a new role beyond the familiar defense contractor. And in "When Chèvre Was Weird: Hippie Taste, Technoscience, and the Revival of American Artisanal Food Making," Heather Paxson reminds us that many types of food that are ubiquitous today—as well as labels like "artisanal" and "organic"—sprang from various people's concerted efforts, back in the 1970s, to embrace a countercultural lifestyle apart from the perceived consumerism and rat race of conventional American middle-class routines.

Underlying all four of these sections is a common theme: consumption. As Thomas Frank's book *The Conquest of Cool* made clear, business goals and counterculture grooviness were hardly polar opposites. Instead, they developed a symbiotic relationship through much of the 1960s and 1970s, as advertising executives appropriated elements of coolness and youthfulness to increase profitability and refine American capitalism.[8] Likewise, the essays in this volume reveal ways in which many people sought to reconcile science, technology, and hipness, melding a certain form of hip consumerism with enthusiasm for science. Many of the personalities who engaged in groovy science were indeed selling something, be it artisan cheese, midwifery manuals, or ideologies of innovation. One person's rebellion, idealism, or self-exploration was quite often another hipster's day job or meal ticket. As Robert Heinlein, one of the counterculture's favorite science fiction writers, put it, "there ain't no such thing as a free lunch."[9]

Understanding the historical contours and contributions of groovy science is important for several reasons. American historians have recently "discovered" the 1970s. A fast-growing literature now seeks to displace the 1960s as "the pivotal decade" for understanding the United States and the world during the years since World War II.[10] While many of the recent books admirably disentangle subtle features of the labor market and financial institutions, political allegiances, religious and intellectual trends, and publishing practices, most pay little attention to science and technology—and those that do tend to revert to Teller's and Roszak's simplistic dichotomy, pitting counterculture against science.[11] Yet the long 1970s were a period of significant change in American science and technology as well. A growing literature by historians of science and technology has begun to reveal colorful and sometimes-unexpected entanglements between the youth movements, counterculture(s), and science and technology over the long 1970s.[12] Building on those exciting studies, our goal for *Groovy Science* is to intervene before American historians' narrative of the 1970s settles into its own grooves, reifying Roszak's attractive yet misleading caricature and overlooking the many ways in which an embrace of science and technology (of a *particular type*) became part of what it meant to be countercultural.

At the same time, we hope that *Groovy Science* will help spur other historians of science, technology, and medicine to focus on the long 1970s. Thanks to the contributions from at least two generations of scholars, we now know a great deal about science and technology during the height of American-Soviet tensions. The story of American science in the twentieth century has come to be dominated by the now-familiar narrative of skyrocketing expansion after World War II, followed by precipitous decline around 1970—that is, by the narrative arc of the High Cold War.[13] Yet science and technology in the United States did not grind to a halt with the onset of détente and stagflation. New approaches, motivations, and institutions emerged, some with clear debts to prior patterns and others striving for genuine novelty. Groovy science represented improvisations, mutations, and hybridizations of what had only lately seemed to be the norm. The recent books on particular instances of what we have dubbed "groovy science" have opened a window on to a swirling, technicolor pastiche of ideas, people, and events. The time has come to chart common trends and themes across the physical, biological, and social sciences—common features that remain obscured if one relies only on a collection of tightly focused monographs.

Two caveats are in order. The first concerns lumping. We are mindful of the fact that there was no single, monolithic entity in the United States that could be called "the counterculture." Instead, disparate, sometimes-inchoate, and ever-changing assemblages of people, organizations, and ideas all passed under the "counterculture" banner. Part of the counterculture, to be sure, was politically oriented, working to change the existing power structure, protest against the ravages of the Vietnam War, and increase minority representation via a variety of means. This group was more inclined to accept hierarchy and organizations and their attendant bureaucracies. In comparison, another wing of the counterculture eschewed formal organization and drifted toward communal living arrangements, social experimentation, and consciousness-expanding activities. Other useful distinctions can be drawn, between (say) the splintering of the New Left and the quirky outcroppings of the New

Age.[14] Folded into this mix was the diverse experience of minorities, gays, and women, groups that were drawn in varying degrees to groovy science yet for whom a great deal of historical exploration still remains to be done. The term "counterculture" retains heuristic power in the singular, though as lived experience for thousands of young people the word conjured a rich and complex ecosystem.

The second caveat addresses splitting—namely, our focus in this volume on people, places, and events that were largely operating within the United States. Certainly there were countercultures and, indeed, instances of groovy science that percolated in other countries. Many of the ideas and individuals presented in this volume had an appeal that spanned national borders, and many groovy-science acolytes were attuned to developments in other parts of the world.[15] Nonetheless, because the essays in this volume range across the physical, biological, and social sciences, and because the historiography of the long 1970s has been an especially active area among American historians recently, we have focused on instantiations of groovy science within the United States for the sake of overall coherence. We hope that the examples gathered here will help galvanize broader, comparative discussions of science, technology, and counterculture.

Groovy science was an amalgam of earnestness and playfulness, an unsteady mix often fueled by a sense that society was on the brink of some new revolution. Roszak also thought that revolution was nigh: marked not by the emergence of a new kind of science but by a gaggle of "Theosophists and fundamentalists, spiritualists and flat-earthers, occultists and satanists," whose "radical rejection of science and technological values" represented the most recent ascendancy of "anti-rational elements in our midst."[16] Yet the countercultural critics had their reasons to turn away from the hulking infrastructure of Cold War science; and their own efforts were hardly antirational. With historical distance, we can discern connections to a deeper past that many "tuned-in" practitioners wished to spurn, just as we can trace linkages to thriving, mainstream pursuits today. In all these ways, thinking about groovy science can help us identify the constant interweaving of science and technology with the fabric of daily life—even when that fabric was an exuberant, paisley-patterned polyester.

Notes

1. Edward Teller, "The Era of Big Science," *Bulletin of the Atomic Scientists* 27, no. 4 (1971): 34–36.

2. Theodore Roszak, *The Making of a Counter Culture: Reflections on the Technocratic Society and Its Youthful Opposition* (Garden City, NY: Anchor / Doubleday, 1969), 50, 215.

3. See, e.g., Don Lattin, *The Harvard Psychedelic Club: How Timothy Leary, Ram Dass, Huston Smith, and Andrew Weil Killed the Fifties and Ushered in a New Age for America* (New York: HarperOne, 2010); Jeffrey Kripal, *Esalen: America and the Religion of No Religion* (Chicago: University of Chicago Press, 2007); David Kaiser, *How the Hippies Saved Physics: Science, Counterculture, and the Quantum Revival* (New York: W. W. Norton, 2011); Fred Turner, *From Counterculture to Cyberculture: Stewart Brand, the Whole Earth Network, and the Rise of Digital Utopianism* (Chicago: University of Chicago Press, 2006); and Andrew Kirk, *Counterculture Green: The "Whole Earth Catalog" and American Environmentalism* (Lawrence: University Press of Kansas, 2007). The remainder of this paragraph is based on David Kaiser, "Consciousness on the Charles," *Historical Studies in the Natural Sciences* 41 (2011): 265–75, esp. 274–75.

4. Kaiser, *How the Hippies Saved Physics*, xviii.

5. Turner, *From Counterculture to Cyberculture*; Kirk, *Counterculture Green*; and Andrew Pickering, *The Cybernetic Brain: Sketches of Another Future* (Chicago: University of Chicago Press, 2010).

6. On the etymology of "groovy," see Neil Hamilton, *The ABC-Clio Companion to the 1960s Counterculture in America* (Santa Barbara, CA: ABC-Clio, 1997), 130. Eugene Landy's *The Underground Dictionary* is quoted in Sam Binkley, *Getting Loose: Lifestyle Consumption in the 1970s* (Durham, NC: Duke University Press, 2007), 1.

7. See, e.g., Paul Forman, "Behind Quantum Electronics: National Security as Basis for Physical Research in the United States, 1940–1960," *Historical Studies in the Physical Sciences* 18 (1987): 149–229; Daniel Kevles, "Cold War and Hot Physics: Science, Security, and the American State, 1945–56," *Historical Studies in the Physical Sciences* 20 (1990): 239–64; Stuart W. Leslie, *The Cold War and American Science: The Military-Industrial-Academic Complex at MIT and Stanford* (New York: Columbia University Press, 1993); Peter Galison, *Image and Logic: A Material Culture of Microphysics* (Chicago: University of Chicago Press, 1997); Jessica Wang, *American Science in an Age of Anxiety: Scientists, Anticommunism, and the Cold War* (Chapel Hill: University of North Carolina Press, 1999); David Kaiser, "Cold War Requisitions, Scientific Manpower, and the Production of American Physicists after World War II," *Historical Studies in the Physical and Biological Sciences* 33 (2002): 131–59; David Kaiser, "The Postwar Suburbanization of American Physics," *American Quarterly* 56 (2004): 851–88; and David Kaiser, "The Atomic Secret in Red Hands? American Suspicions of Theoretical Physicists during the Early Cold War," *Representations* 90 (2005): 28–60.

8. Thomas Frank, *The Conquest of Cool: Business Culture, Counterculture, and the Rise of Hip Consumerism* (Chicago: University of Chicago Press, 1997).

9. Robert Heinlein, *The Moon Is a Harsh Mistress* (New York: Putnam, 1966).

10. See, e.g., Bruce J. Schulman, *The Seventies: The Great Shift in American Culture, Society, and Politics* (New York: Free Press, 2001); Beth Bailey and David Farber, eds., *America in the Seventies* (Lawrence: University Press of Kansas, 2004); Edward D. Berkowitz, *Something Happened: A Political and Cultural Overview of the Seventies* (New York: Columbia University Press, 2006); Binkley, *Getting Loose*; Rick Perlstein, *Nixonland: The Rise of a President and the Fracturing of America* (New York: Scribner, 2008); Bruce Schulman and Julian Zeilzer, eds., *Rightward Bound: Making America Conservative in the 1970s* (Cambridge, MA: Harvard University Press, 2008); Jefferson Cowie, *Stayin' Alive: The 1970s and the Last Days of the Working Class* (New York: New Press, 2010); Niall Ferguson et al., eds., *The Shock of the Global: The 1970s in Perspective* (Cambridge, MA: Harvard University Press, 2010); Judith Stein, *Pivotal Decade: How the United States Traded Factories for Finance in the Seventies* (New Haven, CT: Yale University Press, 2010); and Daniel Rodgers, *Age of Fracture* (Cambridge, MA: Harvard University Press, 2012).

11. See, e.g., Richard Kyle, *The New Age Movement in American Culture* (New York: University Press of America, 1995), chaps. 6–7; Peter N. Carroll, *It Seemed Like Nothing Happened: America in the 1970s*, 2nd ed. (New Brunswick, NJ: Rutgers University Press, 1990), 236–38; Schulman, *The Seventies*, 96–101; and Lattin, *Harvard Psychedelic Club*, 3, 220. A welcome exception to the trend is Bailey and Farber, *America in the Seventies*, which includes Timothy Moy's important essay "Culture, Technology, and the Cult of Tech in the 1970s," 208–27.

12. Turner, *From Counterculture to Cyberculture*; Eric J. Vettel, *Biotech: The Countercultural Origins of an Industry* (Philadelphia: University of Pennsylvania Press, 2006); Kirk, *Counterculture Green*; Kelly Moore, *Disrupting Science: Social Movements, American Scientists, and the Politics of the Military, 1945–1975* (Princeton, NJ: Princeton University Press, 2008); Andrew Pickering, *The Cybernetic Brain: Sketches of Another Future* (Chicago: University of Chicago Press, 2010); Kaiser, *How the Hippies Saved Physics*; Patrick McCray, *The Visioneers: How a Group of Elite Scientists Pursued Space Colonies, Nanotechnologies, and a Limitless Future* (Princeton, NJ: Princeton University Press, 2012); and Michael Gordin, *The Pseudoscience Wars: Immanuel Velikovsky and the Birth of the Modern Fringe* (Chicago: University of Chicago Press, 2012).

13. For a review of recent trends, see Sally Gregory Kohlstedt and David Kaiser, eds., *Science and the American Century: Readings from "Isis"* (Chicago: University of Chicago Press, 2013).

14. See Turner, *From Counterculture to Cyberculture*, 3–9; and Kaiser, *How the Hippies Saved Physics*, xxi–xxiii.

15. Within the recent literature, Andrew Pickering's *Cybernetic Brain*—which focuses mostly on developments in the United Kingdom—provides a helpful indication of cross-national trends, as does Peder Anker, "The Call for a New Ecotheology in Norway," *Journal for the Study of Religion, Nature, and Culture* 7 (2013): 187–207.

16. Roszak, *Making of a Counter Culture*, 50–51.

Part One: Conversion

1

Adult Swim: How John C. Lilly Got Groovy (and Took the Dolphin with Him), 1958–1968

D. Graham Burnett

What did it mean to be "groovy" circa 1970? It meant knowing how to hang, how to float, how to be at one with others, with animals, with the universe itself. I believe we can treat the following text as paradigmatic of the project as a whole:

> I suspect that whales and dolphins quite naturally go in the directions we call spiritual, in that they get into meditative states quite simply and easily. If you go into the sea yourself, with a snorkel and face mask and warm water, you can find that dimension in yourself quite easily. Free floating is entrancing. . . . Now if you combine snorkeling and scuba with a spiritual trip with the right people, you could make the transition to understanding the dolphins and whales very rapidly.[1]

Spiritual cetaceans? Trippy, collective, free-floating ethology of the odontocetes? Where are we?

The short answer is that we are located firmly in the head—the very heady head—of one of the most impor-

FIGURE 1.1 John C. Lilly at work in the lab. Reprinted by permission of the Flip Schulke Archives. © Flip Schulke.

tant, and one of the strangest, scientists of the 1960s: John C. Lilly, the man whose work with brains and behaviors of dolphins had lasting implications for cultural understanding of human beings' nearest aquatic kin (fig. 1.1). A pioneering neurophysiologist, a troubled military psychologist, an apostate cetologist and animal fantasist, ultimately a Pied Piper of whale hugging and cosmonaut of heightened consciousness— John C. Lilly traced a fascinating trajectory across the postwar period. Retracing a parabolic portion of his path (he rose and he fell) will allow us, I think, to catch several striking views of what we may indeed want to call "groovy science."

Lilly and the Cetacean Brain

Born in 1915, Lilly, from a well-to-do family in Saint Paul, Minnesota, took a bachelor of science degree from the California Institute of Technology in 1938 and studied at Dartmouth Medical School for two years before moving to the University of Pennsylvania, where he completed his MD in 1942 and remained on the faculty. There, under the influence of Britton Chance and Detlev Bronk, Lilly pursued research in biophysics, including applied investigations into real-time physiological monitoring—work linked to wartime service in military aviation, where techniques for assaying the respiration of airmen were needed.[2] Lilly had contact through his family with the neurosurgeon Wilder Penfield in the later 1940s and developed an interest in neuroanatomy and the electrophysiology of the brain. By 1953 he had been appointed to the neurophysiology laboratory of the National Institute of Mental Health (NIMH), where he worked under Wade Marshall as part of a joint research program with the National Institute of Neurological Diseases and Blindness.

By the mid-1950s Lilly's lab in Bethesda, Maryland, was performing in vivo electrical stimulation of the brains of macaques—work aimed at cortical mapping by means of correlating point applications of currents at varying thresholds with specific behaviors and reactions in subject animals.[3] Reporting on some of these investigations at a conference on the reticular formation of the brain, held in Detroit in 1957, Lilly would explain,

> The neurophysiologist has been given a powerful investigative tool: the whole animal can be trained to give behavioral signs of what goes on inside. . . . We are in the position of being able to guess with less margin of error what a man might feel and experience if he were stimulated in these regions.[4]

This was, in many ways, unpleasant business, Lilly acknowledged, pointing out that he had "spent a very large fraction of my working time for the last eight years with unanaesthetized monkeys with implanted electrodes." In addressing the nebulous region where neurology, psychology, and animal behavior overlapped, Lilly permitted himself some observations on the affective universe of his scientific subjects:

> When an intact monkey grimaces, shrieks, and obviously tries to escape, one *knows* it is fearful or in pain or both. When one lives day in and day out with one of these monkeys, hurting it and feeding it and caring for it, its experience of pain or fear is so obvious that it is hardly worth mentioning.[5]

It would not be the last time that Lilly would reflect on the inner lives of his experimental animals with considerable confidence. But his experimental animal was about to change. Like a number of American psychology researchers in the mid-1950s—including the echolocation researcher Winthrop Kellogg—Lilly was in the process of leaving monkeys behind for the bottlenose dolphin, *Tursiops truncatus*.

His first brush with the study of cetaceans came in 1949 when, during a visit to a neurosurgeon friend on Cape Cod, Lilly learned that a recent storm had beached a whale on the coast of southern Maine. A plan took shape for an impromptu expedition north, with a view toward collecting a novel brain.[6] As it happened, Lilly was acquainted from his days at the University of Pennsylvania with the Swedish-Norwegian physiologist and oceanographer Per F. "Pete" Scholander, who had also worked with Detlev Bronk in aviation physiology during World War II and had then moved to the Woods Hole Oceanographic Institute.[7] Scholander—something of a daredevil, and fascinated by the physiology of extreme environments—had published research on dive physiology and decompression, and while still living in Scandinavia he had conducted a number of pioneering studies on the deep-diving capabilities of marine mammals, particularly whales.[8] Lilly looked up Scholander and recruited him for the trip, and the three men suited up for a drive to Maine. Shortly after reaching the carcass (a large pilot whale), exposing the skull, and beginning to chip away toward the brain, they were joined by two other researchers who had independently made the drive up from Woods Hole: William Schevill and his wife and collaborator, Barbara Lawrence. They were, reportedly, somewhat miffed to discover that they had been beaten to the punch and were particularly concerned that the hacksaw dissection might have damaged the airways of the upper head, which they had come to examine. In the end, however, the cadaver would be theirs, since Lilly and his partners found that the brain had largely been dissolved through autolysis; the smell alone overpowered them.

Though he headed home with little to show for the trip, Lilly had brushed the shores of cetology, and his curiosity did not dissipate. At a meeting of the International Physiological Congress four years later, in 1953, Lilly and Scholander again crossed paths, and Scholander suggested that Lilly get in touch with a leading expert on captive dolphins, Forrest G. Wood, who at that time handled the animals at Marine Studios. Lilly did, and as a result, he was one of eight investigators to participate in what came to be known informally as the "Johns Hopkins expedition" in the autumn of 1955. Like prewar projects, the expedition featured a mixed crew of physiologists and medical men gearing up to vivisect some bottlenose dolphins, only this time it would be in the carnival environs of a Florida ocean theme park rather than a remote fishing village on a barrier island.[9]

In preparation for this 1955 trip, Lilly spent the summer in correspondence not only with Wood (securing access to a set of dolphins for experimental work) but also with Schevill at Woods Hole (concerning the anatomy of the airways of the common dolphin)[10] and with Scholander (concerning restraint techniques and the respiratory characteristics of the odontocetes).[11] Using this information, Lilly worked up a dolphin respirator that would, it was hoped, permit the surgeons and neuroscientists of the party to expose the brain of an anesthetized animal in order to begin the work of cortical mapping by neurophysiological techniques. It appears that no one in Europe or the United States up to this point had attempted a "surgical" intervention on a dolphin or porpoise.

The Johns Hopkins expedition of 1955 was at best a qualified success. Lilly and the other investigators were unsuccessful with their anesthetics and their respirator, and in the end they euthanized, without dexterity, five dolphins, apparently alienating a number of the Marine Studios personnel in the process.[12] But if the 1955 investigations were not a triumph, they did deepen Lilly's continuing interest in the cetacean brain.[13] Having heard a set of Wood's recordings of bottlenose dolphins at Marine Studios, Lilly was much struck—as were a considerable number of others at this time—by the range and apparent complexity of dolphin phonation. In October 1957 and again in 1958—after a visit with Schevill and Lawrence in Massachusetts, where they were conducting work on the auditory range and echolocatory capabilities of a bottlenose dolphin in a facility near Woods Hole—Lilly returned to Marine Studios. This time he was equipped to undertake investigations of the dolphin brain and behavior using techniques like those he had deployed and refined with macaques at NIMH: namely, percutaneous electrodes, driven by stereotaxis, that could probe the brain tissue of an unanesthetized, living animal.[14] Over the two visits, three more animals were sacrificed, and Lilly experienced a kind of scientific epiphany that would shape his scientific life, even as its reverberations eventually unmade his scientific reputation.[15]

Compressing a complicated encounter that took place over several days—and which continued to draw Lilly's reflections and reconstructions for years—is not easy, but we can summarize Lilly's sense of his findings this way: First, Lilly persuaded himself that, in comparison to his experience with monkeys, the dolphins appeared to learn very rapidly how to press a switch to stimulate a "positive" region in their brains (and to turn off stimulation to a region causing pain).[16] Second, he claimed to have been much struck by the sense that an injured experimental subject, when returned to the tank with other dolphins, "called" to them and received their ministrations, suggesting an intraspecies "language."[17] Third, on reviewing the tapes made of these investigations, Lilly grew increasingly certain that his experimental subjects had been parroting his speech and other human sounds in the laboratory. These three elements—intelligence, an intraspecies language, and (perhaps most significantly) what he took to be fleeting glimpses of an attempt at interspecies communication—left Lilly with a feeling that he was on the cusp of something vast. Reflecting on the work of 1955, 1957, and 1958 in his Lasker Lecture in April 1962, Lilly tried to explain:

> We began to have feelings which I believe are best described by the word "weirdness." The feeling was that we were up against the edge of a vast uncharted region in which we were about to embark with a good deal of mistrust concerning the appropriateness of our own equipment. The feeling of weirdness came on us as the sounds of this small whale seemed more and more to be forming words in our own language.[18]

After hammering his way into hundreds of mammalian brains, Lilly suddenly heard a voice.

Odd as this breakthrough may seem, Lilly was not alone in his sense of the magnitude of what had happened in the Marine Studios laboratory in the late 1950s. One of Lilly's medical friends who had been in attendance in October 1957, during work on dolphin number six, later mused to him in a letter, "I keep thinking of that first moment when the first, clearly purposeful switch-pressing response occurred. This is one of the extraordinary moments in science."[19] Loren Eiseley, the anthropologist who had become the provost of the University of Pennsylvania, wrote publicly that "the import of these discoveries is tremendous, and may not be adequately known for a long time."[20] And in 1961 Lilly would write of the discoveries in still-grander world-historical terms, situating his own research at the cusp of the fourth "great displacement" in the history of science: citing Freud, Lilly explained that although man had, over the last five hundred years, been thrust from the center of the universe, from the center of nature, and finally from the center of his own mind, modern man still thought of himself as the center of all intelligence. This (final?) pillar of human exceptionalism now teetered beside the dolphin tanks. It was his predilection for claims like this—sweeping, visionary, laced with self-aggrandizing enthusiasm—that eventually seriously tried the patience of Lilly's colleagues, who placed increasing pressure on him in the early 1960s to deliver some reproducible scientific results.

Those results were slow in coming. Lilly's first published report on his dolphin experiences appeared in December 1958 in the *American Journal of Psychiatry*, in an article entitled "Some Considerations regarding Basic Mechanisms of Positive and Negative Types of Motivations." It is a deeply strange document, one that opens a window obliquely on to the world of brain research during the Cold War. Because it attracted a spate of articles in newspapers and magazines across the country—and launched the writing project that culminated in *Man and Dolphin* a little over two years later—it is worth examining this initial presentation in some detail.[21]

In view of the paper's reception and impact, it is striking that the discussion of the dolphin work at Marine Studios represents less than one-third of its total length, and that this section is sandwiched in the middle of a wide-ranging discussion of the positive and negative "motivation" regions of the mammalian brain. Lilly's primary concern in this paper was to reflect on the fact that neurophysiological work over the previous five years had established the existence of brain regions that, under stimulation, trigger "negative-painful-stop" responses, whereas other regions trigger "positive-pleasurable-start" responses. At issue, finally, was the balance between the aggregate sizes (and influences) of these two "parts" of the brain. So one reads, for instance, "Of course we like to think that in the total action of the brain, the positive tends to overbalance the negative, and that the intellectual functions might be neutral ones, neither positive nor negative, found in excess of the positive and the negative."[22]

Lilly then posed a question: Would point stimulation within the neocortex and the cerebellum—the regions of "higher" brain function—generate neutral, or perhaps mildly positive, effects? It appeared, strangely in his view, that in monkeys it did neither: strong negative response zones could be found in these sophisticated regions. So what? Well, this was troubling, according to Lilly's analysis. Were the centers of *abstract reasoning* laced with "halt" zones? This seemed to raise a striking possibility: the tractability of sophisticated cognitive functioning by means of electrical stimulation. Enter the specter of "mind control."

But perhaps, Lilly went on to reason, these issues—the "positive" versus the "negative" in higher brain function and, crucially, the plasticity of subjects under electrical stimulation—could only really be addressed in a brain larger and more complex than that of the macaque. As Lilly put it, "May not a larger brain be more impervious to such tampering with its innards? May not the trained, sublimating, and sometimes even sublime human mind resist, and even conquer such artificially evoked crassly primitive impulses?" Human experimentation was the only way to get sure answers to these questions. Acknowledging that this remained too risky to contemplate, Lilly then introduced *Tursiops truncatus*: "So far we have found only one animal that has a brain the size of ours who will cooperate and not frighten me to the point where I can't work with him—this animal is the dolphin." A description of Lilly's work at Marine Studios followed, emphasizing the success he and his collaborators had had in finding both positive and negative regions in this large brain. Two paragraphs at the end of this section became the core of the newspaper articles that followed, so they are quoted here in their entirety:

In this abbreviated account, I cannot convey to you all of the evidence for my feeling that if we are to ever communicate with a non-human species of this planet, the dolphin is probably our best present gamble. In a sense, it is a joke when I fantasy that it may be best to hurry and finish our work on their brains before one of them learns to speak our language—else he will demand equal rights with men for their brains and lives under our ethical and legal codes!

Before our man in the space program becomes too successful, it may be wise to spend some time, talent, and money on research with the dolphins; not only are they a large-brained species living their lives in a situation with attenuated effects of gravity, but they may be a group with whom we can learn basic techniques of communicating with really alien intelligent life forms. I personally hope we do not encounter any such extraterrestrials before we are better prepared than we are now. Too automatically, too soon, too many of us attribute too much negative systems activity to foreign language aliens of strange and unfamiliar appearance and use this as an excuse for increasing our own negative, punishing, attacking activities on them.[23]

With that, Lilly was finished with dolphins, and he returned to the shadow subject of his presentation: "What does all this mean in

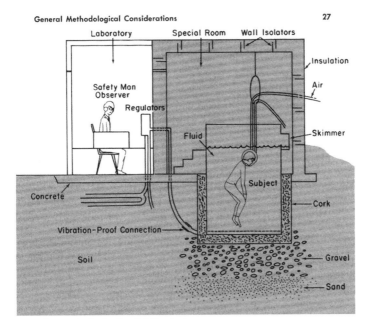

FIGURE 1.2 Sensory deprivation in a flotation tank.

terms of us, our species?" Was it possible, without percutaneous electrodes, to investigate the "positive" and "negative" systems of our own
brains? "Turning inward, examining our minds, their deep and primitive workings, can we see evidence of the actions and inner workings
of the positive, pleasure-like, start, and the negative, pain-fear-like, stop
systems?"[24] Lilly's answer was yes, and the technology for doing so
preoccupied him for the last third of his talk: isolation tanks—large,
temperature-regulated tanks in which neutrally buoyant subjects were
confined, in a breathing hood, in total darkness and without sound or
sensory input, for as long as they could stand it (fig. 1.2).[25]

Military Minds

We may seem to be impossibly remote from the study of whales and
dolphins, and yet the links are tighter and stranger than one might expect. Lilly was the inventor of the water-immersion technique of sensory
deprivation, and he was a significant early contributor to the broader
area of sensory-deprivation research, opening the third major center
for such investigations (after McGill and Princeton) in the world and
spawning a number of labs (most importantly that of Jay Shurley at
the Oklahoma City VA hospital) that built tanks to his specifications.

While working on such systems for several years at NIMH (from 1955–56 forward), Lilly conducted extensive self-experimentation in his tank, developed new equipment for it, advised on the building of similar systems, and lectured around the country on their use.

Such experimental environments seem, in retrospect, macabre, but it is necessary to recall the impetus for this sort of work. The earliest such investigations (at McGill, from 1951 forward) were stimulated by an interest in Russian and Chinese "brainwashing,"[26] and there can be no doubt that Lilly's move into this area was linked to such preoccupations, which were firmly established on the national stage in the United States in the mid-1950s. One of the files in the Lilly papers is labeled "Indoctrination, Forced," and it contains material on solitude, isolation, and "brainwashing." An adjacent file, labeled "Solitude," contains, most interestingly, Lilly's notes on a conversation with "Dr. Sperling" in the Research and Development office of the Surgeon General's office, dated 23 April 1956. The notes read, "called this date, re: brainwashing, etc, mentioned by Dr. Felix before senate appropriations committee two weeks ago." A set of newspaper clippings makes it clear that this exchange took place at the height of *Manchurian Candidate* fears over the practices of Chinese "mind control" scientists.[27] Other correspondence reveals that systems like Lilly's were wanted for two purposes: first, as a tool for screening tests and personality assessments (in order to find individuals particularly resistant to such situations and techniques) and, second, as training instruments to improve the resistance of those who might face sensory-deprivation conditions—not just soldiers, spies, and diplomats but others subjected to the rigors of solitary environments, particularly pilots, astronauts, and those manning remote meteorological or monitoring stations.[28]

Precisely how close Lilly's ties to the world of intelligence operations actually were remains obscure. In his autobiography, he wrote that his decision to leave NIMH in 1958 was motivated by his growing unease with the encroachment of application-oriented, apparently government-linked, investigators seeking information about the work of his Bethesda laboratory.[29] Lilly was involved, well into 1959, in advising members of the security establishment about the potential uses of his neurophysiological investigations, though an unseemly flap over his security clearance in May of that year certainly added friction to those relationships.[30] As of 27 August 1959, Lilly had "Secret" clearance, and documentation in Lilly's FBI records indicates both that J. Edgar Hoover personally attended to his file and that he was to be treated as persona non grata by the bureau after 1960.[31]

What emerges from a close reading of Lilly's 1958 paper in this context is the remarkable way in which his early dolphin investigations were entangled with this set of seemingly remote preoccupations—"brainwashing," "reprogramming," and "mind control"—that were reverberating through the sciences of brain and behavior in the mid-1950s. For instance, the questions Lilly posed after reporting his work with macaques—Can humans "resist or conquer" such situations? Do different people discover different degrees of "egophilic" or "egophobic" affect in isolation tanks?—were questions wholly tied to the pressing problem of the imperviousness and durability of the Cold War human agent.

These very un-cetological matters might be of merely anecdotal value to our understanding of Lilly's work with dolphins if his 1958 paper were simply a salad of his diverse interests. But this is not the case. Rather, the paper offers palpable clues to the early ties between Cold War psychological-neurological investigations aimed at "accessing" the mind of recalcitrant (and taciturn) enemies and Lilly's research program on interspecies communication. Indeed, by the time Lilly had expanded his brief asides on the promise of dolphin-human communication into the popular book *Man and Dolphin* of 1961, the language of "psy-ops" permeated his discourse. For instance, alluding to the "less well controlled" dolphin training "of the past," which used food rewards, Lilly pointed out that as a result of his work and that of others, "we now have push-button control of the experiences of specific emotions by animals in whose brains we have placed wires in the proper places. . . . Using this 'reward stimulation' technique, we demonstrated quite satisfactorily that a dolphin can vocalize in two different ways." One of these, above the surface ("in the air"), suggested that already the animals were being brought into the sphere of their captors.

Another mind control technique was less invasive but also promising—"intraspecies solitude":

> If a human being is isolated from other humans for a month or more, and is confined to a small area geographically and a small range of activities, his interest in his surroundings and its minutiae increases radically. . . . Further, if a confined, isolated human is allowed brief contacts with other humans even without a shared language, he begins to find their presence comforting, and a pleasant relief from the "evenness" of his surroundings. If these humans control his only sources of food as well as his sources of intraspecies stimulation, he may adopt to their de-

mands in subtle and not so subtle ways. He may, given time, learn their language, take on their beliefs, etc.

When we catch a dolphin and put him alone in a small tank, we are imposing similar "solitary confinement" strictures on him. Maybe we can thus capture his loyalty, and his initiative.[32]

Later, in his effort to break through to (or perhaps simply to "break") a dolphin, Lilly would deploy this and other techniques borrowed from the margins of the world of the mind control "spooks"—including a set of extended "chronic-contact" experiments and, finally, experimental psychopharmaceuticals, particularly D-lysergic acid diethylamide (LSD 25), which in the early 1960s was closely linked to the clandestine work of the CIA's "Artichoke" project.[33]

This strange imbrication of the techniques of mind control and animal communication in the late 1950s and early 1960s suggests at least one way in which the isolation tank and sensory-deprivation research fitted with Lilly's program of cetological investigations in this period. But there were others. Among Lilly's papers, in his "Solitude" file, I discovered his copy of the English translation of Jacques-Yves Cousteau's 1953 best seller *The Silent World*, a book (and, in 1956, a film) that introduced Cousteau and deep-sea scuba diving to hundreds of thousands of Americans. Lilly's annotations in the margins of this volume suggest that he read the text with care and took particular interest in Cousteau's reflections on the experience of weightlessness and isolation in the silent suspension of the underwater world. The development of the actual mask and breathing technologies of Lilly's isolation tank must thus be seen in the context of a growing interest in scuba diving and undersea environments.[34] In fact, in describing his efforts to communicate with dolphins in this period, Lilly spent a good deal of time on the need for the experimenter to attempt to commensurate him- or herself—imaginatively, even physically—with the subject. And here, Lilly explained, the flotation/isolation tank could be of considerable use, since it offered a glimpse of the dolphin's perceptual universe.[35] It was for this reason, apparently, that Lilly saw to it that an isolation tank was built at the Saint Thomas lab of his Communication Research Institute (CRI), adjacent to the indoor dolphin tank. In the early 1960s Lilly's isolation technology appeared poised to give him insight into the "mind in the waters" that had become, by 1958, his new experimental subject.[36]

Dolphin Discussions

It was with an eye on dedicating himself to this new experimental subject that Lilly, in 1958, departed from NIMH, in the glow of widespread
interest in what this hard-driving scientist planned to accomplish with
the dolphin. He was jumping several ships at once, in that he simultaneously separated from, and subsequently divorced, his wife of two
decades, mother of two of his children. By 1959 Lilly not only had a
trade contract for a pair of books on dolphins but had also begun work
on opening what would become a dedicated research facility for the
study of these animals. From the start, Lilly—keen to reinvent himself at
some distance from middle-class respectability (and the smell of monkey
cages)—had his eye on the Caribbean, and visits to the Bahamas, Ja-
maica, and, finally, Puerto Rico gave him the lay of the land. He secured a
position at the University of Puerto Rico in pharmacology during 1959–
60 as he continued his reconnaissance of the region, and by the end of
that year he was remarried, to Elisabeth Bjerg, a former fashion model
from Saint Croix. He had also settled on Saint Thomas as the location
for his new institute, found real estate and agents to help him secure it,
and was in a position to entertain visits from two grant makers from the
Office of Naval Research (ONR), including, significantly, Sidney Galler.
ONR announced that it was willing to top up the funding of his recently
incorporated institute (CRI), depending on how the National Science
Foundation (NSF), National Institutes of Health (NIH), and Department
of Defense responded to his grant requests. Meetings in Washington, DC,
followed, and by 13 June, Lilly could write to his friend Orr Reynolds at
NASA that "the Institute has been given a grant by the N.S.F. (with help
from O.N.R. and D.O.D.) to build the world's first laboratory devoted
to the study of the intellectual capacities of the small, boothed [*sic*, for
"toothed"] whales." The institution's budget was set at about $100,000
a year, though expenses went up when, later that year, Lilly decided to
open a second lab in Miami—what he called a "bipedal" arrangement,
allowing the work to be "close to our supply of animals."

Shortly thereafter, NASA money was forthcoming through Reynolds,
under the Office of Space Sciences' Biosciences Program, which sponsored a "behavioral sciences division" dedicated to, among other things,
"investigations on the mechanisms of inter- and intra-species communication of intelligent information, emotional status, and basic drives, in
an attempt to discover the mechanisms which nature has evolved, and
to supplement these mechanisms by technological devices."[37]

FIGURE 1.3 The Communication Research Institute.

Meanwhile, the building itself was well under way. Navy Underwater Demolition Team frogmen excavated the dolphin pool below the laboratory building with the help of six hundred pounds of TNT (fig. 1.3). By March 1960 Lilly was ready to bring his first two dolphins—secured through Wood at Marine Studios—down to Saint Thomas. Media coverage of the transfer included a front-page piece in the *Herald Tribune*, but the enthusiasm was short-lived: both dolphins died in a matter of weeks, and Lilly returned to Florida in what was, at least in part, an effort to learn more about how to maintain dolphins in captivity. To this end, he spent time at Theater of the Sea and the Marine Laboratory of the University of Miami. All the while, Lilly was pulling together the manuscript of *Man and Dolphin*, which would be published in August 1961.

Interestingly, Lilly's talking dolphins beat him into print. In April 1961 the Hungarian physicist Leo Szilard published his biting satire on the arms race, *The Voice of the Dolphins*. This futuristic tale— written in the summer of 1960 (as Szilard fought cancer in Memorial Hospital in New York City) and circulated in various forms among nuclear-armament negotiators in 1960—apparently emerged out of Szilard's conversations with Lilly in the late 1950s at NIH.[38] In the story, Lilly's name is explicitly invoked and linked to the founding of a Soviet-American scientific research institute, located in Vienna, which succeeds—it claims—in "breaking through" to dolphins, which prove, according to the scientists, to be brilliant thinkers. Over the next two decades, the Vienna Institute's dolphins serve as (yes) modern "Delphic" oracles who guide the world, via their scientist mouthpieces, through the moves of a chess-like game of nuclear disarmament (using Szilard's controversial scheme for de-escalation). In the end, their work done, the dolphins all die mysteriously, the Vienna Institute dissolves, its records vanish in a fire, and the world has been pulled back from the brink of nuclear annihilation. Szilard winks at the end, alluding to the doubts of some concerning whether the guiding intelligence behind the whole miraculous process had indeed been the dolphins or merely the crafty community of international scientific adepts. The success of this book, which over the next three years sold over thirty-five thousand copies in the United States alone (it was translated into six other languages, becoming a "minor classic of the nuclear age"), makes it clear that the nebulous idea of dolphins as an alternative intelligence—and as poten-tial interlocutors for human scientists—was very much in the air by the time *Man and Dolphin* appeared, and that these ideas were firmly linked to the name John C. Lilly.[39]

And *Man and Dolphin* delivered. Its opening sentence made an ex-travagant prediction: "Within the next decade or two, the human spe-cies will establish communication with another species: nonhuman, alien, possibly extraterrestrial, more probably marine, but definitely highly intelligent, perhaps even intellectual." Lilly's money (quite liter-ally: he had invested his own resources in CRI) was on the dolphin. In a closing chapter he explored the potential legal, moral, and social problems that would be confronted, going so far as to suggest that in the distant future the world might confront a crisis comparable to that of racialized global human inequality: "For a long time, presumably, they [the educated dolphins] will be in the position of the Negro races who are attempting to become Westernized."[40] At the same time, the encounter with this parallel oceanic intelligence would give human beings "a perspective of which we can be only dimly aware at the present time. Our own communication among ourselves will be enhanced and improved by such contact."[41]

This unstable froth of *Planet of the Apes* futurism and Cold War development theory met with popular enthusiasm (fired in part by an appealing photo-essay in *Life* magazine at the end of July) and a guardedly positive review by the well-known turtle biologist Archie Carr in the *New York Times*. Largely laudatory assessments in a variety of other papers in the United States and the United Kingdom followed. Even the *Quarterly Review of Biology* praised the text for putting "forth in a very readable fashion a study that many people have heard about."[42] Within a year, the initial run of ten thousand copies had sold out, and a second printing was under way.

During this period Lilly completed work on a handful of scientific papers reporting the results of his dolphin researches (up to the publication of *Man and Dolphin*, Lilly's only published scientific articles mentioning dolphins were the 1958 essay in the *American Journal of Psychiatry* and a brief note in *Science* with Alice Miller, "Sounds Emitted by the Bottlenose Dolphin," which appeared in May 1961). He also expanded his research enterprise at CRI, securing more dolphins, cultivating the board of trustees, and writing grants for new research programs and staff. These years, 1961–63, represent the waxing of CRI and Lilly's program. Sending out complementary copies of *Man and Dolphin* to a who's who of old-guard cetology, movie celebrity, and national political power, Lilly soon found himself not only receiving laudatory letters from the respected ethologist-anthropologist Gregory Bateson (who would join CRI shortly thereafter) but also entertaining, in Saint Thomas, President Kennedy's personal physician and a host of other distinguished visitors who wanted to meet the dolphins and the dolphin doctor.[43]

A typescript draft of a program description for CRI, which can be dated to late 1962, captures Lilly's vision for the institute in this heady period, and annotations in his hand indicate his careful efforts to position CRI between psychology, medicine, neurology, and animal behavior:

> The institute is studying intensively one of the unusual creatures of the sea—the bottlenose dolphin. This is a mammal with a brain larger and more complex than the human brain. The unexplored biological territory of the large mammal brain affords

an unequaled opportunity for tests in neurophysiology, brain function, communication, and intelligence.[44]

What follows is a list of benefits and objectives of this work, which Lilly carefully renumbered to accord with his sense of their significance, moving "improved techniques of human brain surgery" from first to last and replacing it with, first and second, "improved understanding of human learning and educational processes" and "a more systematic approach to the measurement of interspecies intelligence."

Writing an entry on interspecies communication in the McGraw-Hill *Yearbook of Science* in 1962, Lilly would briefly mention the work of Catharine Hayes and Keith Hayes with chimpanzees before moving on to say that "insofar as the author knows, there have been no systematic and serious attempts to date on the part of the human to learn to speak with another species in its own tongue." This, and the teaching of human language to dolphins, were the work, Lilly reported, of CRI.

This brief entry captures the combination of dynamic enthusiasm, medicoscientific gravitas, and thinly veiled self- (and CRI) promotion that pervaded Lilly's production in the bustling period that followed the publication of *Man and Dolphin*. During this time Lilly's NSF Career grant came through, and his trips to Europe and Washington were frequent. Lilly appeared to be in the vanguard of a vigorous new area of well-supported (and popular) research.

The air of gravitas and possibility that surrounded dolphin work in these years must be linked to rumors (and realities) of military interest. In *Man and Dolphin*, Lilly himself alluded presciently to these "implications" of future work on communication with the small odontocetes:

> Cetaceans might be helpful in hunting and retrieving nose-cones, satellites, missiles, and similar things that men insist on dropping into the ocean. They might be willing to hunt for mines, torpedoes, submarines, and other artifacts connected with our naval operations. They might also be willing to do scouting and patrol duty for submarines or surface ships, and they might carry their protagonist activities to the point where they can be used around harbors as underwater demolitions team operators.[45]

His speculations on these matters had rapidly reverberated through popular channels as well as through communities of specialists.[46]

In fact, in the summer of 1963 Forrest G. Wood quit his position as the curator at Marineland of Florida (the former Marine Studios) and accepted a new job—as the head of the newly formed Marine Biosciences Facility of the Naval Missile Center (NMC), where he would jointly administer (with Thomas G. Lang, from McLean's Naval Ordnance Test Station, NOTS) the emerging Cetacean Research Facility being constructed at Point Mugu. It was this nascent institution that would grow, through several changes of name and location, into the US Navy's Marine Mammal Program, and it was under the auspices of this program that Wood and his collaborators would succeed, over the next decade, in realizing a not-insignificant number of Lilly's seemingly bizarre ideas about the potential military uses of dolphins and porpoises. As early as 1964 the navy trainers and cetacean researchers (with help from commercial animal-training guru Keller Breland) succeeded in getting a dolphin to work untethered in the open ocean and to return when signaled; in 1965 a navy-trained bottlenose aided the "aquanauts" living in the experimental underwater habitat Sealab II by shuttling matériel and messages to and from the surface; over the next three years several marine-mammal deepwater marking and recovery programs were developed, and at least two of these were operationalized as navy "systems"; and finally, by 1970–71, a number of navy dolphins accompanied a specialized team of navy divers to Cam Ranh Bay, in Vietnam, where they were deployed in a program to stop Vietcong sappers.[47]

Lilly himself was, in the early 1960s, closely connected to this research program. A remarkable memo in Lilly's papers dated 17 October 1961 records Lilly's notes on a long telephone conversation with US Navy physicist William B. McLean, in which the latter sketched the rapidly shifting profile of West Coast cetacean research. Lilly learned of the work of the Lockheed group and discovered that they were at odds with the China Lake / Tom Lang group, who were investigating laminar flow (Lilly noted to himself, "We have a copy of their report," indicating how closely he was connected to this hydrodynamic research). Moreover, McLean gestured at others who were trying to attract research funding for work on porpoises and dolphins: Kenneth Norris at UCLA, Peter Scholander at UCSD, and several people at UCLA's Brain Research Institute (all of whom had separate plans afoot for cetacean research programs), not to mention the momentum gathering behind the proposed new cetacean research facility at Point Mugu, about which Lilly was informed when the whole program was still aborning. In a marginal note one senses Lilly's despair at keeping track of who was who: "There are as many plans as there are groups."[48]

A few months later, Lilly would have an even clearer sense of the situation, because in February 1962 McLean invited him out to NOTS for several lectures. It appears that Lilly made a considerable impression there, since he subsequently received a number of letters from NOTS personnel enthusiastically pressing several very curious lines of inquiry. For instance, T. W. Milburn (the head of the psychology division at NOTS and the director of Project Michelson) followed up in April with some brainstorming:

> Reasoning that the machine is such an integral part of man's culture, and that many men find it enormously pleasant to interact with machines, I have wondered whether it might not be possible to develop some mechanical equipment that a dolphin might use. Again, you will see how indegted [*sic*] I am to you for the short conversations we have had thus far, thinking in terms of the short drive, I wondered whether it might be at all feasible (and I realize that the idea may sound a little fantastic) to arm dolphins with some sort of weapon that would enable them more easily to attack shark.

He continued the letter with further thoughts:

> Bill McLean and I have been discussing the possibility of developing some dolphin toys, large, complex, mechanical devices, that might be of interest to dolphins even in the open seas, that would involve some kinds of buttons to push that would generate running water, perhaps with one trained dolphin teaching others. I would very much appreciate your own reflections as to whether [this] sounds too wild to contemplate.[49]

These missives seem so odd that it is difficult to resist the speculation that they are in fact written in some sort of circumlocutory code—a suspicion enhanced by a single NOTS-NMC Off-Base Authorization Form that appears in this file of Lilly's correspondence, indicating that in the fall of 1962, Lilly was scheduled to receive a registered "Secret" visit at CRI from "Louis R. Padberg," a navy electronics engineer working through the NOTS-NMC life sciences department.[50]

Even as Lilly's network of military and intelligence connections, medical associates, biophysics colleagues from wartime service, and old Caltech friends (several of them in powerful positions within the new institutions of postwar scientific funding) kept him, and CRI, plumped with research contracts and grants in the early 1960s, the established experts in animal communication, on the one hand, and captive-dolphin handlers, on the other, began to undermine his standing and his claims. If they could not deny him the fanfare that often attended his showy science, they could snipe, and they did. Already by the end of 1961, several reviewers had grumbled. James W. Atz (an ichthyologist at the American Museum of Natural History), writing in the bulletin of the New York Zoological Society in December, was at pains to explain—as several hostile reviewers would be—that there was nothing wrong with radical scientific theories and that Lilly, if correct in his central claims, would rightly take his place beside Darwin as "one of the greatest, most creative innovators in biology." Who could criticize originality and breadth of imagination? However, *Man and Dolphin* presented not a "single observation or interpretation" that could withstand scrutiny. Where this was the case, visionary hypothesis crossed the line into irresponsible delusion of the gullible populace: "Scientists and educators who believe that it is important for members of a democratic society to have a rational view of animal life can only wish that Dr. Lilly had not felt called upon to put himself so prominently in the public eye."[51]

The first shot had been fired, and more blasts were to come, particularly as 1962 came and the scientific papers backing up the "discoveries" described in *Man and Dolphin* were not forthcoming. (Lilly alluded there to a number of manuscript articles, but the published papers that eventually emerged did little to buttress his more extravagant claims.) The Cornell-based professor of linguistics and anthropology Charles Hockett, who had written several long essays on the relationship between animal communication systems and the evolution of human language, gave Lilly's book a terse and damning review in *American Anthropologist*, where he acknowledged that dolphins were fascinating but worried out loud that Lilly might have "dealt a body blow to the important program of research in which he wishes to play a part."[52] The most extensive critique appeared simultaneously, in *Natural History*, jointly authored by Margaret C. Tavolga (a specialist in animal behavior in the Biology Department at Fairleigh Dickinson University who had also worked with Wood's dolphins at Marine Studios) and her husband, William N. Tavolga (who, as it happened, had chaired the symposium on marine bioacoustics held at Lerner Marine Laboratory, in the Bahamas, in April 1963). The Tavolgas declared the book a primer for a young scientist seeking guidance on "how not to do scientific research," and they went after Lilly's "unsound and naïve grasp" of terms like "language," "speech," and "communication." Drawing on the very fields Lilly had disregarded, the two animal behavior researchers wrote,

> Dr. Lilly takes the view that "those who speak are those who are capable of learning language." If "one two three" said with very poor intelligibility by a dolphin is indicative of the giant-brained animal's ability to speak, and therefore to learn language, what is to be said of a parrot's clear-cut, if bird-brained, "Polly wants a cracker"? Furthermore, if the parrot is then given a cracker, have we established communication with an alien species?

In their assessment, the director of CRI needed to do some more thinking about "communication": "There is no doubt in the minds of most of us that dogs, dolphins, and many other animals are able to communicate by sounds or other means (to us or to each other) their relatively simple needs and wants, but this remains a far cry from language."[53]

Sex, Drugs, and Dolphins

As these assessments make clear, Lilly increasingly found himself on the defensive in the early 1960s. For instance, though he was one of the organizers of the First International Symposium on Cetacean Research in Washington, DC, in 1963—a watershed gathering of scientists studying whales and dolphins—Lilly's presentation at the gathering was marred by a shout-down kerfuffle in which his work at CRI (and before) was dismissed as trumped up, unethical, and possibly dishonest. His antagonist in this showdown was none other than Forrest Wood, the head of the fast-growing navy program. It was an indication of the way things were going: Lilly was losing friends fast by 1963. Lilly the star was about to become Lilly the shooting star—and his descending streak would trace a significant trajectory across the counterculture.

The arc of Lilly's fall can be concisely traced on a plane defined by two of the era's significant axes: sex and drugs. Let's do sex first. In 1961 *Man and Dolphin* presented the Saint Thomas laboratory of CRI as a kind of scientific *Swiss Family Robinson*, where Lilly and his beautiful wife and their six children (five from their combined previous marriages, one of their own) shared the labor of scientific life with bottlenose dolphins (this sort of thing was already a commonplace in primate work like that of Keith and Catharine Hayes). As the jacket copy on *Man and Dolphin* explained:

> A large fraction of the family home is taken up with the work of the program on the dolphins. The whole family participates: the elder boys have taken motion pictures of the animals in the experimental situations; the younger group have been involved in feeding and in swimming with the animals.
>
> Elisabeth handles a major portion of the administrative load of the institute, and helps with the research and writing.

But this scenario proved more advertising than reportage. Unhappy in their island fastness, Elisabeth and the children had decamped for Miami for good by the end of 1963, and the Saint Thomas lab promptly began its drift from the outer orbit of the military-industrial complex on to the hyperbolic arc of what would come to be called, in a general shorthand, "the sixties."

Absent the straitening forms of domesticity, the CRI facility in the Caribbean began to look less and less like an outtake from a wonkish version of *Flipper* and more and more like a bachelor crib for randy scientists seeking sun and surf.[54] The aging roué Aldous Huxley visited in these years, and Bateson (known for his rakish open shirts and nonconformist sensibility, and having recently embarked on a third marriage, to Lois Cammack, twenty-five years his junior) often presided in the dolphin pool, entertaining an irregular flow of the rich, brilliant, and/or curious. A youthful Carl Sagan made his way down from Harvard, having gotten to know Lilly through their mutual interest in "breaking through" to nonhuman species (he and Lilly together created a semisecret society of SETI researchers who called themselves "The Order of the Dolphins," wore a tie clip–like dolphin insignia, and sent each other coded messages to test their readiness for extraterrestrial contact).[55] Sagan thoroughly enjoyed himself and later claimed that while doing so he made the acquaintance of a young woman named Margaret Howe, then working at one of the resorts on the island.[56] Invited into the loose CRI community, the attractive and tomboyish Howe would eventually be hired by Bateson to "manage" the establishment when he decamped for Hawaii. She would play the crucial role in Lilly's program of "chronic-contact" experiments in 1964 and 1965—experiments (if this is the right term) that would appear prominently in *The Mind of the Dolphin*, Lilly's 1967 sequel to *Man and Dolphin* (fig. 1.4). The published account of this research—which narrates the way Howe, in

FIGURE 1.4 Margaret Howe in the flooded laboratory of CRI.

a skintight leotard, spent several weeks with a male *Tursiops* in a living facility flooded to thigh depth (her mouth brightly painted, to help the dolphin read her lips)—wends its way to a "climax" of sexual contact, offered as something like a proxy for increased interspecies mutuality.[57]

The book, quite possibly because of this frisson of aquatic bestiality, received more public attention than its weaknesses (of organization, clarity, and purpose) perhaps merited, but it was nevertheless not the sort of monograph from which a conventional scientific reputation could easily recover. Not only did the text entertain the notion that dolphins incline toward a "Polynesian" rather than an "American" model of sexuality, but it did so in the context of a sweeping indictment of the impoverishingly limited worldview of the human species, particularly its North American variety. The difficulties of communicating with dolphins are laid firmly at the feet of humanity, who are presented as a sickly lot, driven by fear and violence, inclined to deception and repression.[58]

Our only hope, Lilly suggested, was to listen to the dolphins, since (as he put it in an epigraph), "Through dolphins, we may see us as others see us."[59] There is a lingering sense throughout the book that dolphins offer humans a kind of higher psychotherapy.[60] Dredging up everything from Montaigne to William James to the Kinsey reports, Lilly presented sexual liberation and what he called "wet courage" as the preconditions for transcending the species barrier: "The purpose here is to free up one's own mind to see the new possibilities of feeling and thinking without the dry civilized structures."[61] The process amounted to a kind of ecstatic ethnography, one that made use of "hypnosis, drugs," and technology to conjure up the *Umwelt* of creatures that see with sound and can therefore (?) speak in images.[62] It was necessary to develop our own "latent 'acoustical-spatial thinking,'" Lilly asserted, and to this end he sketched the notion of "a dolphin suit with built-in, three-dimensional, sonic and ultrasonic emitters and receivers" and further imagined a fully synesthetic approach to the translation of dolphin phonation:[63]

> The internal picture which the dolphin can then create while sounding slash calls, the internal picture which he creates of his surroundings in terms of beat frequencies coming stereophonically combined from the two ears, must be a very interesting kind of picture. It is as if to us the nearby objects emitted a reddish light and the further objects emitted a bluish light, with the whole spectrum in between. We might see, for example, a red patch very close by and then a dimmer, blue patch in the distance farther away . . . a blue background downward symbolizing the bottom, a red patch up close meaning a fish nearby, and a large green object swimming between us on the bottom meaning another dolphin.[64]

Lilly—suspended in his flotation tank, listening to the whine of the hydrophones—was drifting into the trippy world of stereophonic psychedelia.

And trippy is meant here in the literal sense. By his own account Lilly first took LSD late in 1963, with Constance Tors (wife of Ivan, director of *Flipper*), on a visit to Los Angeles.[65] He wrote subsequently that this initial exposure consisted of a pair of "classic, high-energy" trips "filled with fantastic personal and transpersonal revelations and terrific intellectual breakthroughs."[66] A less fantastic trip followed (complete with an unhappy psychodrama of his failing second marriage), and then, in May 1964, shortly after addressing the annual meeting of the Acoustical

Society of America, a near-death experience linked to LSD experimentation.[67] Undaunted—indeed, it would seem, fascinated—Lilly apparently
secured additional LSD through his links to NIMH and began a series
of some twenty self-dosing experiments at the CRI laboratory in Saint
Thomas between 1964 and 1965.[68] These seem to have included at least
some injected exposures during which Lilly isolated himself in his flotation tank, Margaret Howe (who refused to try LSD with him) apparently serving as his "safety man [*sic*]." Sources suggest that by January
1965 Lilly was also injecting the dolphins with the drug, nominally to
test whether it had any effect on their vocalizations, though these experiments do not seem to have been closely controlled.[69] In this period he
drafted several versions of research protocols along these lines, which
served as applications to Sandoz Pharmaceuticals for relatively large
shipments of additional LSD-25.

I think it is important to acknowledge the difficulty of entirely freeing these aspects of Lilly's story from a slightly ludicrous taint: LSD?
To *dolphins*? And a woman in a *tank*, giving them *hand jobs*?[70] Having
presented lectures on this material on several occasions, I am familiar with the collective smirk by which an audience reflexively relegates
the endgame of the Lilly saga to the unproblematic category of period
burlesque, and I am myself not wholly immune to this reaction. Nevertheless, it is worth making the effort to see these most extravagant
aspects of Lilly's work without the distorting glaze of hindsight. In the
early 1960s there were quite a few researchers in biology and psychology departments giving LSD-25 to various animals and observing the
effects.[71] There were also, of course, formal experimental investigations
with the drug that made use of human subjects, and self-dosing was
considered by many a necessary preparation for therapeutic or investi-
gative prescription. Furthermore, as I have suggested above, both Lilly's
LSD experiments and the chronic-contact experiments with Howe can
be understood not as madcap "1960s-style" *antitheses* to the buttoned-
down world of military bioscience but rather as their very apotheosis.
LSD was a notable tool in the kit of Cold War scientists of mind and
behavior, and they understood it to be an instrument for reducing the
inhibitions of those with whom they wanted to talk. Lilly wanted to
talk with dolphins; therefore, LSD presented itself as a very plausible
approach. Similarly, "chronic contact" was a recognized technique for
"winning over" recalcitrant or taciturn persons of interest, and careful
management of erotic potentials was a nontrivial element of some of
these protocols.[72]

These contextualizing observations aside, there can be no doubt that

Lilly crossed a set of lines between 1963 and 1966. Were they invisible lines? Perhaps, though a proper answer to this question would thrust us firmly into the middle of some very serious and difficult historical debates about the cultural upheavals of the second half of the 1960s—debates that on the whole have yet to extricate themselves from the political stalemates the era itself did much to define. We do not need to resolve those thorny matters to know that things were changing quickly in these years and that the transgressive passages of figures like Lilly were doing much to clarify new boundaries. Timothy Leary, of course, is paradigmatic here, and his own story parallels (and ultimately intersects) that of Lilly in interesting ways.[73] It was in the spring of 1963, right about the time that Lilly was getting ready to participate in the First International Symposium on Cetacean Research, that Leary and Richard Alpert (later Ram Dass) were dismissed from Harvard under a cloud of opprobrium that had gathered over what appeared to be their overzealous promotion of LSD experiences. Legal restrictions on the drug followed, and by 1966 LSD-25 was so tightly regulated in the United States that there were effectively no legitimate research programs making use of the product. By that time Sandoz had issued a recall for outstanding orders of the drug, and a rising district attorney named G. Gordon Liddy was working overtime to stamp out the Leary-centered experiment in drug-addled alternative living centered at Millbrook, in upstate New York, subject of Tom Wolfe's *The Electric Kool-Aid Acid Test*. Leary himself would soon be jailed. In one sense, the party was over. In another sense, it was just getting going.

And that sums up Lilly's position by 1967 as well. Mounting scandal and critical scientific assessments of CRI's work by increasingly hostile peers (including Kenneth Norris and others in attendance at the First International Symposium on Cetacean Research) backed Lilly into a defensive posture and left him lashing out at his opponents.[74] But to no avail. With the emerging revelations about his unconventional experiments and no stream of peer-reviewed publications to back up his showboating public claims, his allies could no longer defend him. With his NSF monies soon terminated and other federal agencies requesting the return of loaned equipment, financial matters at CRI reached a critical point. By the end of the year, Lilly had been forced to shutter both the Saint Thomas and Miami laboratories, though not before five of his captive dolphins died (Lilly called it the result of a "hunger strike" on their part; neglect seems more likely); three others were later released into Miami Harbor. By May 1968 Lilly, separated from Elisabeth (they would later divorce), wrote to his literary agent in New York that he was "looking for a job." He headed out to the West Coast, for mind-expanding peripatetic trajectory from Berkeley to Esalen and eventually well beyond; he and Leary would meet up on this trip.[75] Interestingly, as he was winging into the California sunset, Lilly stopped by the navy's Point Mugu dolphin laboratory for a quick lunch with an old friend who was working on a vocoder that could "translate" human speech into "dolphinidese" (and vice versa)—Lilly and the navy had both been working on such devices for several years.[76] Wood, discovering him on base, had a cow.[77]

Conclusion

Wood's reaction speaks volumes. He and Lilly were emphatically headed in different directions in May 1968. And in this sense their confrontation back in August 1963, when they traded barbs at the First International Symposium on Cetacean Research, must be understood as the first cracking sound of what rapidly opened into a splitting fissure in cetacean research in the United States in a critical decade. Tectonic plates ground menacingly under that local flare-up of scientific infighting: Lilly, increasingly preoccupied in the mid-1960s with erotic and ecstatic exploration of his own mind and those of his animal subjects, eventually came to feel that the dolphins—sexually liberated, stereophonic, nonmanipulative superintelligences—were leading him to a new kind of self. Turning on, tuning in, and effectively dropping out, John Lilly left the world that made him—the world of the Cold War biosciences—behind and went on to become a major-minor figure of the pacifist, drug-friendly, ecosensitized counterculture. Wood, by contrast, still clipped his hair close, and his well-trained dolphins had serial numbers. They, too, possessed remarkable abilities, but about them he and his fellow scientists and trainers could not speak.

It must be said that the parting of their ways has about it the air of a fairy tale for the era—which is in fact what it promptly became. Right after the First International Symposium on Cetacean Research, Lilly entertained four of the European participants down at CRI in Miami: the two whale ear specialists Reysenbach de Haan and Peter Purves; the Dutchman Dudok van Heel, who was participating in a new effort to keep captive porpoises in the Netherlands; and the French bioacoustics expert René-Guy Busnel. Did conversation among this rump of the symposium in Miami return to Lilly's showdown with Wood? There is no way to be sure, yet it would appear that Busnel, at least, returned to Europe with an acute sense of what the final scene in Washington represented and a view of the matter quite sympathetic to Lilly. Back in France, Busnel would be in communication with the French novelist and English professor Robert Merle (a recent winner of the Prix Goncourt), and in this period Merle would turn his hand to a dramatic tale that pitted a Lilly-like character (named "Dr. Sevilla") against a navy porpoise-training establishment under the direction of a Wood-like "M. D. Morley"; intelligent dolphins are caught in the middle, and World War III is narrowly averted. Published in 1967 in French as *Un animal doué de raison*, and two years later in English as *Day of the Dolphin*, this thriller would go on to become a major motion picture starring George C. Scott. It would, in a way, immortalize the conflict between Lilly and Wood, fixing it as a confrontation between the forces of peace and war, eros and thanatos, as nothing less than a showdown between the age of Aquarius and the age of the hydrogen bomb.[78]

Tremors along this strange new fault line would continue for years, and in the smoke and rattle (the miasma of the Vietnam conflict, the mounting quakes of antiwar protest, and increasingly truculent environmentalism), dolphins, and their larger cousins, would come to symbolize the aspirations of many who hoped to defy the cultures of death. And it was Lilly who had led them, John C. Lilly—the Eisenhower-era, pocket-protector-wearing, right-stuff engineer who emerged from the complex decade that separated 1958 from 1968 as a guru-sage of countercultural enlightenment.

Acknowledgments

The material in this essay is adapted from chapter 6 of *The Sounding of the Whale: Science and Cetaceans in the Twentieth Century* (Chicago: University of Chicago Press, 2012).

Notes

1. Lilly, cited in Joan McIntyre, comp., *Mind in the Waters: A Book to Celebrate the Consciousness of Whales and Dolphins* (New York: Charles Scribner's Sons / Sierra Book Club, 1974), 83.

2. This work was done through the E. R. Johnson Foundation for Medical Physics, which was run by Bronk, and which had contracts with the army and navy air forces through the Committee on Medical Research of the Office of Scientific Research and Development. Interestingly, in light of Lilly's later work on underwater breathing masks at the National Institute of Mental Health, these gas-monitoring technologies were applied, among other things, to detect mask leakage. For a discussion of the nitrogen meter that Lilly ap-

parently helped to develop, see "Curriculum Vitae, John Cunningham Lilly, M.D., 1968," p. 2, file "CRI personnel," box 3C2[D1], Lilly Papers, Stanford University Library (hereafter cited as Lilly Papers). See also John C. Lilly and Thomas F. Anderson, "The Nitrogen Meter: An Instrument for Continuously Recording the Concentration of Nitrogen in Gas Mixtures," Report 299, 28 February 1944, Division of Medical Science, Acting for the Committee on Medical Research of the Office of Scientific Research and Development, Committee on Aviation Medicine, National Research Council. This device used photoelectric monitoring.

3. It is important to emphasize the novelty of this sort of work in the period. It was in 1954 that Olds and Milner demonstrated that a rat could learn to stimulate its own brain, and later investigations by Delgado and others demonstrated similar behavior in cats, as well as the reverse—namely, learning to turn off a current that apparently caused pain/fear/discomfort.

4. John C. Lilly, "Learning Motivated by Subcortical Stimulation: The Start and Stop Patterns of Behavior," in *Reticular Formation of the Brain*, ed. Herbert H. Jasper et al. (Boston: Little, Brown, 1958), 705.

5. Ibid., 719.

6. Lilly recounts this story in *Man and Dolphin* (Garden City, NY: Doubleday, 1961), 40–47.

7. Much can be learned about Scholander's work from his autobiography, *Enjoying a Life in Science: The Autobiography of P. F. Scholander* (Fairbanks: University of Alaska Press, 1990), and from a shorter memoir published earlier, "Rhapsody in Science," *Annual Review of Physiology* 40 (1978): 1–17. In 1963 Scholander attended the First International Symposium on Cetacean Research. His papers are held at Scripps Institution of Oceanography. I have consulted these holdings (five boxes), which contain some interesting material on his work with whales and dolphins, including a set of photographs depicting his visit to Brødrene Saebjørnsen's whaling station in Steinshamn, Norway, in the 1930s. These papers also contain a folder of his notes on the hydrodynamics of dolphin bow riding, work that resulted in a pair of articles in *Science* in 1959: "Wave-Riding Dolphins: How Do They Do It?," *Science* 129, no. 3356 (24 April 1959): 1085–87; and, with Wallace D. Hayes, "Wave-Riding Dolphins," *Science* 130, no. 3389 (11 December 1959): 1657–58.

8. The most substantial early piece of this work was the monograph published in 1940 in *Hvalrådets skrifter*: P. F. Scholander, "Experimental Investigations on the Respiratory Function in Diving Mammals and Birds," *Hvalrådets skrifter* 22 (1940): 5–131. I write about this work in "Self-Recording Seas," in *Oceanomania: Souvenirs of Mysterious Seas*, ed. Mark Dion and Sarina Basta (London: Michael Mack, forthcoming). A valuable discussion of Scholander's research in this area, along with a full bibliography, can be found in John W. Kanwisher and Gunnar Sundnes, eds., *Essays in Marine Physiology, Presented to P. F. Scholander in Honor of His Sixtieth Birthday*, *Hvalrådets skrifter* 48 (Oslo: Universitetsforlaget, 1965). Some of the early experiments involved the use of pressure gauges affixed to whaling harpoons.

9. The investigators, in addition to Lilly, were J. Rose, V. Mountcastle, and L. Kruger from Johns Hopkins Medical School; C. Woolsey and J. Hind,

University of Wisconsin; Karl Pribam, Institute for Living, Hartford, CT; and Leonard Malis, Mount Sinai Hospital. The full records of this work can be found in box 6A1–B1, Lilly Papers.

10. Ibid.

11. As early as 1940 Scholander had done respiratory analysis on several restrained and submerged *Phocoena*.

12. Ibid. These records include minute-by-minute logs of each operation and phonograph disks recording the interactions of the scientists during each intervention. Given the broad disagreements that erupted later over this work, closer attention to these materials might prove interesting.

13. The 1955 investigations also set in motion the research that would lead, almost a decade later, to the first successful techniques for major surgery on the small whales. See E. L. Nagel, P. J. Morgane, and W. L. McFarland, "Anesthesia for the Bottlenose Dolphin, *Tursiops truncatus*," *Science* 146, no. 3651 (18 December 1964): 1591–93.

14. See John Cunningham Lilly, John R. Hughes, Ellsworth C. Alvord Jr., and Thelma W. Galkin, "Brief, Non-injurious Electric Waveform for Stimulation of the Brain," *Science* 121, no. 3144 (1 April 1955): 468–69; and John C. Lilly, "Electrode and Cannulae Implantation in the Brain by a Simple Percutaneous Method," *Science* 127, no. 3307 (16 May 1958): 1181–82. Note that Lilly alleged that Schevill and Lawrence were working in a navy facility; William Watkins, a navy bioacoustics researcher close with both men (interview by the author, 9 August 2003), insisted that the work was done in a private pool on Nonamesset Island, owned by the Forbes family.

15. It was also on these trips that Lilly became interested, through Wood, in the apparent ability of these animals to control the direction of their sound. Using an early AMPEX stereo tape recorder (on loan), Lilly and Wood were able to hear clearly that the click trains emitted by captive dolphins had directional specificity. Wood discusses this finding in his *Marine Mammals and Man: The Navy's Porpoises and Sea Lions* (Washington, DC: Robert B. Luce, 1973). See also Gregg Mitman, *Reel Nature: America's Romance with Wildlife on Film* (Cambridge, MA: Harvard University Press, 1999), 248.

16. This was done by means of a switch, placed within reach of the animal's beak. While I have never seen a reference to this problem, it must be asked whether contact with the switch could have been a product of convulsions and/or efforts by the animal to escape its constraints. Lilly's repeated emphasis on the "purposive" could perhaps be read as special pleading.

17. This issue of the "distress call" was central to later disputes; trainers and animal handlers were well aware of "epimeletic," or caregiving, behavior among these animals. Wood, and before him McBride and Hebb, had raised the subject of the "language" value—"language in the sense that a dog's barking or growling is a language"—of these whistlings. See F. G. Wood, "Underwater Sound Production and Concurrent Behavior of Captive Porpoises, *Tursiops truncatus* and *Stenella plagiodon*," *Bulletin of Marine Science of the Gulf and Caribbean* 3, no. 2 (March 1953): 124–25.

18. McIntyre, *Mind in the Waters*, 71.

19. Lawrence S. Kubie to Lilly, 31 July 1961, file "Kubie, Lawrence S.," box 3A1–C1, Lilly Papers.

20. Loren Eiseley, "The Long Loneliness: Man and the Porpoise; The Solitary Destinies," *American Scholar* 30, no. 1 (Winter 1960–61): 58.

21. For a discussion of the flurry of headlines prompted by the original presentation (at the San Francisco meeting of the American Psychiatric Association in May 1958), see Wood, *Marine Mammals and Man*, 3, 12n1. William Evans also takes up some of this publicity in *Fifty Years of Flukes and Flippers: A Little History and Personal Adventures with Dolphins, Whales, and Sea Lions, 1958–2007* (Sofia, Bulgaria: Pensoft, 2008).

22. John C. Lilly, "Some Considerations regarding Basic Mechanisms of Positive and Negative Types of Motivations," *American Journal of Psychiatry* 115 (1958): 499. Note that, beginning in the 1940s, Lilly "undertook psychoanalytic training as a student in the Philadelphia Association for Psychoanalysis," where he worked with Robert Waelder, a student of Freud's. (See Lilly typescript 1968, file "CRI personnel," box 3C2[D1], Lilly Papers.) Lilly invested eight years in psychoanalysis, a period that overlapped with his neurophysiological and isolation studies. Exactly what role these experiences may have played in the ease with which he moved from electrophysiology of the brain to questions of personality is worth consideration.

23. Lilly, "Some Considerations regarding Basic Mechanisms of Positive and Negative Types of Motivations," 501.

24. Ibid.

25. The best review of this extraordinary research enterprise is John P. Zubek, ed., *Sensory Deprivation: Fifteen Years of Research* (New York: Appleton-Century-Crofts, 1969). See also Leo Goldberger, review of *Sensory Deprivation*, ed. John P. Zubek, *Science* 168, no. 3932 (8 May 1970): 709–11.

26. See Zubek, *Sensory Deprivation*, 9, for a discussion.

27. Relevant material in these files includes a typescript by Robert J. Lifton (in the neuropsychiatry division of the Walter Reed Army Institute of Research) entitled "Chinese Communist 'Thought Reform': 'Confession' and 'Re-education' in Penal Institutions," and an essay by Edgar H. Schein (at the Army Medical Service Graduate School) entitled "Chinese Brainwashing." Published materials represented as clippings include several *New York Times* pieces, including "New Evils Seen in Brain Washing," 4 September 1956, and "Two Challenge Views on Brainwashing," 22 September 1956. See files "Solitude" and "Indoctrination, Forced," box 5A1, Lilly Papers.

28. Sensory deprivation screening was used, for instance, in the selection of the astronauts for Project Mercury. In the Lilly Papers, I discovered that he was for several years in this period a dues-paying member of the Slocum Society (founded in 1955) and received its newsletter. Joshua Slocum (1844–1909), a New England captain, became an international celebrity at the turn of the century after he successfully completed a solo circumnavigation in 1895–98, the first such exploit recorded. He was lost at sea, alone, in 1909. His popular book on his successful voyage, *Sailing Alone around the World* (New York: Century, 1900), recounts several hallucinatory intervals during long crossings

(though it is difficult to assess the tone of these passages, which have a comical quality). The Slocum Society was founded (perhaps paradoxically) to create a community of solitaires, particularly those dedicated to long solo voyages.

29. See Francis Jeffrey and John Cunningham Lilly, *John Lilly, So Far . . .* (Los Angeles: Jeremy P. Tarcher, 1990), 82–100. See also a brief discussion of Lilly's situation and CIA interest in his work in (the more reliable) John D. Marks, *The Search for the "Manchurian Candidate": The CIA and Mind Control* (New York: Times Books, 1978).

30. "Dr. Lilly's problem concerned a meeting held at the Pentagon in May 1959. This meeting was called in order that ranking officers of the Office of Naval Research, the Air Force, and the Army could hear a briefing by Dr. Lilly on his work on the brain of dolphins, Dr. Lilly explained that the military was interested in this field [TEXT CENSORED] inasmuch as research by himself and other scientists had established that by the use of electrodes placed in the brains of animals and humans the will could be controlled by an outside force. He explained that if an electrode were placed in the brain of a subject. He [*sic*] could make the subject experience great extremes of joy or depression, for example. Dr. Lilly stated that the potential of this technique in 'brain washing' or interrogation or in the field of controlling the actions of humans and animals is almost limitless. He stated that our officials are aware that the Soviets are intensely interested in this field and that they are conducting extensive experiments and that their progress has roughly paralleled that of ours." "John Cunningham Lilly (Dr.)," memorandum, Jones to DeLoach, p. 2, FBI personal file. Deletion in category b2, "solely related to the internal personnel rules and practices of an agency."

31. Ibid., p. 4. Lilly apparently found himself caught in a wrangle between the security establishments of the Defense Department and the FBI in late 1959. Having been asked to leave a Pentagon briefing (as noted above) because of a security "problem" with his clearance, Lilly made a set of inquiries and learned (from an unnamed informant) that the problem had originated with the FBI. He followed up, only to be told that this was not the case, and that the FBI wanted to know who had told him this. He refused to divulge his source, despite several visits from agents, both in Miami and in San Juan. The FBI appears to have learned the identity of the source on its own and marked Lilly as "uncooperative."

32. This and the quotations above are from Lilly, *Man and Dolphin*, 190–91. Similar discussions of the importance of "isolation" and "confinement" run through Lilly's early articles on dolphins in *Science*. See, e.g., John C. Lilly and Alice M. Miller, "Vocal Exchanges between Dolphins," *Science* 134, no. 3493 (8 December 1961): 1873–76.

33. See file "Sandoz," box 3D2, Lilly Papers. On Artichoke, see Marks, *Search for the "Manchurian Candidate."* More generally on the history of LSD in this period, see Jay Stevens, *Storming Heaven: LSD and the American Dream* (New York: Grove Press, 1998).

34. The edition of Cousteau's *The Silent World* is the fourth printing, from 1961, so Lilly must have read the book in that year or later. In 1961 Lilly and Jay Shurley published their essay "Experiments in Maximum Achievable Physi-

cal Isolation with Water Suspension of Intact Healthy Persons" in *Psycho-physiological Aspects of Space Flight*, ed. B. E. Flaherty (New York: Columbia University Press, 1961), 238–47. It was Lilly's last article in this area. I think it likely that the annotations in *Silent World* were made in that year, particularly as several of them deal with the rubber "Furney goggles" Cousteau describes. Lilly and Shurley discussed the form of similar latex masks in their correspondence concerning the flotation tank. Other annotations to the text include small marks next to "I turned over and hung on my back" (p. 5), and "As we submerged, the water liberated us from weight" (p. 78). For a discussion of the film version of *The Silent World*, as well as a brief treatment of Cousteau's broader importance in the growing American fascination with the undersea world in the 1950s, see Mitman, *Reel Nature*.

35. "In the course of some experiments I conducted from 1954 through 1956 I was suspended in water for several hours at a time, and I noticed that my skin gradually became more and more sensitive to tactile stimuli and an intense sense of pleasure resulted. However, if the stimulation was carried too far it became intensely irritating, I reasoned that the dolphin is suspended in water all of his life, twenty four hours a day, and possibly had developed an intensely sensitive skin" (Lilly, *Man and Dolphin*, 172). The issue of "commensurating" with the dolphin appears in many places in Lilly's published and unpublished work; see, e.g., ibid., 209. Mitman has explored this idea of experimental commensuration in the study of animal behavior in "Pachyderm Personalities: The Media of Science, Politics and Conservation," in *Thinking with Animals: New Perspectives on Anthropomorphism*, ed. Gregg Mitman and Lorraine Daston (New York: Columbia University Press, 2005), 175–95. Mitman and Daston together take up the problem in their introduction to this same volume.

36. There remains yet another link between sensory-deprivation research and dolphin study in this period: as it turns out, the pioneering figure in sensory-deprivation work in 1951 was Professor D. O. Hebb, at McGill University. Hebb, who had a simultaneous appointment at the Yerkes Laboratory of Primate Biology in Orange Park, Florida, was the very same Hebb who coauthored the foundational 1948 article (with Arthur McBride) "Behavior of the Captive Bottle-Nose Dolphin, *Tursiops truncatus*," *Journal of Comparative Physiological Psychology* 41 (1948): 111–23. This was really the first scientific paper to document behavioral observations on the captive marine mammals of the recently reopened Marine Studios. How Hebb, too, bridged the universes of sensory deprivation and cetology is not absolutely clear. One possibility is that work with captive primates in this period encouraged exploration of the behavioral ramifications of prolonged isolation and boredom, since monkeys respond rapidly and markedly to these conditions; this observation might explain both Lilly's and Hebb's early curiosity. Of the unsavory aspects of some of Hebb's other Cold War work there can be little doubt; see, e.g., Alfred W. McCoy, "Science in Dachau's Shadow: Hebb, Beecher, and the Development of CIA Psychological Torture and Modern Medical Ethics," *Journal of the History of the Behavioral Sciences* 43, no. 4 (Fall 2007): 401–17. I have found no evidence of a link between Lilly and Hebb, but it seems likely they knew each other through sensory-deprivation work, and it is surely possible that Hebb stimu-

lated Lilly's early dolphin interests. Whatever the case may be, the unlikely ties between isolation studies and dolphin studies in this period demand a revised reading of Loren Eiseley's curious and moving essay on Lilly's work, tellingly entitled "The Long Loneliness: Man and the Porpoise; The Solitary Destinies."

37. "Behavioral Biology Program—Biosciences Programs—Office of Space Sciences," sec. 3, file "Reynolds, Dr. Orr. E., NASA," box 3D2, Lilly Papers.

38. A version of the story was circulated among Soviet and American scientists at the Pugwash Conference in 1960.

39. See Barton J. Bernstein's introduction to Leo Szilard, *The Voice of the Dolphins, and Other Stories*, Nuclear Age Series (Stanford, CA: Stanford University Press, 1992), 4 ("minor classic"). See also William Lanouette and Bela A. Silard, *Genius in the Shadows: A Biography of Leo Szilard, the Man behind the Bomb* (New York: Charles Scribner's Sons, 1992). Lanouette interviewed Lilly before his death about his interactions with Szilard.

40. Lilly, *Man and Dolphin*, 217. Consider, as context, Pierre Boulle's *La planète des singes* (Paris: Juilliard, 1963) and its reception in the United States. For a study of the way race played out in the American film world, see Eric Greene, *Planet of the Apes as American Myth: Race and Politics in the Films and Television Series* (Jefferson, NC: McFarland, 1996). By 1965 Lilly was having his CRI staff read and comment on an English translation of Boulle's novel. There is a larger story to be told about the relationship between Lilly's work and the world of science fiction in these years.

41. Lilly, *Man and Dolphin*, 223.

42. Bryan P. Glass, review of *Man and Dolphin*, by John C. Lilly, *Quarterly Review of Biology* 36, no. 4 (December 1961): 311. See also "He Barks and Buzzes, He Ticks and Whistles, but Can the Dolphin Learn to Talk?," *Life* 51, no. 4 (28 July 1961): 61–66; Archie Carr, "Have We Been Ignoring a Deep Thinker?," review of *Man and Dolphin*, by John C. Lilly, *New York Times Book Review*, 3 September 1961, 3; Ted Hughes, "Man and Superbeast," review of *The Nerve of Some Animals*, by Robert Froman, and *Man and Dolphin*, by John C. Lilly, *New Statesman* 53, no. 1619 (23 March 1962): 420–21; B. A. Young, "Placid and Self-Contained," review of *Man and Dolphin*, by John C. Lilly, and *The Nerve of Some Animals*, by Robert Froman, *Punch*, 14 March 1962, 443; unsigned review of *Man and Dolphin*, by John C. Lilly, *New Yorker*, 16 September 1961, 178; and Robert C. Cowen, "Can We Converse?," review of *Man and Dolphin*, by John C. Lilly, *Dolphins: The Myth and the Mammal*, by Antony Alpers, and *Porpoises and Sonar*, by Winthrop N. Kellogg, *Christian Science Monitor*, 14 December 1961, 11.

43. For a taste of this, see Lilly's file of correspondence with Dr. Janet Travell, box 3D2 (where there is also a file of replies to and letters on *Man and Dolphin*), Lilly Papers.

44. "The Programs of the Communication Research Institute," typescript, file "Worcester Foundation," box 3D2, Lilly Papers.

45. Lilly, *Man and Dolphin*, 219.

46. In fact, a newspaper article in the *Staten Island Advance* of 13 March 1962 had already "broken" the story that the navy was working with dolphins at Point Mugu.

47. See, e.g., Blair Irvine, "Conditioning Marine Mammals to Work in the Sea," *Marine Technology Journal* 4, no. 3 (1970): 47–52. See also Sam H. Ridgway, *Dolphin Doctor: A Pioneering Veterinarian and Scientist Remembers the Extraordinary Dolphin That Inspired His Career* (San Diego, CA: Dolphin Science Press, 1987); and Wood, *Marine Mammals and Man*. The actual deployment in Vietnam (a seven-month tour of duty, apparently not very successful) remains clouded in rumor. The minimal (but presumably reliable) information available appears in a historical essay by the public affairs officer at the Space and Naval Warfare Systems Center in San Diego, Tom LaPuzza, "SSC San Diego Historical Overview," accessed 3 March 2011, http://www .spawar.navy.mil/sandiego/anniversary.

48. Memorandum of 17 October 1961, file "China Lake—Dr. W. B. McLean," box 3D2, Lilly Papers.

49. T. W. Milburn to Lilly, 10 April 1962, in ibid. Note that Milburn, who would later write about the psychological dimensions of deterrence, appears to have attended the First International Symposium on Cetacean Research; in other correspondence he expressed interest in Lilly's isolation tanks. Project Michelson was an integrated research program on strategic deterrence. For a sense of some of the concerns, see Louis D. Higgs and Robert G. Weinland, *Project Michelson Preliminary Report*, Technical Progress Report 309 (China Lake, CA: US Naval Ordnance Test Station, 1963).

50. File "China Lake—Dr. W. B. McLean," box 3D2, Lilly Papers. There is, among these materials, a reference to additional documents in a "vault"; I was not able to locate those materials in the Lilly Papers, and it seems likely they were destroyed. Padberg was among the attendees of the Lerner Marine Laboratory bioacoustics symposium held in April 1963.

51. This and the quotation above are from James W. Atz, review of *Man and Dolphin*, by John C. Lilly, *Animal Kingdom: Bulletin of the New York Zoological Society* 64, no. 6 (December 1961): 190. Atz was an observer at the Washington symposium in 1963. See Symposium Program, folder 46, box 54, Hubbs Papers, Scripps Archive.

52. Charles F. Hockett, review of *Man and Dolphin*, by John C. Lilly, *American Anthropologist* 65, no. 1 (February 1963): 176–77. Hockett was the author of "Logical Considerations in the Study of Animal Communication," an invited conclusion to *Animal Sounds and Communication*, ed. W. E. Lanyon and William N. Tavolga (Washington, DC: American Institute of Biological Sciences, 1960), 392–430.

53. This and the quotations above are from Margaret C. Tavolga and William N. Tavolga, review of *Man and Dolphin*, by John C. Lilly, *Natural History* 71, no. 1 (January 1962): 7.

54. It is interesting to think about this transformation in the context of a growing literature on the question of domesticity and the laboratory. I have been inspired here in part by Deborah Harkness, "Managing an Experimental Household: The Dees of Mortlake and the Practice of Experimental Philosophy," *Isis* 88, no. 2 (1997): 247–62.

55. This would be wonderful material to pursue, and a point of departure would be the file entitled "Order of the Dolphins" in box 3C2[D1], Lilly

Papers. This group (born of the first Green Bank conference in November 1961) eventually included most of the leading figures in what would become the discipline of exobiology (Frank Drake, Melvin Calvin, J. B. S. Haldane, and others). These gentlemen (and a few women, too) entertained each other by circulating encoded messages like those "that might be received from another civilization in space" and generally mused about the possibilities for extraterrestrial life. For some context on all this, consider James E. Strick, "Creating a Cosmic Discipline: The Crystallization and Consolidation of Exobiology, 1957–1973," *Journal of the History of Biology* 37, no. 1 (2004): 131–80; and Steven J. Dick, *The Biological Universe: The Twentieth Century Extraterrestrial Life Debate and the Limits of Science* (Cambridge: Cambridge University Press, 1999).

56. Lilly himself says that Bateson "discovered" Howe (Jeffrey and Lilly, *John Lilly, So Far . . .* , 118), but Sagan's role is outlined in William Poundstone's biography *Carl Sagan: A Life in the Cosmos* (New York: Henry Holt, 1999). Margaret Howe Lovatt denies Poundstone's account (which is based, according to Poundstone, on Sagan's own writings and, Poundstone says, confirmed by an interview with Lilly) and says that she heard about the CRI work while working at a hotel on the island and went there on her own initiative, where it was Bateson who generously folded her into the research program on a volunteer basis. Margaret Howe Lovatt, interview by the author, 25 August 2009.

57. See chap. 14 of John C. Lilly, *The Mind of the Dolphin: A Nonhuman Intelligence* (Garden City, NY: Doubleday, 1967). It is interesting to note that Lilly had Howe read *Planet of the Apes* (by Pierre Boulle, trans. Xan Fielding [New York: Vanguard Press, 1963]) to prepare her for her chronic-contact work. This fact, as well as the observations about lipstick and so forth, can be confirmed in the remarkable manuscript files of this work in file "1965, St. Thomas," box 6A1–B1, Lilly Papers.

58. Lilly, *Mind of the Dolphin*, 128.

59. Ibid., xvii.

60. In its original formulation, this idea came from Bateson, who wrote to Lilly shortly after reading *Man and Dolphin* to propose that, if Lilly was right about dolphin intelligence, there was reason to think that these animals had evolved to apply the bulk of their cognitive capacity to the social world rather than to the material world (roughly speaking, because they had no hands). Bateson suggested that this might mean that the dolphins would make, if we could speak to them, ideal psychotherapists for humanity, so obscenely obsessed with things and so inept in relationships. See Bateson to Lilly, 16 October 1961, file of replies to and letters on *Man and Dolphin*, box 3D2, Lilly Papers.

61. Lilly, *Mind of the Dolphin*, 170.

62. Ibid., 91.

63. Ibid., 135.

64. Ibid., 152.

65. Lilly had Hollywood links. In 1961 the glamorous actress Celeste Holm, fascinated by news reports of Lilly and his talking dolphins, sought

him out while performing at the neighboring Coconut Grove. Later, one of her sons, Ted Nelson, would spend a year working at CRI, before becoming one of the leading figures at the intersection of information technology and the counterculture. For these links, see Lawrence S. Kubie to Elisabeth and John Lilly, 14 September 1961, and Elisabeth Lilly to L. S. Kubie, 6 December 1962, file "Kubie, Lawrence S.," box 3A1–C1, Lilly Papers.

66. Jeffrey and Lilly, *John Lilly, So Far . . .* , 134.

67. In his autobiography (ibid., 135) this episode (which involved a coma, hospitalization, and, apparently, some small permanent damage to Lilly's eyesight) is blamed on an improperly washed syringe.

68. The number twenty is Lilly's own (ibid., 139); Margaret Howe Lovatt recalls many fewer (interview by the author, 25 August 2009).

69. Margaret Howe Lovatt recalls one occasion on which Lilly used a jackhammer on the rock wall of the pond in which an LSD-dosed dolphin was swimming, apparently to try to get a rise out of the animal, which was otherwise not behaving in a particularly striking way (interview by the author, 25 August 2009). Lilly himself cites a "project report" on this work that I have not been able to find in its original form (*The Human Biocomputer: Programming and Metaprogramming [Theory and Experiments with LSD-25]*, Scientific Report CRI0167 [Miami: CRI, 1967]), but it is reasonable to assume that much, if not all, of the content of this document appears in Lilly's later published writings under essentially the same title: *Programming and Metaprogramming in the Human Biocomputer: Theory and Experiments*, available in a second edition (New York: Three Rivers Press, 1987).

70. There is, of course, a fictionalized version of these events: Ted Mooney's novel *Easy Travel to Other Planets* (New York: Farrar, Straus, and Giroux, 1981).

71. Lilly corresponded with some of these scientists. See H. A. Abramson to Lilly, 23 November 1964, file "Abramson, H. A.," box 3C2[D1], Lilly Papers.

72. It should be noted that Margaret Howe Lovatt claims that she herself was the primary proponent of the chronic-contact work (interview by the author, 25 August 2009).

73. Lilly followed the Leary story with some care. See Lilly to Frederick G. Worden, 12 December 1967, with clipping, file "Worden, Frederick G., Brain Research Institute, Los Angeles, California," box 3D2, Lilly Papers.

74. Consider the suite of very hostile peer reviews he received, in file "National Institutes of Health," box 3D2–D1, Lilly Papers. Also relevant is Scott McVay's account of overhearing Norris's dismissal of the work of CRI during an on-site evaluation circa 1964: "You see what sort of cockamamie stuff is going on here?" Scott McVay, interview by the author, 7 July 2003.

75. Lilly to Peter Matson, 2 May 1968, file "Matson, Mr. Peter H.," box 3A1–C1, Lilly Papers.

76. There is a history to be done on these devices, which were the holy grail of dolphin research in the 1960s. One of them features prominently in the navy film *Dolphins That Joined the Navy* of 1964, in which a navy researcher is shown speaking *Hawaiian* into the converter device. It was apparently believed that this language was particularly well suited to dolphin

communication. For a technical account of both the (unclassified) navy work and that of Lilly, see the cover story by Richard Einhorn, "Dolphins Challenge the Designer," *Electronic Design* 15, no. 25 (6 December 1967): 49–64. Also useful is the discussion by Wood in chap. 5 of his *Marine Mammals and Man*. The great controversy here, at least among conspiracy theorists, involves the untimely death of the navy's main researcher on this vocoder project, Dwight W. Batteau. I bought (from a collector of such things) a copy of the one-hundred-plus-page report filed by Batteau's collaborator Peter R. Markey shortly after his partner was found dead in a lagoon in Hawaii: Batteau and Markey, "Man/Dolphin Communication: Final Report, 15 December 1966–13 December 1967, Prepared for U.S. Naval Ordnance Test Station, China Lake, California, Contract No. N00123-67-C-1103, Listening, Incorporated, 6 Garden Street, Arlington, MA." For a somewhat-histrionic whirl through the history of the vocoder, consider Dave Tompkins, *How to Wreck a Nice Beach: The Vocoder from World War II to Hip Hop; The Machine Speaks* (New York: Stop Smiling, 2010).

77. This is Lilly's account: Jeffrey and Lilly, *John Lilly, So Far . . .* , 150.

78. NB: Robert Merle's *The Day of the Dolphin*, trans. Helen Weaver (New York: Simon and Schuster, 1969), is dedicated to Busnel. For an expanded sense of the way this split in the field looked to a participant, consider the memoir of one of the trainer-divers who became increasingly radicalized across these years: Richard O'Barry and Keith Coulbourn, *Behind the Dolphin Smile* (Chapel Hill, NC: Algonquin Books, 1988).

2 Blowing Foam and Blowing Minds: Better Surfing through Chemistry

Peter Neushul and Peter Westwick

Surfing seems the archetype of a counterculture. It appears to offer no productive benefit to society, besides the fun it affords the individual rider. Surfing's countercultural image defined the groovy sixties and seventies for many commentators. Alvin Toffler, in *Future Shock*, famously declared that surfers were "a signpost pointing to the future."[1] To limn sixties youth, Tom Wolfe insinuated himself with the Pump House Gang at Windansea, where insolent teenagers mocking the squares perfectly expressed the generation gap.[2] Surfers were even cooler than Wolfe knew: the Windansea crew had immediately pegged him, in his wingtips, bespoke suit, and natty tie, as a kook himself. That's why he ending up hanging around some kids a block down the beach at the pump house; the real surfers were under Windansea's traditional palm shack fronting the reef. After his piece appeared, the locals graffitied the pump house: "Tom Wolfe is a dork."[3]

And, of course, there was Timothy Leary, whose cosmic mantra—"turn on, tune in, drop out"—defined the sixties counterculture. Surfing entranced the guru. Leary gave popular talks titled "The Evolutionary Surfer" and proposed that surfers "are truly advanced people. . . . You could almost say that surfers are mutants, throw-aheads of the human race." Surfers, Leary enthused, rode nature's energy bands, expressing their individuality through pure contact with nature. He concluded, "Surfing is the spiritual aesthetic style of the liberated self."[4]

To people like Leary, surfing expressed the romantic impulses of sixties counterculture, a movement that seemed to shun modern technological, industrial culture for passion, risk, adventure, and engagement with nature.[5] The editor of *Surfer* magazine intoned in 1970 that "surfing by itself is clean and basic enough to transcend this era of anarchy and unrest. Because surfing, in its pure form, deals with an equilibrium involvement between man and his nature, it becomes, almost by definition, an ecologically pure undertaking. Perhaps even an undertaking so basic that it performs an evolutionary function." Another surf writer similarly offered surfing as a solution to the problems of the modern world: "Surfers are a new race. . . . Of all men, the surfer, almost alone, strives to live in harmony with nature and the world around him. . . . When everyone does it, there will be no time for wars and killing, no time for stupid rules made by stupid old men. Everyone will be too busy surfing and getting it on."[6] This view of surfing as a transcendental, personal communion with nature, however, obscured surfing's fundamental connections to modern industrial technology.

In the late 1960s and early 1970s surfers launched what became known as the shortboard revolution. This performance revolution, inspired in part by psychedelic drugs, introduced radically new surfboard designs and shifted the prevailing mass-production model in the surfboard industry to a backyard-craftsman ideal. Instead of surfboards knocked out by the thousands in a few standard models, surfboard shapers now worked alone with a surfer to develop a custom design tailored to the individual's style. This craftsman ideal, however, was underpinned by the industrial-scale, highly toxic process chemistry used to make polyurethane foam, polyester resin, and fiberglass, the main ingredients in these surfboards.

The Shortboard Revolution

Surfing in the early sixties rode a wave of popular interest, propelled by postwar affluence and baby boom demographics and reflected in the "Gidget" and "Beach Party" movies and the Beach Boys. The surf boom drove increasing demand for surfboards and that supply came from mass production. Starting in the late 1950s, major surfboard makers in Southern California scaled up production lines, and a handful of large labels, named after individual surfer/shapers—Velzy, Hobie, Weber, and Noll—exerted a virtual monopoly, shipping cheap, mass-produced boards across the West and East Coasts and to Hawaii.[7] Hobie made a hundred boards a week by the late fifties, and by the midsixties Dewey Weber was making three hundred a week. Greg Noll knocked out two hundred a week, thanks to his seventy employees in a twenty-thousand-square-foot factory that featured high-tech ventilators, temperature controls, and separate facilities for making foam, shaping boards, laminating, and sanding.[8] Surfboard production, originally a one-at-a-time craft, was vertically integrated by the midsixties.

At this time manufacturers also tried the so-called pop-out approach, using carefully molded foam blanks ("popped" out of molds, hence the name) that eliminated the process of planing the foam by hand. For a time in the early sixties pop-outs outproduced hand-shaped boards.[9] Problems with foam strength and consistency undermined pop-outs, but the main problem was price: pop-outs were not cheaper than shaped boards, since shapers like Hobie or Noll could hire eighteen-year-old kids for close to nothing to crank out boards all day, in exchange for the cachet of working for a big-name in the industry.

Mass production meant disciplined management, as the casual surfers who started these companies found themselves racing to meet orders and cracking down on teenaged surfer-employees who blew off work when the waves were good. Greg Noll, whose public image combined big-wave heroics with beer-soaked debauchery, made a management poster featuring himself in a serape and sombrero, wielding a machete, with the caption: "Watcheth, for ye know not when the master co-meth."[10] These shapers, in short, became The Man.

The mass-production paradigm reflected surfing's prevailing image. In the early 1960s surfing enjoyed a mainstream, middle-class reputation. Gidget's dad, the picture of 1950s middle-class fatherhood, when informed that his young daughter had taken up surfing, declared that he was glad to see her pursue a healthy outdoor sport. (And Gidget in the end chose the frat-boy surfer Moondoggie, not the beach-bum Kahuna, as her beau.) The surfers in *The Endless Summer*, the surf-movie-turned-crossover-hit in 1966, strolled through airports with short haircuts and dressed in suits and ties. Surfing soon underwent its own sixties revolution, however, known as the shortboard revolution, which plugged deeply into counterculture currents.

The shortboard revolution occurred between 1967 and 1970. Arguments still rage among Australians, Californians, and Hawaiians over who and what constituted the revolution; it will suffice here to say that, within a couple years, the average board length shrank by a third, or three feet: from a ten-foot board, with rounded nose and tail and blocky fin, to six or seven feet, drawn to narrow points at nose and tail, with a raked and foiled fin. The "shortboard revolution" changed the way surfers rode waves, the waves they chose to ride, and—most important for our purposes here—the way surfboards were built.

Surfboard manufacturing shifted from mass production to a craft mode. Instead of major manufacturers like Hobie, Weber, or Noll churning out thousands of standard models from their factories, a plethora of surfboard shapers turned to custom boards shaped in backyards and garages. This was in part because shorter boards were more sensitive to an individual surfer's size and style. A six-foot board that could float a 140-pound surfer sank under a 200-pound rider. These two surfers, though, could happily swap waves on the same ten-foot board without noticing too much difference. With shorter boards, instead of buying a standard model off a surf-shop rack, a surfer would design a custom board after long conversations with the shaper about his or her ability, riding style, preferred wave type, and so on. This individual relationship between surfer and shaper underpinned the surfboard production process.

The shift to craft production reflected the ethos of the sixties. Counterculture rebellion against large-scale production and technology produced a do-it-yourself movement, represented by Stewart Brand and the *Whole Earth Catalog* and by the emerging "small technology" or "appropriate technology" movements.[11] In this age of backyard tinkerers making their own gear, surfboard shapers joined the turn to individuality and away from mass production, providing an oceangoing version of Yvon Chouinard forging his own rock-climbing hardware or Burt Rutan building airplanes in his garage.

Drugs played a large part in the shortboard revolution. Tales abound of shapers taking bong hits or dropping acid and then turning out radically new designs. Gerry Lopez described a late 1967 meeting of shortboard pioneers Dick Brewer (fig. 2.1) and Bob McTavish: "A little chemical stimulation helped pique Brewer's interest until he and McTavish were so deeply engrossed in discussing surfboard design that the other Australians finally got in their car, started it up and began beeping the horn before they got Bob to leave. Meanwhile Brewer was all fired up, we followed him over to the shaping room where my blank was sitting on the rack, and he immediately took the saw and cut two feet off the tail."[12]

Allan Weisbecker similarly recalled, or rather didn't recall, "There

FIGURE 2.1 The shaper as guru. Dick Brewer (center) with Hawaiian surfers Gerry Lopez (left) and Reno Abellira (right), circa 1967. Photograph by David Darling.

are large parts of '69 I don't even remember. But I do remember buying blanks from a guy living in a tree house for a baggy of pot, then I'd give Brewer another little baggy to shape it, take it to this guy Wolfman to finish. The whole thing would cost me like 80 bucks. For a Brewer, man! The Holy Grail of boards." Weisbecker concluded, "It's not a coincidence that the acid movement and the revolution in surfing came at the same time. . . . We were so high we would have tried riding a barn door."[13]

This drug use reflected the general atmosphere of upheaval and experimentation we now associate with the sixties. It is no accident the new surfboard designs emerged around 1968, the year youth culture in general declared a revolution. Drew Kampion, editor at *Surfer* in these years, in a bit of generational hyperbole called the shortboard revolution the "greatest conceptual shift in surfing history."[14]

Surfers, however, were not content to just sample drugs; they played a central role in the creation of the sixties drug culture. Weisbecker, for example, took a break from surfing the North Shore to visit the source of his Moroccan hash. As he discovered there, "You could buy 40 bucks worth and sell it for $1000 stateside. You didn't have to have an MBA to figure out you could make some serious money doing that." Opting for a more accessible source, Weisbecker and a surfer friend soon had a sailboat running bales of pot from Colombia, with one run of twenty thousand pounds netting them each a million dollars.[15]

A group of Laguna Beach surfers raised this technique to the peak of sophistication in an enterprise that mushroomed, so to speak, into a global drug-smuggling ring worth $200 million, responsible for half the LSD and hash in the United States. The group was known as the Brotherhood of Eternal Love (fig. 2.2). Its founders had made a couple of keen observations: surf travel was beginning to flourish, much of it to places with good drugs; and hollowed-out surfboards provided a handy way to sneak drugs through customs. Surfers thus had a natural excuse to visit drug sources and a way to get drugs back. They started on a small scale, smuggling dope from Mexico on their way back from Baja surf trips. They incorporated formally as a tax-exempt organization in Laguna in 1966 and soon developed a system of secret bank accounts, false identities, chemical labs cranking out millions of LSD

FIGURE 2.2 David Nuuhiwa, Brotherhood of Eternal Love founder John Gale, and an unidentified youth in Laguna Canyon, 1971, the year Nuuhiwa won his second US surfing championship. Photograph by Jeff Divine.

doses, and smuggling routes from South America, Hawaii, and Central and Southeast Asia.[16]

The brotherhood's network brought in hash and pot measured not by the ounce or the pound but by the ton. The feds refused to believe that a bunch of hippie surfers were capable of running such a complicated operation. The US government belatedly recognized it in the 1970s as "one of the largest and most complex drug systems in the history of this country's narcotic law enforcement efforts" and added that the drug ring had made Laguna Beach "the psychedelic drug capital of the world."[17] This drug network is what first connected Leary to surfers—and hanging around the Laguna surfers is what got Leary busted by the local police.

The brotherhood gave away its smuggling secret in *Rainbow Bridge*, a 1972 film about a commune on Maui and their trek to a Jimi Hendrix concert on Mount Haleakala. The film centered on surfers as cosmic messengers, the advance guard of an enlightened extraterrestrial civilization, and it featured footage of several leading surfers—including Mike Hynson, the clean-cut star of *The Endless Summer*, who had now grown out his hair and taken up with the brotherhood. Hynson was smuggling drugs in hollow boards while living on High Avenue, literally as well as psychochemically, in La Jolla. One scene showed surfers cracking open a hollow board and pulling out big bags of hash, which they then joyfully sampled. The film may have helped open the eyes of the feds. An interagency drug task force the following year recommended that "all surfboards coming in from Hawaii should be broken. The indications have been that they do contain hash."[18]

The influence of surfing's countercultural sixties carried on to 1970, when the first "Expression Session" was held in Hawaii without judges, scores, winners or losers. In a soulful "anticontest" format proposed by *Surfer* editor Drew Kampion, the Expression Session pitted the surfer against himself rather than other contestants. Petty rivalries trumped soul, however. At the inaugural Expression Session, Hawaiian surfers protesting their exclusion from the session in favor of "chickenshit California surfers" sparked a brawl at the opening "good karma party."[19]

By the 1970s the romantic image of drug culture began to fray, revealing a darker fabric. Top surfers spiraled into heroin addiction, surfer-smugglers found themselves incarcerated in horrific foreign jails or shot dead in bungled drug deals, and surf towns from Australia's Gold Coast to Oahu's Westside to the original "Surf City," Huntington Beach, suffered the scourge of drugs. The emergence of professional surfing in the 1970s tempered drug use, as a new generation of pro surfers challenged drug associations in their argument for surfing as an elite, competitive sport. Mark Richards, an eventual four-time world champion, attacked the general view of surfers as "drug addict dole blodgers" and promoted instead a clean-cut image that was more appealing to sponsors.[20]

Foam and Fiberglass

Chemistry did not influence surfing just through psychedelics. It provided the raw materials that drove surfing's popularity. Fiberglass and foam, the staples of the modern surfboard, were developed by the prewar chemical industry and made their way into surfing through the aerospace industry, which intersected the surf community in Southern California. They remain the preferred material for almost all surfboards today.

Ancient Hawaiians had shaped six-foot *paipo* and *alaia* boards from koa wood along with the fifteen-foot-plus *olo* boards made from wili wili wood. After the revival of surfing at the turn of the century, surfers fashioned plank boards out of imported redwood. These boards could weigh well over a hundred pounds, deterring all but the most physically fit from taking up surfing, and surfers perennially searched for light but strong materials for surfboards. They found them in the chemical and defense industries.[21]

The origins of polyurethane can be traced to nineteenth-century German isocyanate chemistry and to the 1930s, when Bayer Chemical Company, the maker of aspirin and sulfa drugs, developed new urethane foams. In a laboratory at Leverkusen on the banks of the Rhine, Otto Bayer (no relation to the company founder) discovered that the reaction of two alcohol groups, isocyanates and polyols, created urethane. He also found that specific polyols such as esters and ureas could be used to produce foamed urethane polymers. The rigid foams combined high strength with a wide range of elasticity. DuPont obtained basic patents for urethane in the late 1930s, but the materials did not attract major interest until the postwar years.[22]

When the Allies entered Germany at the end of World War II, military intelligence reports described progress in urethane production. Shortly thereafter, the US Army Air Force expressed interest in the rigid foam. Bayer licensed its production technology to US firms, and DuPont, Monsanto, Princeton University, Goodyear Aircraft, and Lockheed Aircraft shifted their own urethane research into high gear. Lockheed chemists patented a "foamed-in-place" rigid polyurethane method and, by the 1950s, licensed American Latex Products to produce "Lockfoam" using their technique. Polyurethane foams had hundreds of commercial applications in the postwar era, especially for insulation in the burgeoning refrigeration industry. In the early days, however, the US military was one of the largest consumers.[23]

The Electric Boat Division of General Dynamics worked closely with the National Oil Products Chemical Company of Arlington, New Jersey (Nopco), during the late 1950s to develop a method for using Lockfoam polyurethane foam to fill voids in their new generation of nuclear submarines. Nopco's Lockfoam could be foamed in place, flowing into and bonding to the walls of cavities. Foam cut the submarine's weight by over eight tons. Lockfoam could be blown in different weights; heavier foams contributed structural strength to pressurized parts of the submarines. In applying foam in submarine stabilizers, engineers at the Philadelphia Navy Shipyard struggled to get rid of bubbles in the foam and shrinkage. The navy and Nopco found that by heating the outside skin to 75 degrees Fahrenheit, problems with shrinkage disappeared. Nuclear submarines like the *Skipjack* contained up to twenty thousand pounds of Lockfoam.[24] In aircraft, foamed-in-place urethanes increased the rigidity of control surfaces by filling voids in ailerons and rudders, and since they were buoyant, they could also help keep a downed aircraft afloat.[25] Nopco created a foam plastics division in 1955 and built two plants, one in New Jersey and one in Los Angeles, with a combined annual capacity of two million pounds of urethane and vinyl foams. In the early 1960s the company shifted its focus to isocyanate raw materials, building a plant in Linden, New Jersey, with a production capacity of up to twenty million pounds a year.[26]

Meanwhile, another material emerged from the chemical and defense industries. Games Slayter, a chemist at Owens Illinois Glass Company, invented fiberglass (brand name "Fiberglas") in 1931 as a new form of insulation. The key to his new material was a machine that could draw liquid glass into extremely fine strands. Slayter accomplished this by heating glass marbles in an electric furnace with two hundred holes in the base. High-speed winding mechanisms pulled the molten glass through the holes at over a mile a minute. The strands were then twisted and woven to make a new, fiberglass textile. Owens Illinois Glass and Corning Glass both participated in the development of fiberglass and, in 1938, formed a new company, "Owens Corning Fiberglas," to mass-produce and market the new product as an insulator and textile. By the 1950s Owens Corning refined fiberglass to a point where the flexible strands were 1/15 the thickness of a human hair and had a greater tensile strength than steel.[27]

Fiberglass was lighter than cotton, as flexible as silk, and strong. During World War II, the aircraft industry used fiberglass yarn as cloth for parachute flare shades and in wing liners for aircraft. After the war, the aerospace industry supported much of fiberglass production—for instance, as ablatives in rocket nozzles and reentry vehicles. Surfers began using fiberglass together with polyester resin as a strong, waterproof coating on their boards soon after the war, and in the 1960s they began experimenting with more advanced weaves, such as twist weaves. Fiberglass may have had an unconscious attraction for surfers: it was basically spun glass, or sand; surfers in a way were still lying on the beach, even when out in the water.[28]

Meanwhile, surfboard shapers continued to seek lighter-weight materials for the buoyant core of their boards. California shapers first began using fiberglass over balsa wood, but balsa was in short supply and notoriously porous if the fiberglass coating was dinged. Surfers instead turned to the wealth of synthetic foams emerging from industry in the postwar era. Bob Simmons was the first to experiment with foam in his quest for a lighter board. A Caltech mechanical engineering student who moonlighted at Douglas Aircraft during the war (and whose brother invented the strain gage at Caltech, a staple of aviation research), Simmons intersected the aircraft industry from several directions. He sandwiched Styrofoam between wood veneers and created light, buoyant boards that rode faster than the existing planks and hollow boards. Styrofoam was readily available in the postwar era, and Simmons experimented with his own formula, blowing his foam in a backyard mold. The drawback to Styrofoam was that it was nearly impossible to shape and dissolved when combined with the new fiberglass and resin coating.[29]

Polyurethane foam entered the world of surfing in 1956, twenty years after Otto Bayer created the synthetic material at his Leverkusen laboratory. Dave Sweet began shaping balsa-core "Malibu Chips" on the beach in Malibu during the early 1950s. Sweet purchased his first surfboard from Bob Simmons and, following Simmons, began experimenting with Styrofoam and managed to produce an epoxy-sealed prototype. His direction changed in 1954, when a friend showed Sweet a block of polyurethane foam. Polyurethane foam was compatible with polyester resin, which meant Sweet could apply the same fiberglass/resin coating used on balsa-core boards. Finding no commercial polyurethane available in the size required to shape a surfboard, Sweet set out to make his own, meanwhile continuing to build balsa boards as a way to finance his quest for a synthetic substitute. He used his savings to build an experimental clamshell mold in his basement room at a Hollywood boarding house. Learning by trial and error, Sweet began by purchasing chemicals from Nopco in small amounts and buying his first one-ton, steel-and-fiberglass mold from Techniform, an LA-area aerospace machine shop. At one point he shared his mold with his younger brother and Cliff Robertson—yes, the actor who played Kahuna in *Gidget* and who surfed in real life—to make molded pop-outs for Robertson/Sweet Surfboards.[30]

Meanwhile, in consultation with Reichhold Chemical, Dave Sweet continued to refine his foam-blowing process to the point where he produced blanks that looked like finished surfboards. He then incorporated inserts into the mold in order to create specific shapes. Sweet routed wooden stringers into both sides of his boards but also offered the option of a through-hull stringer. Annual production in a good year was around eight hundred boards. Sweet opened a store on Fourteenth and Olympic in Los Angeles and served a steady stream of customers until the late 1960s, when the shortboard revolution cut sales drastically and forced him out of business.[31]

Sweet introduced polyurethane to surfing but had no desire to scale up production. This step was accomplished by Hobart "Hobie" Alter and Gordon Clark in Laguna. Hobie did not know of Sweet's previous work with polyurethane; he obtained a sample of polyurethane from Kent Doolittle, a Reichhold resin salesman, and was impressed by the dense chunk of foam. He learned that you could sand polyurethane and that the material worked with fiberglass and resin. Hobie Surfboards had just hired Gordon Clark as a glasser, after he left his job at the Huntington Beach oil fields. Clark had a bachelor's degree in engineering from Pomona College, where he studied math, physics, and chemistry and also took up surfing. He picked up the nickname "Grubby" because he would leave work in the oil fields and drive straight to Dana Point, where he slept in the Hobie parking lot before going surfing at Trestles. Alter invited Grubby to join him in experimenting with foam. The secretive Clark told him to stop talking to outsiders, and the two began mixing their own foam in empty ice cream containers lined with wax paper. Clark overcame problems with heat and humidity to come up with a formula that was strong and stable enough for surfboards.[32]

Using capital from Hobie Surfboards, the two rented a building in Laguna Canyon where Clark focused exclusively on foam. Like Sweet, they started with test molds and found the process extremely messy. By the middle of 1958, Alter and Clark invested in a steel and concrete mold large enough to blow foam for half a surfboard blank. Unlike

Sweet, Alter and Clark did not attempt to produce blanks for entire surfboards. Instead, they blew the blanks in halves and glued the pieces together with a wooden stringer in the middle. Once poured, a polyurethane blank took forty minutes to mix, expand, and cool. Where a balsa blank required laborious laminating and chipping, a power planer cut through foam like butter, and shapers could ready several boards for glassing in a day's work.[33]

Air bubbles remained a problem, pocketing the finished blanks. Refined chemistry and experimentation with mold linings eventually eliminated bubbles and produced a blank that could be clear-coated, striped, or finished with whatever pattern the customer desired. Alter pushed foam boards hard during the summer of 1958, when he and Clark had three molds turning out more than enough blanks to supply his shop.[34]

A year later, Clark bought out Alter, retooled, and started Clark Foam. Hobie continued to make surfboards but also shifted to the mass production of lightweight shot-foam catamarans called "Hobie Cats" that revolutionized sailing. Unlike Hobie and Sweet, Clark did not try to shape and sell boards himself; he just concentrated on blowing foam and selling the blanks to shapers. Hobie and Sweet had blown foam to supply their own surfboard business, not sell it to others. A few competitors followed Clark into the foam blank business, including Chuck Foss, Harold Walker, Ron Haydu, and Rogers Foam. Gordon Clark continued to scale up production and squeeze out competitors; by the 1970s Harold Walker's foam factory in Wilmington, California, was the only alternative source of blanks, but it never posed a significant challenge to the Clark Foam juggernaut.[35] Clark wooed shapers by incorporating their preferred rocker into his wide range of blanks, which cut down on shaping time, and he also varied the size, type, and shape of stringers available in his blanks.

Above all, Clark won market share by keeping prices low. This ruthlessly undercut competitors but also nobly supported small-scale shapers; Clark pointedly refused to give large-scale producers a price break. He could do so because of the quality of his products and the scale of his operation. The secretive Clark did not divulge his production figures, but rumor had it that he shipped a thousand foam blanks a day, several times more than his closest competitor at Walker.[36] He bought raw chemicals by the ton, built massive, many-ton concrete molds and reactor vessels to blend foams and resins, and invested in high-tech ventilation and environmental-control systems. Clark's cheap foam in turn depended on the supply of raw chemicals from a global chemical industry underpinned by military applications.

With raw materials for polyurethane production readily available, Clark was soon selling foam at a fraction of the cost of balsa. His process used large volumes of toluene di-isocyanate (TDI) available from Bayer, FMC, Wyandotte Chemicals, and Allied Chemical. Clark combined the isocyanate with a polyester polyol blend produced in Texas by Celanese Corporation to meet the growing demand at the time for polyurethanes. TDI is a hazardous material but its pervasiveness eased acceptance.[37]

In short, the proliferation of small-scale, backyard shapers in the late 1960s and early 1970s depended on the supply of cheap foam blanks from Clark and on the cheap raw chemicals supplied by Bayer and other major firms in the global chemical industry, whose production runs were in turn supported by the defense industry. Like the idyllic tropical reefs that surfers loved, the do-it-yourself ideal in surfboard making rested on a vast, unnoticed infrastructure. Surfers, however, certainly encountered this substrate when they wiped out and hit the reef, and shapers eventually recognized their own infrastructure on a day, decades later, when Grubby Clark wiped them out.

By 2000 Clark had an iron grip on 90 percent of the $200 million US market and 60 percent of the world market for polyurethane blanks. Three years earlier, *Surfer* magazine heralded Clark as the second-most powerful man in surfing (after Bob McKnight, CEO of Quiksilver). The article included a photo of Grubby with both middle fingers raised in a double-barreled salute to the surfing community.[38]

Over the years Clark had regaled his customers with annual letters describing the state of the surfboard industry, on topics ranging from the introduction of shaping machines during the late 1970s to the need to ventilate resin fumes from glassing shops in the 1990s. In December 2005 surfboard shapers received a rambling fax notifying them that Clark Foam was closing immediately. The global supply of blanks disappeared in a day, and surfboard shapers desperately scrambled to find foam. Dubbed "Blank Monday," Clark's abrupt decision shocked the surfing world.

Why did he quit? His fax claimed that government regulators and health-related lawsuits by employees threatened to ruin his company: "I may be looking at very large fines, civil lawsuits, and even time in prison." Clark's missive perplexed industry insiders. There was no cluster of lawsuits attacking Clark Foam, and while he had visits from state and federal environmental agencies and the local fire marshal, none had taken action. Nevertheless, after over thirty years in the blank business, Clark dismantled his plant and auctioned off much of his equipment.

He ordered his workers to smash his eighty-odd concrete master blank molds. The shattered molds, source of numerous world-championship boards, became a site for pilgrimages by surfers.[39]

The pressure on Clark may have been more economic than environmental. Clark's annual letter the year before had warned of the emergence of offshore surfboard manufacturers, especially in Asia. A company in Thailand called Cobra International had recently opened a new factory outside Bangkok that dwarfed Clark's operation; the surf industry began speaking with awe about "the Cobra factory," which cost $20 million to build, employed 2,800 people, and could crank out 250,000 boards a year.[40] Chinese manufacturers were similarly tooling up large factories. Clark may have seen the handwriting on the wall and jumped out of the business before Asian competitors pushed him out.

Conclusion

The rise of offshore blank manufacturers, together with computerized shaping machines, sparked fears among surfers of the demise of the surfboard craftsman and of the individual relationship between a surfer and a shaper. Some of these fears are being realized, as surfers today can buy Chinese-made boards at Costco for less than half the price of a custom board. Surfers conveniently forget that the backyard-craftsman ideal just removed one step—shaping—from mass production. They also forget that the backyard craftsman has not always been the model, and that the current shift to mass production is perhaps more a return to the model of the early sixties. The backyard craftsman was the product of a particular historical moment starting in the late sixties, propelled by do-it-yourself ideals, psychedelic drugs, and new technologies.

This is not a simple spinoff argument, or technological determinism. The military-industrial complex in this case did not foster mass production and consumerism; rather, industrial chemicals enabled the backyard-craftsman, individualized model of surfboard production—and that in turn reflected the context of sixties do-it-yourself, small-technology ideals.

The sixties counterculture is supposed to represent a backlash against science and technology, which had become associated with the military, environmental pollution, and technocratic challenges to democracy. In the usual telling, sixties youth viewed technology as the source of society's problems, not the solution. But science and technology were not necessarily opposed to the counterculture. Science and technology re-

flect their social context, and for the sixties and seventies that included
the counterculture, as the chapters in this volume make clear.[41]

The role of high technology in surfing extends beyond surfboards
to wetsuits (another product of military research and industrial chemistry), urethane leashes, synthetic surfwear, and even the waves themselves, thanks to artificial reefs and wave pools. But the romantic culture surrounding surfing, a premodern Polynesian pastime, discouraged
surfers from thinking too much about such high-tech associations. In
the twentieth century surfing became a highly technological pursuit that
involved industrial petrochemicals: polyether and polyester polyols and
toluene di-isocyanate in foam, and styrene monomer resin catalyzed by
methyl ethyl ketone peroxide (MEKP). Surfers, whom one might think
of as environmentally conscious, also ignored the environmental implications of their technology. This toxic soup of volatile chemicals, once
turned into a surfboard, is chemically inert (unless tossed in a fire, which
surfers occasionally do as a ritual "sacrifice" to attract waves during
flat spells, usually after a few beers); but this inertness means that surfboards will last for centuries when tossed in the local landfill.

Surfing's countercultural image was just that, an image. Surfing
survived for centuries because of its connections to the mainstream,
whether in ancient Hawaiian society or the modern world. Yet surfing
managed to remain a counterculture for decades, if not centuries, and it
is still seen that way today even though it is a $10 billion global indus-
try.[42] The developments we describe here bolstered surfing's counterculture credibility. The shortboard revolution undercut the Gidget-era surf
boom: the big factories went out of business, and that made it harder
for a beginner to buy a board. Then, when you did buy a board from
your local backyard shaper, it was shorter, which made learning how to
surf much harder. So the shortboard revolution acted like a brake and
kept surfing—for a time—from becoming too mainstream.

But there was a contrary impulse: surfers want to make a living
off surfing, and that often meant attracting more people to the sport.
This economic imperative sometimes overlaid a democratizing impulse.
Thus, Tom Morey, a former composites engineer for Douglas Aircraft
and then Avco (where he designed ablative nozzles for the Nike-Zeus
missile), in the early 1970s invented a cheap, do-it-yourself, entry-level
board for beginners that he explicitly hoped would democratize surfing.[43] What did he call his invention? Of course: the "boogie board,"
which indeed eventually introduced millions of people to surfing. The
surf industry meanwhile finessed this problem by making money, not

by selling surfboards to surfers, but by selling surfwear to people who don't surf.

Surfing remains the prototype of a counterculture, reinforced by popular images of Jeff Spicoli–type burnouts. This traces back to the groovy sixties, when surfers grew out their hair, smuggled drugs, and moved to Maui communes. The counterculture transformed surfing, in particular influencing surfboard design, but surfing also influenced the sixties counterculture, helping to create the drug economy and inspiring such gurus as Timothy Leary with the transcendent surfer lifestyle. Surfers were no longer clean-cut kids enjoying a little wholesome outdoor recreation; they were dope-smoking rebels giving American society and middle-class culture the finger. Surfboard manufacturers were no longer large employers with high-tech, high-volume facilities and responsible management; they were backyard operators trading shaping jobs for bags of pot. This image added to surfing's groovy appeal; for all those young baby boomers looking to escape boring middle-class culture, what better way than to load a few surfboards into your VW van and head for the coast?

Notes

1. Alvin Toffler, *Future Shock* (New York: Random House, 1970), 364, 288.

2. Tom Wolfe, *The Pump House Gang* (New York: Farrar, Straus and Giroux, 1968).

3. Steve Barilotti, "Pump House Redux," *Surfer*, January 1995.

4. Steve Pezman, "The Evolutionary Surfer: Dr. Timothy Leary Interview," *Surfer*, January 1978.

5. E.g., Jacques Barzun, *From Dawn to Decadence* (New York: Harper-Collins, 2000), 465–91, on 474. It is no coincidence that the eighteenth-century Romantic movement emerged just as European explorers were reaching Hawaii and other Polynesian islands—and encountering surfing. The exotic sport, the pure pursuit of pleasure, inspired Europeans on a continent wracked by revolution and war. Andy Martin, "Surfing the Revolution: The Fatal Impact of the Pacific on Europe," *Eighteenth-Century Studies* 41, no. 2 (2008): 141–47.

6. Drew Kampion, "I, II, & III," *Surfer* 11, no. 5 (1970); and Paul Witzig, "When Everyone Surfs," *Surfer* 12, no. 3 (1971); both reprinted in Sam George, ed., *The Perfect Day: 40 Years of "Surfer" Magazine* (San Francisco: Chronicle Books, 2001), 52, 55. Kampion was the editor of *Surfer* at the time; Witzig, an Australian surf filmmaker.

7. Drew Kampion, *Stoked: A History of Surf Culture* (Salt Lake City, UT: Gibbs Smith, 2003), 103. See also Greg Noll, *Surfer's Journal* 6, no. 2 (1997): 45–46; and Mike Diffenderfer, *Surfer's Journal* 6, no. 3 (1997): 31.

8. Nat Young, *History of Surfing*, rev. ed. (Angourie, NSW: Palm Beach Press, 1994), 84; and Greg Noll and Andrea Gabbard, *Da Bull: Life over the Edge* (Berkeley: North Atlantic Books, 1989).

9. Gordon Clark, "History of Surfboard and Sailboard Construction," in *Essential Surfing*, by George Orbelian (San Francisco: Orbelian Arts, 1987), 174.

10. Noll and Gabbard, *Da Bull*, 114.

11. Fred Turner, *From Counterculture to Cyberculture: Stewart Brand, the Whole Earth Network, and the Rise of Digital Utopianism* (Chicago: University of Chicago Press, 2008); and Andrew Kirk, *Counterculture Green: The "Whole Earth Catalog" and American Environmentalism* (Lawrence: University Press of Kansas, 2007). On appropriate technology, see E. F. Schumacher, *Small Is Beautiful: Economics as if People Mattered* (New York: Perennial Library / Harper and Row, 1975), 190–201; and Carroll Pursell, "The Rise and Fall of the Appropriate Technology Movement in the United States, 1965–1985," *Technology and Culture* 34 (1993): 629–37.

12. Gerry Lopez, "Prodigy," *Surfer's Journal* 6, no. 1 (1997): 22.

13. Chris Mauro, "The Surfer Interview: Allan Weisbecker," http://surfermag.com/magazine/archivedissues/allan/index.html.

14. Quoted in David Rensin, *All for a Few Perfect Waves: The Audacious Life and Legend of Rebel Surfer Miki Dora* (New York: Harper Entertainment, 2008), 183.

15. Mauro, "Interview: Weisbecker"; and Allan Weisbecker, *In Search of Captain Zero: A Surfer's Road Trip beyond the End of the Road* (New York: Jeremy P. Tarcher / Putnam, 2001). See also Peter Maguire and Mike Ritter, *Thai Stick: Surfers, Scammers, and the Untold Story of the Marijuana Trade* (New York: Columbia University Press, 2014).

16. Joe Eszterhas, "The Strange Case of the Hippie Mafia," *Rolling Stone*, 21 December 1972; and Nicholas Schou, *Orange Sunshine: The Brotherhood of Eternal Love and Its Quest to Spread Peace, Love, and Acid to the World* (New York: St. Martin's, 2010).

17. US Senate Committee on the Judiciary, "Hashish Smuggling and Passport Fraud: 'The Brotherhood of Eternal Love,'" 93rd Congress, 1st Sess. (3 October 1973).

18. Mike Hynson, *Transcendental Memories of a Surf Rebel* (Dana Point, CA: Endless Dreams Publishing, 2009), 177, 235; and Joe Eszterhas, "The Strange Case of the Hippie Mafia," *Rolling Stone*, 7 December 1972.

19. Drew Kampion, "The Death of All Contests," *Surfer*, September 1970.

20. Peter Westwick and Peter Neushul, *The World in the Curl: An Unconventional History of Surfing* (New York: Crown Books, 2013), 129–30, 159. For a contemporary perspective on surfing and drugs, see Chas Smith, *Welcome to Paradise, Now Go to Hell: A True Story of Violence, Corruption, and the Soul of Surfing* (New York: HarperCollins, 2013).

21. Ben R. Finney and James D. Houston, *Surfing: The Sport of Hawaiian Kings* (Rutland, VT: Tuttle, 1966); and Tom Blake, *Hawaiian Surfboard* (Honolulu: Paradise of the Pacific Press, 1935).

22. Otto Bayer, "Das Di-Isocyanat-Polyadditionsverfahren (Polyurethane),"

Angewandte Chemie 59 (1947): 257–72. See also German Patent 728.981 (1937), I. G. Farben. For a general treatment of the topic, see Raymond B. Seymour and George B. Kauffman, "Polyurethanes: A Class of Modern Versatile Materials," *Journal of Chemical Education* 69 (1992): 909.

23. "Urethane Plastics—Polymers of Tomorrow," *Industrial and Engineering Chemistry* 48 (September 1956): 1383–91.

24. Benjamin S. Collins, "Foamed Plastic Replaces 50-Year-Old Method of Filling Submarine Voids," *Marine Engineering* 63 (1958): 82–83.

25. "Urethane Plastics—Polymers of Tomorrow," 1383–91.

26. "Nopco to Build Plastics Plants," *New York Times*, 30 June 1955, 41; "Expansion Planned by Nopco Chemical," *New York Times*, 25 August 1955, 31; and "Nopco to Build in Jersey," *New York Times*, 14 February 1961, 57.

27. Games Slayter, "Fiberglas: A New Basic Raw Material," *Industrial and Engineering Chemistry* 32, no. 12 (December 1940): 1568–71. See also John A. Morgan and Leon E. McDuff, "Fiber Glass in the Space Age," *Glass Industry*, June 1960.

28. P. H. Kemmer, "Development of Glass-Reinforced Plastics for Aircraft," *Modern Plastics* 21 (1944): 89–93; "Army Experimental Plane Has Glass Reinforced Fuselage," *Automotive and Aviation Industries* 90 (1944): 29; and N. B. Miller and E. L. Strauss, "Heat-Blast Erosion Effects on Reinforced Plastic Laminates," *Society of Plastics Engineers—Journal* 14 (1958): 37–40, 69.

29. For the Simmons story, see Westwick and Neushul, *World in the Curl*, 97–102; and Peter Westwick and Peter Neushul, "Aerospace and Surfing: Connecting Two California Keynotes," in *Where Minds and Matters Meet: Technology in California and the West*, ed. Volker Janssen (Berkeley: University of California Press / Huntington Library, 2012).

30. Mark Fragale, "Dave Sweet: First in Foam," *Longboard Magazine*, September/October 2000.

31. Ibid. For Grubby Clark's perspective on Sweet, see Orbelian, *Essential Surfing*, 174.

32. Andrew Rusnak, "Fun in the Son: Shaping Hobie Alter Part II," *Composites Fabrication Magazine*, June 2001, 2–9.

33. Clark, "History of Surfboard and Sailboard Construction," 164–98.

34. Ibid.

35. Steve Tepper, interview by the authors, 14 October 2010, Goleta, CA. Tepper poured blanks for Walker during the 1970s.

36. William Finnegan, "Blank Monday: Could Grubby Clark Destroy Surfing?," *New Yorker*, 24 August 2006.

37. TDI is used in, for example, car upholstery and carpets. According to Australian blank manufacturer Midget Farrelly: "TDI is not going to disappear. We just have to learn to handle it better." Quoted in Donna Dawson, "Waves of Change: From Shock to Opportunity," *Composites Technology*, June 2006, www.compositesworld.com/articles/waves-of-change-from-shock -to-opportunity.

38. Finnegan, "Blank Monday."

39. Ibid.

40. Amanda Jacob, "Cobra Strikes Out into Industrial Markets," *Re-*

inforced Plastics, May 2004, 24–29. See also Tim Baker, "Made in Thailand," *Surfing*, January 2003.

41. See also David Kaiser, *How the Hippies Saved Physics: Science, Counterculture, and the Quantum Revival* (New York: W. W. Norton, 2012).

42. Thomas Frank, *The Conquest of Cool: Business Culture, Counter-culture, and the Rise of Hip Consumerism* (Chicago: University of Chicago Press, 1997).

43. Tom Morey, telephone interview by the authors, 14 April 2009.

3

Santa Barbara Physicists in the Vietnam Era

Cyrus C. M. Mody

The 1970s: Hiccup or Preview?

For many scientists in the United States (and even for many historians of American science), the 1970s have a reputation as a lost decade. Compared with both the early Cold War and the Reagan buildup of the 1980s, research budgets of the '70s were dismal. As a result, mega-projects such as new particle accelerators and giant telescopes had to be put on hold, while bench-top scientists lacked new tools and facilities.[1] Those tight budgetary times were partly a result of federal spending on Vietnam and on social programs, but many scientists also blamed the public—and especially the youth counterculture—for declining appreciation for Establishment scientists' expertise and patriotic contributions. Some scientists and their institutions responded to the Vietnam-era protest movement with angry denunciation, but many also experimented with a new, more civilian-minded kind of research—only to close down those experiments after a few, unsuccessful years.[2] All in all, the '70s seem to have been a trough of malaise, cultural contestation, tight budgets, and failed idealism in which nothing of much importance was achieved—a period so different from what came before and what came after that it is hard to find a place for it in the historical record.

There's a sense in which that account of the '70s is correct, of course. American scientists at the time worried about their profession's course, and many were relieved when the Reagan administration ramped up federal research budgets and (after some prompting) championed a return to basic science.[3] Military-industrial research institutions, such as the Jet Propulsion Laboratory, that had pursued civilian projects in mass transit, alternative energy, and environmental monitoring in the '70s returned to their favored national-security patrons in the '80s. Moreover, the emergence of the biotech and personal-computer industries helped restore Americans' faith in both the achievements and the economic value of science after a decade in which many of the country's original science-based industries (steel, chemicals, electrical and electronic products, aviation) struggled to overcome their complicity in the Southeast Asian conflict and rising global economic competition. In certain respects, then, the '80s were a restoration following the turbulent interregnum of the '70s.

Yet as the other chapters in this volume show, there was still plenty of good science done in the United States in the '70s, much of it conducted by, in partnership with, or in response to members of the youth counterculture and various protest movements. Tight budgets, declining student enrollments in the sciences, and congressional pressure for civilian-minded research encouraged entrepreneurial ventures, interdisciplinary collaborations, and pedagogical experiments by midcareer scientists that, for better and for worse, have become commonplace in the twenty-first century. Those experiments might never have materialized, though, if not for the relaxed mores, questing spirituality, holism, idealism, and opposition to militarism associated with the counterculture and protest movement. Especially in the physical and engineering sciences, young people in the '70s showed much greater enthusiasm for research topics related to environmental problems, biomedicine, and disability technologies than the previous generation had. American physical and engineering scientists' embrace of such topics has persisted to the present, even as the counterculture and the crises that fostered it have faded. To a great extent, the post–Cold War research enterprise first emerged in the lost decade of the '70s.

From Think Tank to Start-up

To see how the crises around 1970 played out at the microlevel—and continue to resonate at the macrolevel—this chapter offers a case study of a small group of physicists in Santa Barbara, California. The physicists who figure most prominently in my case study—Philip Wyatt, David Phillips, and Virgil Elings—were in their thirties in 1970. None of these three were full participants in the youth counterculture, yet each responded to the changes in American society that the counterculture encouraged: for example, growing environmental awareness, interest in new forms of spirituality, growing skepticism of convention and authority, and opposition to the Southeast Asian conflict and the military-industrial complex. Elings, Phillips, and Wyatt individually attended to some of these currents more than others. Though their relations with each other—and with the young people they worked alongside—were sometimes contentious, as a whole the network surrounding this trio exemplifies the diversity of American physicists' strategies for navigating through the 1970s.

The limitations of the case study approach should be obvious; in particular, it is difficult to show whether the actors in a case study are the exception or the rule. Certainly, it would be a mistake to think that physicists—and especially California physicists—were representative of American science in the '70s. Yet, in some ways, that is the point—the actors in this chapter were the exception, and they both suffered and benefited from that exceptionalism. The problems they faced in 1970 were similar to those faced by much of their discipline, yet their responses were idiosyncratic and unusual. They themselves sometimes complained about lack of support and interest from colleagues, both locally and nationally. Some of this group's innovations failed or petered out because of that lack of support. Over time, though, some of their innovations came to seem entirely unexceptional.

Even so, Santa Barbara might seem an unpromising place to explore how physicists navigated through the Vietnam era. The city's physics establishment—in both industry and academia—was decidedly second tier or lower in the late '60s and early '70s. Santa Barbara's physicists were not subject to nearly the same degree of protest that their counterparts in Palo Alto or Cambridge were, nor were their voices nearly as influential in their discipline's response to war, unrest, and economic malaise. Geographically and intellectually, Santa Barbara's proximity to both Los Angeles and the Bay Area allowed its physicists to take advantage of industrial and countercultural spillover, but the city was not a leading site for either the High Cold War military-industrial complex or the countercultural reaction to the excesses of the Cold War state.

Indeed, Santa Barbara's relative isolation was its main selling point for one kind of postwar physicist—the defense think tank researcher. In

1956 General Electric spun off a defense think tank, TEMPO, in Santa Barbara that soon became the nucleus of a small cluster of such firms. Proximity to the Southern California aerospace industry (and Camp Cooke—now Vandenberg Air Force Base—north of town) was surely one rationale for TEMPO's site, but so was the idyllic, even sleepy, milieu. As a TEMPO pamphlet put it, "the Santa Barbara location was chosen to provide isolation from the day-to-day interchange with engineering and manufacturing functions of General Electric and to encourage independent and objective studies by the technical staff."[4]

Many of Santa Barbara's defense think tanks were West Coast arms of East Coast firms. Like TEMPO, many of these outposts were located in Santa Barbara for both pragmatic and hedonic reasons. For instance, EG&G (cofounded by MIT's Harold "Doc" Edgerton) established an office in Santa Barbara because the "logic was, if I [EG&G executive Sandy Sigoloff] move to Santa Barbara, I can hire really great people because people really like to live in Santa Barbara."[5]

One such firm was Defense Research Corporation, a Virginia-headquartered think tank that cultivated a portfolio of research on counterinsurgency, computer-aided literature searching, and antiballistic missile technology. In the early '60s, Philip Wyatt, a young theoretical physicist, joined Defense Research's antiballistic missile group and started exploring ways to use lasers to track missiles. That, in turn, led Wyatt to think about other applications of lasers, and by 1967 he had come up with a scheme for laser detection of pathogens for, among other things, biological-warfare defense.[6] After writing a report on the use of laser inverse-scattering methods to detect pathogens in water, Wyatt secured US Army funding to develop the technology.

Friction with Defense Research management, however, led Wyatt to move over to EG&G. That firm's bread-and-butter contracts were with the Atomic Energy Commission, for whom it recorded data from nuclear-weapons tests. Thus, in Wyatt's words, they welcomed his biowarfare research contract, because "they'd love to get out of the nuclear business and go into something more interesting."[7] During his brief time there, EG&G gave Wyatt the opportunity to study with microbiologists at UCSB and UCLA and also gave him some management training. Both proved essential when, in 1968, Wyatt formed his own start-up company, Science Spectrum, to commercialize laser inverse-scattering particle detection.[8]

Science Spectrum's founding came just at the start of—and in some ways exemplified—a period in which significant portions of the American research enterprise civilianized and moved away from their High

Cold War roots. As Jennifer Light has shown, national-security-oriented think tanks like Defense Research, EG&G, and TEMPO always had some interest in civilian matters, but their search for civilian applications of Cold War expertise intensified when defense funding slowed in the late '60s.[9] Many firms also saw that association with the military was becoming a stigma; Defense Research, for instance, became the more anodyne "General Research Corporation" in this period.

At the same time that defense research funding was being cut, some civilian agencies were expanding and/or shifting their priorities in ways that made them more attractive to outfits like Science Spectrum. After 1970 the National Science Foundation (NSF), for instance, began moving well beyond its traditional role as a small funder of basic research by individual physical and life scientists; that year the Nixon administration funneled an extra $100 million into NSF's budget to pay for new programs in engineering science and "research applied to [civilian] national needs."[10] The Environmental Protection Agency (EPA) formed in 1970 as well, partly as a result of outcry following the previous year's disastrous oil spill in Santa Barbara; the EPA quickly became another promising funder of physics research. The National Institutes of Health (NIH)—and biomedical funding generally—also began a long upward trend, especially after Nixon's declaration of the "war on cancer" in 1971.

These funding trends drew many physicists toward environmental or biomedical applications that they would have ignored in the early Cold War. For some, this shift in orientation was no doubt merely a pragmatic retooling to chase limited funds. For others, it may well have coincided with a sincere reconsideration of personal priorities in a time of ferment. For many, pragmatic and idealistic considerations were probably difficult to separate. Whatever their reasons, at the end of the '60s physicists such as Wyatt began expressing much more interest in civilian biomedical and environmental applications of their work than they would have just a few years earlier.

Thus, while Wyatt's idea for a dynamic light-scattering (DLS) instrument was conceived in a military-industrial context, in the early 1970s Science Spectrum focused much more on civilian applications and agencies. For instance, Wyatt tried to capitalize on the environmental research boom bolstered by the 1969 Santa Barbara oil spill: as he put it in 1970, "It seems that Santa Barbara is becoming a center for the study of ways to fight pollution, and this company is in an excellent position to help in that research" by using DLS "to study smog and haze particles in the atmosphere and find out what causes this pollution."[11]

Indeed, Wyatt described Science Spectrum as just one of many small tech start-ups that could address environmental issues if only policy makers would help them. As he told the *Santa Barbara News-Press*, "the federal government should subsidize companies that can solve pollution problems, thus creating new jobs and making the effort to dispose of waste a stabilizing influence in the economy."[12]

Wyatt had somewhat more success pitching DLS's relevance for biomedical research, perhaps because of the technique's origins in bio-warfare defense. By 1972 Science Spectrum was field-testing its product line at Santa Barbara's Cottage Hospital as a means for assaying "the relative efficiency of a panel of antibiotics against patient bacterial specimens."[13] In 1975 the company received a grant from the Food and Drug Administration (FDA) to develop a technique for detecting "veterinary drug residues in food producing animals" and animal products—research that Wyatt and his employees conducted in collaboration with both the FDA and the US Department of Agriculture.[14] Wyatt et al. also worked with researchers at the Kettering-Meyer Laboratories and the Southern Research Institute to apply DLS to detection of chemotherapy compounds in cancer patients' blood.[15]

With its slate of biomedical applications and customers, Science Spectrum might well have come to be seen as an early player in the biotechnology industry had it been located in the Bay Area instead of Santa Barbara. Like Cetus—one of the first and most important San Francisco biotech firms—Science Spectrum's founder was a physicist looking to strike up collaborations with the life sciences.[16] And like other biotech entrepreneurs, Wyatt was a maverick and showman: while getting his doctoral degree he was written up in the *New York Times* for advancing unorthodox theories about terrestrial antimatter impacts, and while at Defense Research he had been one of the final fifteen candidates in NASA's first scientist-astronaut program.[17] Staying in the public eye through such unconventional tactics aided Wyatt because, like the early biotech entrepreneurs, he had to make his small, offbeat company visible to venture capitalists (VCs). And, in fact, Wyatt had some success in reaching VCs. Tom Perkins—a legendary investor in both microelectronics and biotech—was on one of Science Spectrum's early boards. Another VC outfit, the Value Line Development Capital Corporation, invested $500,000 in Wyatt's company in 1973 and handpicked a new president for the start-up, Harry Brown.[18]

Science Spectrum's chances of landing venture capital were significantly boosted in 1972 when its Differential II product won an Industrial Research 100 Award. That award was made possible in part by another characteristic Wyatt shared with early biotech entrepreneurs: the ability to recruit university faculty members to join his strange little start-up. In particular, Wyatt spent several months in 1969–70 trying—eventually successfully—to persuade David Phillips, an assistant professor of physics at UCSB to join Science Spectrum and develop the Differential II into a marketable product.

The hiring of Phillips seems to have made all the difference in adapting Science Spectrum's products for use by nonphysicists. As the firm had discovered when it hired a microbiologist from UCLA, that discipline's "inexperience with light scattering apparatus emphasized the need for a reliable, easy to use instrument instead of the maze of wire, tape, switches, and dials that had enveloped the original photometer" (see fig. 3.1).[19] A good measure of how important Phillips was in making Science Spectrum's products more user-friendly is that *Industrial Research* gave him top billing, as "principal scientist," in its citation of the five-man team that won the Industrial Research 100 Award.[20] Another indicator is that the main conference room at Wyatt Technology, Science Spectrum's successor company, is even today named after Phillips. Yet another sign is Wyatt Technology's own description of its history: "[Philip] Wyatt began his investigations of the practical applications of the inverse scattering problem in 1967 with studies of means to differ-

FIGURE 3.1 Employees of Science Spectrum with their first multi-angle light-scattering instrument in 1970. The eye patches enhanced contrast when viewing through the instrument's monocular. Back row, left to right: Victorr A. Herma, Herman H. Brooks, Chelcie B. Liu, David T. Phillips, Richard M. Berkman. Front row, left to right: Philip J. Wyatt, James E. Hawes. Reprinted by permission of Wyatt Technology.

entiate bacterial species from one another. Together with his colleagues, most importantly Dr. David T. Phillips, he modified a traditional light scattering photometer to incorporate a laser light source."[21]

Instruments for Parapsychology

My point is that, by a number of measures, Phillips was a competent and serious scientist who made important contributions to the commercialization of DLS instrumentation. There are indications, however, that his UCSB colleagues had their doubts about Phillips's competence, and that these qualms drove Phillips to Science Spectrum. To be sure, leaving academia to join an unproven start-up carried significant financial risk for Phillips and his young family. As it turned out, Science Spectrum was always a marginal endeavor and seems not to have provided Phillips with steady employment for very long. Thus, he continued doing adjunct teaching at UCSB until about 1976, and he generated various sidelines to bring in some extra cash.

Phillips's departure from UCSB was probably not entirely by choice. The UCSB physics department was at the time in the middle of a major, long-term strategic reorganization and an acute budgetary crisis, both of which made Phillips's chances of getting tenure very unlikely. Not long after its founding in 1960, the department had embarked on a campaign to raise its status by shifting its focus away from undergraduate teaching to PhD training, research, and grantsmanship.[22] Phillips had been one of the early hires in that campaign, but by 1970 he had supervised only one PhD dissertation, and his publication record was undistinguished.[23] In fact, he was significantly more productive—in terms of publications—*after* he moved to Science Spectrum than before. Moreover, the kinds of publications Phillips produced at UCSB may not have met with much departmental approval. For instance, he was probably proudest of an article (for a journal focused on undergraduate pedagogy) describing how to build "the poor man's nitrogen laser"—a topic that his senior colleagues probably saw as insignificant and unhelpful for a department trying to break away from its collegiate past and into the big time of lavish federal research grants.[24]

It's quite possible, then, that Phillips saw that his tenure case would be difficult, and/or that he realized his interests did not align with his department's. That view would have been reinforced when Vincent Jaccarino, the chair of the department from 1969 to 1972, made it clear that tenure reviews were going to become more difficult during his reign. As a departmental history put it later, "Largely as a result of

Vince's leadership, there was a significant change in the department's attitude regarding appointments and promotions—particularly for assistant professors. The department realized it had attained the position that allowed it to attract and promote only those physicists that met the highest standards. This meant making some difficult decisions but the department has been the stronger for it."[25]

One way in which Phillips's interests departed significantly from those of most of his colleagues—thereby jeopardizing his tenure prospects—was his growing attention to parapsychology and countercultural forms of mystical experience. Phillips's incorporation of parapsychology into his scientific and personal outlook probably took off only after he moved to Santa Barbara in 1966, but by the early '70s he was sufficiently ensconced in the area's parapsychology community to serve as the research director of the Southern California Society for Psychical Research. He was also acquainted with, and occasionally appeared in public alongside, many of the leading lights of early '70s parapsychology. For instance, he was at one point part of a discussion circle led by Charles Tart at UC–Davis that included Stanford Research Institute (later SRI International) psychokinesis investigator Hal Puthoff, artist and remote viewer Ingo Swann, and Elizabeth Rauscher, a leading member of Berkeley's Fundamental Fysiks group.[26] Phillips also spoke at an Esalen workshop on parapsychology alongside another Fundamental Fysiks researcher, Nick Herbert; headlining that workshop was the Israeli showman and spoon bender Uri Geller.[27]

Phillips was well aware that his psi pursuits did little to advance his departmental standing. As he put it, at UCSB there were "[s]everal distinguished physicists [including Vincent Jaccarino] who are not sympathetic to psychical research. Involvement has psychological risks, these are strong willed people."[28] Nevertheless, Phillips persisted in his participation in the psi community, in part simply from a strong personal desire for "spiritual development" and greater "intuitive awareness."[29] But Phillips's persistence also stemmed from a professional belief that psi phenomena merited scientific investigation, and that some mystical traditions' descriptions of the world converged on those conveyed by quantum mechanics and scientific cosmology.[30]

In other words, Phillips aspired to a scientific approach to psi phenomena that cut between credulity and skepticism. On the one hand, he quite clearly believed in the existence of some phenomena that would usually be labeled paranormal; on the other hand, he expressed dismay at "the everyday occult mish-mash" and "the foolish trickery that abounds in southern California."[31] His philosophy is perhaps best

summed up in his resignation letter to the Southern California Society
for Psychical Research's board of directors in 1974:

> My employment situation has remained unsettled and several
> psychic interests have developed here in Santa Barbara and I
> find that I just don't have time for real involvement in the soci-
> eties [*sic*] activities in Los Angeles. . . . Uncritical belief in any
> and all claims can be reserved for religious groups, systematic
> examination of minute details of particular phenomena may be
> left to the scientific journals, but a good down to earth examina-
> tion of the local psychic folk can do a lot to put the whole field
> in perspective for society members and the man on the street.[32]

Despite the shadow it cast over his career, Phillips's interest in parapsychology lasted the rest of his life. As his son Glen (front man for Toad
the Wet Sprocket) put it in a song memorializing his father, "Dad's in
the garage / place of legend and fable / . . . 43 boxes of textbooks on
physics / 20-odd more on parapsychology / Well there's DTP's papers,
DTP's projects."[33]

With his employment "unsettled," Phillips looked both to his garage
projects and to his interest in parapsychology to earn a little extra cash.
Thus, at about the same time he joined Science Spectrum, he founded
his own start-up, Glendan Company. Glendan was, in many ways, what
management scholars call a "lifestyle company"—the line between
its products and Phillips's own hobbies was often blurry, and Phillips
sometimes offered Glendan's services to the local parapsychology community with little regard for profit.[34] One of the projects Phillips was
most keenly interested in, for instance, was an amplifier he built for
Bill Welch, a devotee of so-called "voice phenomena" (i.e., intelligible
sounds of paranormal origin found in tape recorder hiss and other electronic noise). His correspondence with and about Welch speaks of Phillips's desire to know what the voice phenomena are and how to make
them clearer, with no mention of turning his amplifier into a commercial
product or attempting to derive any revenue from his work.[35]

However, Phillips clearly had bigger ambitions for Glendan than just
building kits for friends. To that end, he sought out at least two angel investors, Winifred Babcock (a New Age author) and Irving Laucks
(a prominent chemist, industrialist, and local peace activist).[36] His approach to Laucks even resulted in a formal "Agreement regarding ESP
Teaching Devices," in which Phillips agreed to pay Laucks a royalty on
Glendan's sale of an ESP testing device.

LEARN ACUPUNCTURE!
THE ACUPOINTER ACCURATELY LOCATES ACUPUNCTURE POINTS!
Electronic Proof! Shows acupuncture points are real by
electrical resistance measurement.
Educational! Essential tool for studying acupuncture.
Effective! Novel circuit* works for almost all points.
Handy! Pocket sized, no loose wires or electrodes.
All Solid State! Integrated circuit, LED indicator light.
Safe! Battery operation, no power line connection.
Economical! $24.95 including acupuncture wall chart. 5% tax in
Calif. Prepaid orders only. Immediate delivery.
No Risk! Return within 10 days for full refund.
*Pat. Pend.
Glendan Company, 64 Warwick Place, Goleta, California 93017

FIGURE 3.2 Advertisement for the "Acupointer," a device for locating acupuncture points, marketed by David Phillips's Glendan Company. (A similar device, the Tobiscope, was marketed by Virgil Elings's firm, Santa Barbara Technology.) Ad appeared in *Today's Chiropractic* 2, no. 5 (October/November 1973): 32. Reprinted by permission of Linda Phillips.

Phillips also wasn't above drumming up business like any small-time entrepreneur. When he saw articles in the *Los Angeles Times* and *Psychic* magazine on Soviet experiments with acupuncture and detection of "bioplasm," for instance, he developed and then marketed a gadget based on the Soviet research.[37] He advertised this hand-held transistorized pen, the Acupointer (fig. 3.2), in several New Age and chiropractic journals in 1973 and even fired off a letter (on Glendan stationery) to the editors of the *Journal of the American Medical Association* chiding them for dismissing acupuncture and then announcing that "[t]o meet the need for an effective, economical device to locate these [acupuncture] points, Glendan scientists have developed the ACUPOINTER."[38]

Like the early biotech entrepreneurs of a few years later, Phillips amplified his start-up's resources and potential customer base by leveraging his continuing connection to academia. Of particular relevance for his parapsychology work was Phillips's collaboration with Robert Morris, a well-known parapsychologist housed in UCSB's Tutorial Program. For several years, Morris and Phillips cotaught a course on parapsychology; they also collaborated on research, to the point that they cowrote a proposal for NSF funding for investigations into "non-causal correlation" (i.e., psychokinesis or remote viewing) and even submitted an article entitled "Observer Influence upon Measurements of an Electrical System" to *Science*.[39]

In presenting his parapsychology work, Phillips tended to list his affiliation with UCSB rather than Science Spectrum. For instance, when he hosted an "audience participation clairvoyance test" in Los Angeles, he told reporters that he was a "physicist and research associate at UCSB."[40] Likewise, when prominent figures in parapsychology came to town, Phillips made the most of his campus connections. For instance, in 1973 he brought the Dutch psychic detective Gerard Croiset into the laboratory of a new hire in physics, Paul Hansma, to try to mentally influence Hansma's experimental equipment (the kind of effect Phillips and Morris investigated in the article they submitted to *Science*). Afterward, Phillips wrote the following to a friend:

> My experiments with Croiset turned out rather poorly. The change in the apparatus we observed when he was in the lab turned out to be a familiar consequence of drift in the bias current in the detector to an unstable point. The designer of the experiment, Prof. Hansma, said that it was nothing unusual. Unfortunately, no one was present at the time who understood the apparatus thouroughly [*sic*].[41]

Phillips also tried to get one of his angel investors, Winifred Babcock, to donate a grant or a gift to UCSB so that he could use the university's facilities to pursue parapsychology research. As we will see, he also worked with a few UCSB students to build parapsychology experimental apparatus.

A Pedagogical Experiment

In addition to his work with Morris, Phillips had another campus connection that also fed into his off-campus entrepreneurship. At the same time that he founded Glendan and joined Science Spectrum, Phillips became involved in a remarkable pedagogical experiment in the UCSB physics department. In 1971 an assistant professor named Virgil Elings took charge of a new degree program offered by that department: a master of scientific instrumentation (MSI). Others had a hand in starting the program, and Phillips was one of its instructors for the first few years, but among tenure-track UCSB faculty Elings was almost alone in carrying the program forward. As an external review committee put it in 1974, the MSI program's

> major strength is clearly the interest and dedication of its faculty director, Dr. Virgil Elings. . . . Herein, however, lies the major weakness of the program. It is entirely dependent upon the continued interest and activity of Dr. Elings. If he should become

> disabled, find another position, or take a sabbatical leave there
> is no one to continue the necessary leadership.[42]

According to the review committee, other tenure-track faculty took little interest in the program, even though the whole department benefited from it: "The other 19 ladder [tenure-track] faculty members [besides Elings] supervise 26 graduate students. The Physics Department is being credited for 20 graduate students (MSI) who are being handled by one ladder faculty member [Elings], thereby giving the other faculty members an exceptionally light load."[43]

Those twenty MSI students were a much-needed boon to the department, since without them the department's graduate enrollments would have flat-lined for almost a decade at about two-thirds of their 1969 peak. Indeed, Elings today indicates that the MSI program was created "because we thought the Ph.D. program was going to die."[44] Yet in taking ownership of the MSI program, Elings abandoned the kind of high-energy physics research that he had originally been hired to do (forsaking the PhD students and federal grants that that kind of research attracted). As a result, Elings risked lowering himself in his colleagues' eyes—particularly since his new career course centered on low-status, terminal-master's-degree students. Thus, the 1974 external review expressed alarm about "the serious threat to [Elings's] career in academic physics if he stays with the MSI program for more than just a few years . . . [rather than] go back to the mainstream of physics."[45] The warning was well taken; whereas Elings spent the full eight years moving from the assistant to associate rank, and another eight moving from associate to full professor, junior colleagues who worked with PhD students were promoted much faster. For instance, Paul Hansma, the electron tunneling spectroscopist mentioned above who was hired in 1974 (and whose fortunes would later be closely tied to Elings's), spent only three years as an assistant and another three as an associate.[46]

Elings's new direction was such a threat to his career in part because it meant he was no longer contributing to his department's long-range strategy to achieve "a sudden rise to international distinction in a period of only a few years."[47] That strategy was constructed in imitation of peer departments at Stanford, MIT, and elsewhere that had successfully adapted to the postwar funding landscape. Those departments prioritized training doctoral (rather than undergraduate and especially terminal-master's-degree) students in a narrow set of basic research fields that were highly valued by federal funders.[48] Prestige subfields such as high-energy physics that could attract the best doctoral students and generate large grants from the Atomic Energy Commission and other national-security funders were an obvious stepping-stone into the top tier. Hence, in the mid-'60s UCSB gradually began assembling faculty clusters in high-energy research and other basic research fields linked to the military-industrial complex in order to compete better for students and grants.

Several trends combined in the late 1960s to undermine that strategy. First, federal defense and space (closely related to defense) R&D budgets began dropping as a percentage of total US R&D in 1965 and only began rising again in 1981.[49] Total federal R&D (defense and non-defense) funding reached a peak of 3 percent of GDP in 1964 and declined steadily until the 1980s.[50] During that period, Congress and both the Johnson and Nixon administrations also shifted away from the early Cold War funding model and toward research that was deemed "relevant" to civilian social problems.[51] For example, the Mansfield Amendment curtailed the ability of military funding agencies to sponsor basic research (such as most academic high-energy physics) that was not directly applicable to battlefield technologies, while the NSF began placing much more emphasis on engineering and on research that would address civilian needs. Perhaps most importantly, though not easily foreseeable at the time, federal funding for the life sciences began a steady upward climb (in constant dollars) while funding for the physical sciences remained flat for almost three decades. According to the NSF, in 1971 federal funding for the life sciences barely surpassed that for the physical sciences; by 1998 the life sciences were attracting more than twice as much federal funding as the physical sciences.[52]

The second, related trend affecting physics departments was the souring job market. In 1970 alone, "sixteen-thousand scientists and engineers, each holding advanced degrees, lost their aerospace-industry jobs. . . . The number of academic [physics] jobs on offer at the [April American Physical Society] meeting dropped . . . [by] a factor of four" from 1968 to 1971.[53] With fewer grants to support stipends, and dimmer job prospects for graduates, enrollment in physics PhD programs waned. As David Kaiser has shown, the number of physics doctorates awarded annually by American universities peaked in 1971, then dropped by a quarter by 1974 and by more than a third by 1978.[54]

These trends significantly hindered the UCSB physics department's move toward the top tier. As a UCSB Office of Research report complained, the department's postwar strategy focusing on basic research fields was increasingly out of touch with federal priorities:

Shortly after World War II, basic research flourished, however, in recent years a new emphasis has evolved toward applied research. In fact, there is a growing concern that even the National Science Foundation may be failing in its original mission and may in fact assume the role of directing instead of supporting research. This change in trend for NSF is evident in its RANN program (Research Applied to National Needs) and in other new approaches that are in their proposed budget. On this campus over 90% of the 313 active projects [over all departments] continue to be directed toward the study of basic research. The increased competition for the shrinking basic research funds has undoubtedly contributed to the 22% decrease in the value of awards and an 8% decrease in the number of awards recorded for this reporting period.[55]

In tandem with that funding shortfall, PhD enrollments in the UCSB physics program dropped precipitously between 1969 and 1970 (see fig. 3.3).

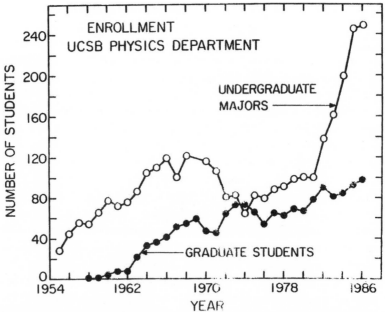

FIGURE 3.3 Number of UCSB physics graduate and undergraduate students, 1955–86. Figure reprinted from Martin Kenney and David C. Mowery, eds., *Public Universities and Regional Growth: Insights from the University of California* (Stanford: Stanford University Press, 2014), courtesy of Stanford University Press. Figure originally appeared in Paul H. Barrett's departmental history manuscript, 1987, UCSB Department of Physics records.

Figure 3.3 also offers evidence of a third trend affecting academic physics departments in the Vietnam era: disaffection arising from the war and other cultural changes. Undergraduates who a few years earlier had seen a physics degree as a path to patriotic public service, stable employment, and deep understanding of nature now stayed away. At UCSB, undergraduate enrollments in physics, according to former department chair Paul Barrett, "peaked at 2.35% [of total UCSB undergraduate population] in 1959 (the post-Sputnik bulge), fell to a low of 0.64% in 1976 (post-Vietnam reaction), and rose to only 0.74% in 1981."[56] According to Barrett, even the recovery after 1981 was more an artifact of UCSB's engineering departments' newly stringent admissions standards than of physics' overcoming its legitimacy deficit.

As Bill Leslie, Matt Wisnioski, and others have documented, student disaffection stemmed from the perception that physical and engineering science departments were too narrowly focused on research funded by national-security agencies, too technocratic and elitist, and thereby too insulated from civil society's concerns.[57] Student activists at these schools opposed not only applied, defense-oriented research but also basic research on the grounds that both did little to solve civilian problems such as pollution.[58] I have explained elsewhere that campus activists believed that applied, defense-related research tended to be more interdisciplinary than basic research and therefore argued that research applied to civilian issues should also be interdisciplinary.[59] At the same time, budgetary pressures encouraged university researchers to cross disciplinary lines to find new funders. Thus, many academic physical and engineering scientists in the early '70s forged much wider-ranging collaborations than were typical in the early Cold War, especially with life science and medical school departments and with humanistic disciplines such as philosophy and music.

Members of the UCSB physics faculty were keenly aware that a student body that had absorbed the values of the counterculture and protest movements wanted their professors to move toward civilian, applied topics and a more interdisciplinary, humanistic outlook. Department brochures from the '70s complained about the counterculture's and the general public's

> serious misconceptions about science. Science is not, as many believe, a mere collection of facts. Scientists are not neutral and colorless professionals. Science is not a set of special tricks for elite minds. . . . [P]hysics is also practical. It serves as a fundamental science in preparing the base for all technological ad-

vances. In the same sense, it serves as one of the most important guides in making difficult technological decisions.

Physics is far more than a technical study. It is part of the totality of our human experience. The influence and interconnections physics has had on our history, on our religions, on our philosophies, on our languages and literatures are deep and broad. To study physics is to begin to see these interconnections and, eventually, to give ourselves a more complete picture of our place in our intellectual heritage.[60]

Note the appeal to a more holistic worldview ("interconnections") and the nod to skepticism of technocracy ("scientists are not neutral and colorless") and authority ("science is not [just] for elite minds [and] is also practical"). Other parts of the brochure appealed to the questing spirituality of the era in language similar to Fritjof Capra's *Tao of Physics*: "Physics is for anyone who enjoys recreating, through understanding and appreciation, those discoveries which give both unity and wonder to the apparent complexity of our physical universe." At the same time, it obliquely defended physicists from association with the conflict in Southeast Asia and argued for their role in overcoming democratic deficits: "physics now seems essential for an informed citizenry. Only then can technologies be encouraged which serve our most important human concerns. At the same time, we will be able to watch out with uncompromising attention for flagrant misuses of our present and future knowledge."

Of course, brochures aside and despite evident student dissatisfaction, most physics faculty members remained focused on basic, relatively monodisciplinary research. Still, UCSB's physics department experimented with various ways to address its budgetary and legitimacy crises. For instance, Melvyn Manalis (a 1970 UCSB PhD, and then a research physicist and lecturer) taught a course on "environmental physics," which was intended to show that "physics is very relevant to modern environmental topics."[61] Similarly, Herb Broida—for whom the current physics building is named—established a Quantum Institute in 1969 to serve as "a problem oriented, interdisciplinary research unit . . . designed to respond rapidly to requests for assistance in the solution of problems important to governmental agencies, private industry, and research foundations."[62]

The department also created a Physics Learning Center (PLC) to make experimental equipment available to "students from all disciplines. With hundreds of demonstrations maintained in an atmosphere of free investigation, the PLC has established itself . . . in the service to local schools and the community."[63] After all, one of the recurring demands of student protesters was that science departments become more involved in their local communities. Interestingly, the PLC was largely the brainchild of Tony Korda, one of the technicians in the department's workshop. Like Phillips, Korda was deeply interested in parapsychology and New Age spiritualism. Indeed, he told Phillips that he possessed a "powerful 'energy' or Kundalini experiences . . . including the flame and many others" and that the inspiration for the PLC had been a Kundalini-esque "experience . . . of poetic and artistic inspiration, together with strong feelings of love for others."[64]

It's in the light of the PLC and Manalis's environmental physics course that we should probably understand Elings's MSI program. That is, these new departmental initiatives were able to be created, and even prosper, because they offered to alleviate the department's budgetary, demographic, and cultural legitimacy crises. Each was, as an external review described the MSI program, "an imaginative response to what seems to be a real need."[65] Yet in each case, the response to crisis was placed in the hands of someone at the margins of the department (Manalis, Korda, Phillips) or someone moving toward the margins (Elings) because of his involvement in the initiative. The rest of the department, meanwhile, continued along much the same trajectory that it had before, if under more straitened circumstances.

The Esoteric and the Tangible

So much for Elings's and his colleagues' understanding of the MSI program. What did MSI students think of it? It's clear that, in the first few years of the program, Elings's recruitment strategy attempted to tap into potential students' desire to do some good in the world in an interdisciplinary fashion, while learning marketable skills to help find scarce jobs. The original program proposal to the NSF, for instance, stated that "the program will satisfy a long felt need for interdisciplinary training."[66] Ads told potential applicants: "The UCSB Master's Program in Scientific Instrumentation is looking for Creative, Hard-working Bachelor's Degree Scientists who want to Solve Real Problems. . . . The one-year program, supported by the National Science Foundation, provides opportunity and experience in interdisciplinary problem solving . . . on campus and in nearby hospitals and industrial laboratories."[67] Or, as Elings told a local reporter, the MSI program was to be "a leader in the growing movement to involve students directly in interdisciplinary research by tackling problems in areas outside their formal training."[68]

The projects that early MSI students worked on reflected that desire for interdisciplinary, civilian, problem-oriented research. These included "an image stabilizer to aid in early detection of cancer in internal organs; a color television speech display system to teach the deaf to speak [fig. 3.4]; a digital heart rate monitor for use by physiologists";[69] a tactile sound mixer built for a blind audiophile undergraduate;[70] ocean-monitoring instrumentation packages;[71] and "a portable device to detect and measure the quantity of lead in gasoline."[72] The origins of the speech display system (known as the "Chromophone") were typical of the program's interdisciplinary reach:

> The need for such a device was voiced by the staff of the Speech Communication Research Laboratory of Santa Barbara in a conversation with Prof. Glen J. Culler of the department of elec-

FIGURE 3.4 David Rosales, a five-year-old hard-of-hearing child, testing the Chromophone, a device for converting sounds into visual patterns on a television. The Chromophone was a product of the UCSB Physics Department's master of scientific instrumentation program. Later MSI students developed it into the SPOT (speech-optical trainer), which was sold by both David Phillips's and Virgil Elings's start-up companies. Photograph by Wilfrid Swalling. From folder "Physics—Scientific Instrumentation," box 79, UArch 12, Office of Public Information Subject Files, University of California, Santa Barbara, Special Collections. Reprinted by permission of the UCSB Special Collections.

trical engineering. . . . One evening he mentioned the project to physicist David Phillips, and the next day Dr. Phillips and Donald Holly, a graduate student in the scientific instrumentation program, were at work with some discarded color TV sets.[73]

Students also acquired interdisciplinary skills and benefited the university by building free experimental apparatus for researchers in many UCSB science departments, especially physics. In lean budgetary times, free equipment must have been highly desirable; during the first year of the program, "the number of cooperating labs on campus increased and soon there were more labs requesting students than there were students."[74]

Many students also used their master's projects to build bridges between the university and surrounding communities. Organizations such as the Santa Barbara Cancer Foundation, the VA hospital in San Francisco, St. Christopher's Hospital for Children in Philadelphia, Cedars Hospital and the John Tracy Clinic in Los Angeles, Goleta Valley Community Hospital, and the Santa Barbara Heart and Lung Institute hosted MSI students and/or field-tested their projects. Elings and Phillips were aware that working with MSI students taxed these organizations, but

> [t]he experience also provides the companies an opportunity to get acquainted with possible employees. . . . Such experience will benefit the student by letting him [*sic*] see what is going on "outside." Also, we feel that the ties which are built between the university and local industry in this way will provide valuable research ideas for both the campus and industry and future employment opportunities for students.[75]

In fact, the MSI program seems to have done quite well in cultivating "future employment opportunities" for graduates. As the UCSB press office proudly noted in 1972, "in contrast to the experiences of students in some other scientific fields, every one of the program's continuing students found employment in instrumentation last year."[76] *Physics Today*, too, lauded the program for "preparing students for physics-related jobs."[77]

Thus, for students, the program offered both an outlet for baby boomer idealism *and* a pragmatic solution to the problem of a poor science and engineering job market. As Mike Buchin, a mid-'70s MSI student who helped invent both the SPOT (speech-optical translator, a variant of the Chromophone) and a thermodilution method for measuring blood flow, put it:

> I got involved in this project mainly out of my own interest in medically-oriented projects. Also, I like getting involved with projects that will help people—applied project work. In the long run, my work has paid off tangibly as well as esoterically (?). I am going to be working as a Research Physicist for Searle Analytical and Radiographics, a firm in Des Plaines, Illinois deeply involved in nuclear medicine.[78]

Countercultural Curriculum

The MSI program was created to meet the needs of a very specific moment in the history of American science and academia. When those conditions receded, the program began to drift and eventually disappeared. In its early years, though, it attracted large cohorts of apparently very capable and enthusiastic students who brought certain aspects of the counterculture into the classroom and, in turn, reshaped Elings's pedagogy and research.

In a minor way, David Phillips was a conduit for some of those countercultural preoccupations. Throughout the MSI program's run, Elings relied on a junior partner to give lectures and advise on projects. For the first five years or so, Phillips was that partner. Unsurprisingly, he brought his parapsychology interests along. For instance, Phillips wrote to the education director of the American Society for Psychical Research in 1972 to say that "I wish to include some ESP and OOB [out-of-body] experiments in the curriculum of the Scintific [sic] Instrumentation Program at UCSB."[79] At least one MSI student, John Placer, built an apparatus for parapsychology research for his master's project (and coauthored an article with Phillips and Robert Morris).[80] At least one other, George Jahn, helped Phillips and an undergraduate, Agi Ban, build a plethysmograph—a standard medical device for measuring lung or blood volume, though Ban and Phillips were following other parapsychologists in using theirs for telepathy experiments.[81]

Of more lasting significance were the countercultural aspects of the program's pedagogical approach. In an era of profound undergraduate distrust of authority, when a top demand of many campus protesters was for student-run courses, Elings and Phillips quickly learned that traditional lectures would neither hold MSI students' attention nor help them with their projects. "The too familiar pattern of factual presenta-

tions of material chosen by the instructor was more tolerated than appreciated. . . . One year's experience indicates that experimental science
students learn more from doing their 'own problem' than from passive
reading and exercises."[82] As a result, Elings became a lifelong advocate
of learning-by-doing and an increasingly harsh critic of conventional
pedagogy. As he described it later, the MSI program was so unconventional in insisting that students learn things for themselves that for some
students it made for "a rude awakening from the spoon-feeding of most
undergraduate experiences."[83]

Over time, Elings's experience with the first few cohorts of MSI students seems to have inspired, or at least amplified, his populist distrust
of authority and pedigree. That, in turn, fed back into the program's
curriculum and the technological approaches students absorbed. Here,
again, the national economic and cultural dislocations of the Vietnam
era played an indirect role. As Christophe Lécuyer and others have
shown, Robert McNamara's attempts to trim military budgeting in the
mid-'60s led to a flattening of semiconductor firms' profits known col-
loquially as the "McNamara Depression."[84] In response, some firms
began pursuing consumer markets much more aggressively. Doing so,
however, required manufacturers to dramatically lower the cost per
chip. By the early '70s, Intel and a few of its peers had succeeded to the
point that electronics hobbyists could easily buy chips that only a few
years earlier would have been beyond their reach.[85] From there, it was
only a small step for hobbyists to begin building "personal computers"
of the kind that Ted Holm and other counterculture icons had been advocating for years—a story now thoroughly ingrained in the supposedly
subversive mythography of Apple, the Whole Earth 'Lectronic Link, and
other Silicon Valley icons.[86]

Less noticed in both popular memory and academic history, however, is that the sudden availability of cheap digital components also
made certain kinds of scientific instrumentation much easier to build.[87]
Elings and Phillips were both quick to embrace cheap digital circuits
for their own projects; note, for instance, in figure 3.2 that Phillips
advertised the Acupointer as "All Solid State! Integrated circuit, LED
indicator light." Phillips and Elings also soon concluded that cheap
integrated circuits made possible a new pedagogical approach to the
MSI program. Since the new, cheap microprocessors could be repro-
grammed for a variety of different tasks, students could be taught the
basics of microprocessor programming as a universal solution to any
instrumentation problem. As Elings put it a few years later, "one exciting aspect of microelectronics is that the resulting 'intelligent' instru-

ments may be easily redesigned to meet altered specifications, often by simple software changes."[88]

Learning to design digital circuits therefore gave MSI students skills that they could adapt to almost any employer's needs—since microprocessor technology provided a flexible tool for problem solving not just in instrumentation but in many other industries as well. Thus, "students now find that a little background in applying electronics to measurement problems opens doors to many jobs that they would otherwise not have access to."[89] As Elings began exposing MSI students to the advantages of digital circuits, however, he came to see that such circuits cast the necessity of a university education in doubt. Anyone, it seemed, could learn to design digital circuits: "one no longer needs a degree in electrical engineering to design state-of-the-art electronic instruments. . . . It isn't true that one needs a course in differential equations or Fourier analysis in order to design top-notch instruments."[90]

If that was the case, then was a degree in electrical engineering worth it? Elings started to think not. As he put it in 1995, "the areas that [MSI] students had done undergraduate work in made little difference in their ability to design instruments. . . . All those esoteric courses made little difference."[91] Some of Elings's best students came in with very unlikely backgrounds: one whom Elings still refers to often and who contributed greatly to his later endeavors entered the MSI program having done his undergraduate work at UCSB in psychology and religious studies. In a sense, Elings's growing disregard for academic credentialing, grounded in his experiences with the MSI program, aligned him surprisingly well with populist science groups of the era such as Science for the People and the New Alchemists.[92]

Inching toward the Marketplace

Over the short term, then, the MSI program trained students for employment, helped UCSB faculty cheaply build research equipment, and provided local firms and hospitals with cheap labor and fresh insight on technical problems. Over the longer term, however, MSI students' idealistic inventions were of little benefit to anyone without some sustained technology transfer out of the university. For that, Elings and Phillips began to take further steps toward entrepreneurship.

In some ways, Elings and Phillips must have been an odd pair: Elings the cynical high-energy physicist with a penchant for blunt, even shocking, language; and Phillips the earnest, spiritual seeker and bench-top experimentalist. Yet the two also had a great deal in common: both grew up in Iowa, graduated from Iowa State in the early '60s, went off to big-name physics PhD programs (Elings, to MIT; Phillips, to Berkeley), and were hired at UCSB in 1966.[93] Both had worked in the trenches of the military-industrial research complex before taking faculty positions: Phillips at the Naval Research Lab and Elings at Union Carbide Nuclear (where he helped "in the design of a nuclear reactor") and Autonetics (where he performed "an analysis on the gyroscope used in the Minuteman missile").[94]

When they got to UCSB, neither contributed to their department's PhD program to quite the extent that their colleagues probably had hoped: Phillips supervised one dissertation; Elings, none. Perhaps they were distracted by outside pursuits: Phillips had his parapsychology and Glendan Company; Elings also had a garage start-up, Santa Barbara Technology (SBT). Indeed, there must have been some cross-marketing or other cooperation between their two start-ups. For instance, in 1973 Elings advertised a product virtually identical to Glendan's Acupointer under the brand name "Tobiscope."[95] Similarly, speech therapists, schools for the deaf, and foreign-language instructors could order SPOT units from SBT as early as 1973, and from Glendan as late as 1981.[96]

Like Glendan, SBT's earliest products were not associated with the MSI program. In addition to the Tobiscope, SBT was perhaps best known for a toy consisting of two parabolic mirrors stacked with their concave sides facing each other, such that an object placed on the bottom mirror appeared to float above the top one. According to legend, this effect was discovered by accident by a UCSB student (in some versions, a "janitor" or "maintenance man") doing some housekeeping in a lab.[97] More certain is that Elings and Caliste John Landry, a PhD student in electrical engineering, patented the device together.[98] Eventually, Elings licensed the patent to another start-up, Optigone, which makes most of the versions of this product that are seen in science museum gift shops today.

Both Glendan and SBT, however, also marketed products based on MSI projects, such as the SPOT and Buchin's thermodilution cardiac-output computer. Elings's later detractors and competitors sometimes spin this as an early example of Elings's questionable ethics regarding intellectual property. That is a reasonable interpretation, since, for example, Elings and Phillips (but not Mike Buchin) took out a patent for the thermodilution cardiac-output computer *after* it had been the subject of Buchin's MSI project.[99] However, it is important to note that one of the critical components of Buchin's version of the cardiac computer was a Wheatstone bridge (a Victorian-era device for measuring electrical resistance), whereas Elings and Phillips's patent ostentatiously states, "A Wheatstone bridge is not used."

Yet given the ad hoc nature of both the early MSI program and university intellectual property practices at the time, Elings and Phillips were probably the only people in a position to make MSI students' projects available to organizations such as hospitals and schools for the deaf. Glendan and SBT may also have served as career models for graduates of the program. Certainly, a fair number of graduates became freelance technicians who did small contract projects for local high-tech companies, either full-time or to supplement their income—work that resembled both what they had done in the MSI program and what Elings and Phillips were doing with their garage companies.[100]

In any case, SBT sold cardiac computers until the early 1980s, and biomedical researchers reported using them in journal articles published well into the 1990s. SBT's other major product in the late '70s was a "dermofluorometer" used to determine whether injured skin flaps were healing properly. The cardiac computer and dermofluorometer marked a new phase in Elings's career in which the MSI program, his own on-campus research, and his off-campus entrepreneurship began to converge. As Elings's departmental biography noted, once he began directing the MSI program "he shifted his research interest to medical instrumentation. Many MSI students gained valuable experience working on some of these projects. Among them was a lung water measurement system, and a light-scattering immunoassay instrument."[101] Today (2012), Elings ascribes this shift to dissatisfaction with the anonymous teamwork typical of high-energy physics—a reason given by other high-energy physicists who moved toward biomedical entrepreneurship in the '70s, such as Cetus cofounder Donald Glaser.[102]

Of course, Elings may also have seen that turning toward biomedical research opened up new funding possibilities: unlike most of his departmental colleagues, he could now solicit funding from places like the American Heart Association. That was the case for a project he began with George Jahn, an MSI student who left UCSB to become a cardiologist, and John Vogel, a doctor at the Santa Barbara Heart and Lung Institute who had previously mentored MSI students such as Mike Buchin. That project led, in turn, to the dermofluorometer line that SBT began selling.[103]

About the same time, Elings also began collaborating on biomedical instrumentation with a new UCSB physics assistant professor, David Nicoli. Upon joining the physics department, probably in the 1976–77 academic year, Nicoli seems to have taken over David Phillips's role as junior partner in the MSI program. In 1978 Elings and Nicoli formed a start-up company, Nicomp, to sell an instrument that they had developed on campus for determining the size and morphology of particles suspended in solution. Well before the Bayh-Dole Act and Genentech's record-setting initial public offering (both in 1980), Elings was on his way to becoming a serial academic entrepreneur.

Into the Reagan Era

The MSI program, meanwhile, continued on, but perhaps with less passion and certainly with less publicity. In the program's earliest days, the best students had been treated as Elings's near peers intellectually and socially. As Mike Buchin puts it, Elings "wasn't a hippie, but he was relaxed" and happy to party with students on occasion.[104] He also coauthored with at least four of the early graduates, boosting his own publication productivity while giving them a useful credential. By 1974, though, the department's external reviewers observed that the MSI program was no longer attracting as high a caliber of student and warned against "the danger of attempting to maintain constant size, rather than setting a floor on incoming qualifications and letting the class size be determined by the number of qualified applicants."[105]

David Phillips and Philip Wyatt also drifted somewhat in the late '70s, as Science Spectrum's cash flow problems forced cutbacks even among its senior staff. As a result, Phillips picked up consulting jobs and, for a few stretches, employment at other local firms such as Renco, a small electronics manufacturer.[106] Wyatt, meanwhile, began winding his company down. In late 1980 or early 1981, he invited his staff to toast the dissolution of Science Spectrum. This being Santa Barbara in the early '80s, their late-afternoon celebration turned into an open-door wine tasting and then into a laboratory experiment when someone decided to test the wines with the company's light-scattering instruments.[107] As a last hurrah, Wyatt sent a paper to *Science* comparing the output of the light-scattering measurements with the ratings of a "nonprofessional consumer panel" consisting of a "physicist, mathematician, electrical engineer, office manager, x-ray technician, coin dealer, electromechanical assembler, and attorney."[108]

Then, to hear Wyatt tell it, something of a miracle happened. His letter to *Science* was picked up in various outlets, resulting in overtures from soft drink companies to perform analyses of their products with his light-scattering instruments. Those requests then led to further articles and publicity, snowballing into even more interest from the military and other customers.[109] In 1982 Wyatt was able to reincorporate, this time under the name Wyatt Technology. Before long, David Phillips was consulting for Wyatt Technology and eventually joined the firm. Interestingly, Phillips's notebooks show that he was still dabbling in parapsychology, but with a 1980s twist. There, right next to computer code for Wyatt instruments and notes on meetings between Wyatt engineers and representatives from the US Army, are data Phillips collected from possible ESP adepts trying to predict the value of the next day's Dow Jones index and the price of silver![110]

For Elings, the early '80s were less eventful but, by extension, less exciting. It wasn't until 1986 that the professional path he had embarked on in 1970 was closed down and a new path opened up. This was a two-step process. First, at a department talk by a Spanish physicist, Nico García, Elings learned the details of the scanning tunneling microscope (STM), a technique that García and Elings's departmental colleague Paul Hansma had each been working on since 1983. During the talk, Elings came to the conclusion that his expertise with digital circuits would allow him to build and sell STMs that were as good as the homemade instruments Hansma and García had. After attending one of the first international STM conferences, he realized that STM was hot enough—its inventors having just received the Nobel Prize— that there was a sizable market for a commercial instrument. Within a few months, he had sold his share of Nicomp and invested the funds in cofounding an STM company, Digital Instruments (DI), with John "Gus" Gurley, one of his best MSI students.

The second shove in a new direction that Elings received was the— apparently quite sudden—cancellation of the MSI program. With the mid-'80s rebound in PhD and undergraduate enrollments and federal research funding, the exigencies that led to the program's creation had disappeared. Long-standing qualms about how a terminal-master's-degree program affected the department's reputation now became more urgent. With Nicoli having left for Nicomp, Elings was essentially alone in running and championing the MSI program. So when a new department chair took over in 1986, the ax fell.

I've written extensively elsewhere about what happened next.[111] DI turned a profit in every year of its existence by supplying approximately half the world's scanning probe microscopes. Elings and Hansma, despite some ups and downs, established an extraordinarily lucrative (for all parties) technology transfer procedure between DI and UCSB. Yet in commercializing a fellow professor's inventions, Elings strained relations with UCSB in ways that he had not when commercializing MSI students' projects—leading Elings to quit his faculty position in 1990. Over the next decade, Elings turned his firm's attention toward industrial customers, leading to DI's "merger" with a semiconductor process equipment company, Veeco, in 1998 for approximately $150 million. Since then, Santa Barbara has become home to a cluster of nanoscience instrumentation companies founded by veterans of DI, the Hansma group, and other UCSB research labs.

Since leaving Veeco, Elings has become a prominent local philanthropist and occasional speaker on entrepreneurship and innovation. In both these guises, he often refers to the MSI program as a source of DI's success and a model for the future of science and engineering education and high-tech commerce. As well he should. The MSI program gave DI some of its key personnel, a successful technological and managerial philosophy, and—through SBT and Nicomp—much-needed seed funding. What is striking about at least some of Elings's philanthropy, though, is that it extends the populist contestation of conventional pedagogy and academic credentialing that arose from the MSI program's more countercultural aspects. For instance, Elings has donated a million dollars to Dos Pueblos High School's Engineering Academy to support a problem-based, learn-by-doing approach that closely resembles the pedagogical philosophy of the MSI program. The academy's founder, Amir Abo-Shaeer, takes much the same populist stance regarding conventional teaching as Elings—though it is perhaps an indication of the extent to which the "counterculture's" ideas about education now suffuse elite opinions on education "reform" that Abo-Shaeer has won a MacArthur Fellowship and is the subject of a best-selling nonfiction book entitled *The New Cool*.[112]

Conclusion

Cool or not, the type of science practiced by Phillips, Wyatt, and Elings did become the new normal. High-tech start-ups—especially those founded by or closely connected to university faculty—had become icons of economic gain and social liberalization by the time Elings left Digital Instruments.[113] With the end of the Cold War, the reorientation of the US research enterprise toward more civilian, applied research came to be seen as a critical part of the nation's "peace dividend."[114] As a result, the kind of research that Science Spectrum and the students in the MSI program had been doing in the early '70s took on a new life. At the same time, over the course of the '90s the dramatic increase in federal funding for biomedical research meant that more and more academic physicists were collaborating with clinical researchers and applying for funding from the NIH. Thus, the biomedical instrumentation projects and start-ups that had made Virgil Elings so different from his colleagues in the early '70s came to be a routine, even mandatory, part of many university physics departments' research portfolios. The overt parapsychological research championed by David Phillips receded in the '80s, but as David Kaiser has shown, that was in part because the mainstream of physics had internalized topics such as Bell's Theorem that had originally been associated with parapsychology.[115]

When historians, sociologists, and economists of science have looked at these changes in the American research enterprise, the role of the counterculture and the budgetary and cultural upheavals of the Vietnam era have largely been ignored. On the one side, those who lament these changes, such as Phil Mirowski and Paul Forman, have generally traced their origins to the success of a conspiracy of neoliberal economists around 1980.[116] On the other side, those who take a more positive or neutral stance have variously chalked up these changes to the effects of the Bayh-Dole Act, the biotech industry, professionalization of university technology transfer officers, and so forth.[117] The fact that these events were all taking place in the shadow of student protest, countercultural ferment, and a hot war in Southeast Asia is treated, if at all, as mere coincidence. Long-term processes of globalization and deindustrialization are depicted in such studies as more important factors in the transformation of the US research enterprise than the strains Vietnam placed on the economy and federal budgets.

Conversely, among historians who have examined the institutional effects of Vietnam-era protest, the story has often been one of momentary disruption, of more humane paths not followed, and a return to a kind of science that serves the state and big business.[118] The idea that the counterculture—though not the war that so much of the counterculture directed its attention toward—might have incubated the high-tech industries of the '80s has been broached most influentially by Fred Turner for computing and "New Media" and by Eric Vettel for biotechnology.[119] Still, we don't yet have a good sense of what scientists (especially bench-top experimentalists) did to respond to the straitened and disputatious conditions of the '70s—for instance, how they secured equipment despite declining budgets and attracted personnel despite rising disaffection with science.

What I've tried to show here is that, for at least some scientists in the late '60s and early '70s, the logical course was to diversify their interdisciplinary collaborations; to follow the shifting winds of civil-

ian priorities into new lines of research, especially related to biomedical, environmental, and disability technologies; to experiment with new
ways of reaching young people; and to forge new ties among industrial,
academic, government, and civil-society organizations. One vehicle for
all these strategies was to found their own start-up companies—hardly
unprecedented in the history of American science but also not common
or valorized in the early Cold War, especially among university faculty.
Since the early '90s, the start-up company has become a commonsense,
much lauded vehicle for translating a more interdisciplinary, civilian-
minded kind of science out of the laboratory (especially the academic
laboratory) and into contact with both the public and a variety of relevant organizations. There is plenty of room for debate as to whether
that is (or can be made to be) a beneficial state of affairs. But if we want
to understand where some of those lofty, transformative visions for
high-tech start-ups originated, we should look to the Vietnam era's disruptions and the unorthodox responses fashioned by the likes of Wyatt,
Phillips, Elings, and their students and collaborators.

Notes

1. For two megaprojects that were slowed by the '70s budgetary morass,
see Lillian Hoddeson, Adrienne W. Kolb, and Catherine Westfall, *Fermilab:
Physics, the Frontier, and Megascience* (Chicago: University of Chicago Press,
2008); and W. Patrick McCray, *Giant Telescopes: Astronomical Ambition and
the Promise of Technology* (Cambridge, MA: Harvard University Press, 2004).
The "facilities crisis" has not yet been addressed by historians of science, even
though it was considered a pressing policy issue at the time and probably
affected many more scientists than the slowdown in megaprojects. See *An
Assessment of the Needs for Equipment, Instrumentation, and Facilities for
University Research in Science and Engineering* (Washington, DC: National
Academy of Sciences, 1971).

2. See, e.g., Peter J. Westwick, *Into the Black: JPL and the American
Space Program, 1976–2004* (New Haven, CT: Yale University Press, 2007),
esp. chap. 5.

3. For a case study of the return to "normal" budgets, and of scientists
cajoling the Reagan administration to back basic research, see Catherine
Westfall, "Retooling for the Future: Launching the Advanced Light Source at
Lawrence's Laboratory, 1980–1986," *Historical Studies in the Natural Sciences*
38, no. 4 (2008): 569–609.

4. The quotation appears to date from 1961 and may be found in *Guide
to the G.E. Tempo Collection, 1957–1969*, MSS 139, G.E. Tempo Collection,
1957–1969, UCSB Library Special Collections (hereafter cited as UCSB LSC).

5. Peter Zavattaro, interview by Mary Palevsky, 31 May 2005, Las
Vegas, NV.

6. "Finalists vie for APS Industrial Physics Prize," *APS* [American Physical Society] *News* 17, no. 7 (2008): 1, 5.

7. Philip Wyatt, interview by the author, 13 October 2011, Santa Barbara, CA.

8. David T. Phillips, "Evolution of a Light Scattering Photometer," *BioScience* 26, no. 16 (15 August 1971): 865–67.

9. Jennifer S. Light, *From Warfare to Welfare: Defense Intellectuals and Urban Problems in Cold War America* (Baltimore, MD: Johns Hopkins University Press, 2003).

10. Dian O. Belanger, *Enabling American Innovation: Engineering and the National Science Foundation* (West Lafayette, IN: Purdue University Press, 1998).

11. Chet Holcombe, "Science Spectrum Receives Contract," *Santa Barbara News-Press*, 26 May 1970; and "Subsidies Are Urged in Pollution Fight," *Santa Barbara News-Press*, 15 January 1971.

12. "Subsidies Are Urged."

13. "Harry Brown President of Science Spectrum," *Santa Barbara News-Press*, 14 July 1972.

14. Philip J. Wyatt, David T. Phillips, and Edward H. Allen, "Laser Light Scattering Bioassay for Veterinary Drug Residues in Food Producing Animals: 1, Dose-Response Results for Milk, Serum, Urine, and Bile," *Journal of Agricultural and Food Chemistry* 24, no. 5 (1976): 984–88.

15. Philip J. Wyatt, Robert F. Pittillo, Louise S. Rice, Carolyn Woolley, and L. B. Mellett, "Laser Differential Light-Scattering Bioassay for Methotrexate (NSC-740)," *Cancer Treatment Reports* 60 (1976): 225–33.

16. Eric J. Vettel, *Biotech: The Countercultural Origins of an Industry* (Philadelphia: University of Pennsylvania Press, 2006).

17. John Hillaby, "Anti-matter Rain on Earth Hinted: Florida Physicist Suggests Some Craters Resulted from Such Assaults," *New York Times*, 26 April 1958; and David J. Shayler and Colin Burgess, *NASA Scientist-Astronauts* (Berlin: Springer, 2007).

18. Philip Wyatt, interview by the author, 13 October 2011, Santa Barbara, CA; "Harry Brown President of Science Spectrum"; Value Line Development Capital Corporation, *Annual Report 1973*, folder 3, box 29, Elgin Groseclose Papers, University of Oregon Special Collections and University Archives.

19. Phillips, "Evolution," 865.

20. See the special feature on the Industrial Research 100 competition and awards in *Industrial Research* 14 (August 1972): 28.

21. Wyatt Technology, *2009 American Physical Society Prize for Industrial Application of Physics*, accessed 22 May 2012, http://www.wyatt.com/events/news/2009-american-physical-society-prize-for-industrial-application-of-physics.html.

22. Paul H. Barrett, departmental history manuscript, 1987, found in UCSB Department of Physics basement "archive."

23. Jong-Jean Kim, "Raman Scattering from Ferroelectric Crystals of KDP Family" (PhD diss., University of California–Santa Barbara, 1970).

24. David T. Phillips and John West, "The Poor Man's Nitrogen Laser," *American Journal of Physics* 38, no. 5 (1970): 655–57.

25. Barrett, departmental history, 30.

26. Charles Tart to Lila Gatlin et al., 2 November 1976, folder 20, box 1, MSS 48, American Religions Collection, Santa Barbara Parapsychology Collection, UCSB LSC (hereafter cited as SBPC).

27. "Esalen Presents Psychic Phenomena," flyer for symposium starting 7 May 1974, folder 20, box 4, SBPC.

28. David Phillips, memo to Robert Morris, re: "Independent Evaluation of Research Program and Technique," 1975?, folder 32, box 1, SBPC.

29. David Phillips to Anne Francis, 16 January 1983, stapled into notebook "Audio Circuits 1982 to June 1984," box 1, SBPC.

30. Linda and Dan Phillips, interview by the author, 26 June 2013, Goleta, CA.

31. David Phillips to Arthur Hastings, 11 August 1974, folder 16, box 1, SBPC; and David Phillips to Barry Singer, 20 October 1980, folder 19, box 1, SBPC.

32. David Phillips to Zepha Bogert, folder 13, box 1, SBPC.

33. DTP = David T. Phillips. "Garage Lyrics," accessed 6 August 2012, http://www.stlyrics.com/songs/g/glenphillips25809/garage2036108.html.

34. Sonali Shah, "From Innovation to Firm Formation: Contributions by Sports Enthusiasts to the Windsurfing, Snowboarding, and Skateboarding Industries," in *The Engineering of Sport* 6, ed. E. Fozzy and S. Haake (New York: Springer, 2006), 29–34.

35. David Phillips to Davis M. Peck, research chairman, Spiritual Frontiers Fellowship, Rockville, MD, 4 September 1974, folder 18, box 1, SBPC; and David Phillips to Bill Welch, Encino, CA, 5 July 1974, folder 20, box 1, SBPC.

36. David Phillips to Winifred Babcock, Santa Barbara, CA, 6 February 1974, PK folder 13, box 1, SBPC; and Robert Morris and David Phillips to Irving Laucks, Santa Barbara, CA, 11 June 1975, folder 17, box 1, SBPC.

37. Harry Nelson, "Psychic Medicine—Is It Real Science or Is It Hokum? 'New Type of Energy' Attracting Attention of Experts World Over," *Los Angeles Times*, 30 July 1972; and J. W. White, "Acupuncture: The World's Oldest System of Medicine," *Psychic* 3, no. 6 (1972): 12–18.

38. David Phillips, letter to editors of *JAMA*, December 1972, folder 16, box 1, SBPC. Advertisements for the Acupointer ran in *Today's Chiropractic*, *Anesthesia and Analgesia*, and *Digest of Chiropractic Economics*.

39. David Phillips and Robert Morris, "Internal State Variables in Noncausal Correlation Studies," draft of NSF proposal, 1975; Robert Morris, letter to editors of *Science*, 27 September 1978; both in folder 32, box 1, SBPC.

40. "Clairvoyance Test Planned Saturday," *Santa Barbara News-Press*, 28 February 1973.

41. David Phillips to Walter Uphoff, Oregon, WI, 1 April 1973, folder 21, box 1, SBPC.

42. I. Fatt, E. L. Hahn, and J. D. Jackson, "Evaluation of the Graduate Program of the Department of Physics, University of California, Santa Barbara," 11 June 1974, folder 9 "Doctoral Program Evaluation Final Report, Gradu-

ate Council, August 1, 1977," box 88, UArch 13, Academic Senate Records, UCSB LSC.

43. Ibid.

44. UCSB Technology Management Program, 2008, interview with Virgil Elings, accessed 16 July 2012, http://www.youtube.com/watch?v=H9aQBF7rIg8. See also Virgil Elings, interview by the author, 9 May 2012, Santa Barbara, CA.

45. Fatt, Hahn, and Jackson, "Evaluation."

46. Information here and elsewhere on the dates of UCSB physicists' employment and promotions is taken from inspection of university directories available in the Special Collections of the UCSB Library.

47. Barrett, departmental history, 4.

48. Stuart W. Leslie, *The Cold War and American Science* (New York: Columbia University Press, 1993); and Rebecca S. Lowen, *Creating the Cold War University: The Transformation of Stanford* (Berkeley: University of California Press, 1997).

49. National Science Foundation, *National Patterns of R&D Resources: 1994* (Washington, DC: National Science Foundation, 1994).

50. H. A. Neal, T. L. Smith, and J. B. McCormick, *Beyond Sputnik: U.S. Science Policy in the 21st Century* (Ann Arbor: University of Michigan Press, 2008).

51. See Vettel, *Biotech*; and Belanger, *Enabling American Innovation*.

52. National Science Foundation, *Federal Funds for Research and Development FY 2002, 2003, and 2004* (Washington, DC: National Science Foundation, 2004).

53. D. Kaiser, "Cold War Requisitions, Scientific Manpower, and the Production of American Physicists after World War II," *Historical Studies in the Physical Sciences* 33 (2002): 131–59 (quotation on 152).

54. Ibid.

55. UCSB Contracts and Grants Office, "Annual Report on Activities Financed from Extramural Funds," 1972, folder "Annual Report on Activities Financed by Extramural Funds, Sept. 1972," box 1, UArch 87, Office of Research Collection, UCSB LSC.

56. Barrett, departmental history.

57. Stuart W. Leslie, "'Time of Troubles' for the Special Laboratories," in *Becoming MIT: Moments of Decision*, ed. D. Kaiser (Cambridge, MA: MIT Press, 2010), 123–44; and Matthew Wisnioski, "Inside 'the System': Engineers, Scientists, and the Boundaries of Social Protest in the Long 1960s," *History and Technology* 19 (2003): 313–33.

58. Vettel, *Biotech*.

59. Cyrus C. M. Mody, "Conversions: Sound and Sight, Military and Civilian," in *Oxford Handbook of Sound Studies*, ed. Trevor Pinch and Karin Bijsterveld (New York: Oxford University Press, 2012), 224–48; and Cyrus C. M. Mody and Andrew J. Nelson, "'A Towering Virtue of Necessity': Computer Music at Vietnam-Era Stanford," *Osiris* 28 (2013): 254–77.

60. UCSB Physics, undergraduate brochure, undated (late 1970s or very early 1980s), folder "Brochures, c. 1970s," box 1, UArch 122, UCSB Department of Physics Collection, UCSB LSC.

61. UCSB Physics, Physics 13, Spring 1976, "Environmental Physics," folder "Course Flyers, 1976–1990," box 1, UArch 122, UCSB Department of Physics Collection, UCSB LSC.

62. UCSB Office of Research and Development, "Organized Research Units," folder "Organized Research Units, UCSB 1973–1974, Nov. 1973," box 1, UArch 87, Office of Research Collection, UCSB LSC.

63. UCSB Physics, undergraduate brochure (see n. 60 above).

64. David Phillips to Anne Francis, director of La Bergerie / Inner Space Foundation, Santa Barbara, CA, 16 January 1983, folder 3, box 2, SBPC.

65. Fatt, Hahn, and Jackson, "Evaluation," 21.

66. "Proposal for a Master of Science Degree Program," UCSB Department of Physics basement archive. The same phrase was used to describe the MSI program in several years' "Graduate Study in Physics" booklets: folder "'Graduate Study in Physics' Booklets, c. 1970s," box 1, UArch 122, UCSB Department of Physics Collection, UCSB LSC.

67. Advertisement, undated but probably around 1973, UCSB Department of Physics basement archive. I have no information as to where this ad was published, but the context indicates it was probably in UC–Berkeley's *Daily Californian*, UC–Riverside's *Highlander*, California State University–Los Angeles's *University Times*, or California Polytechnic State University's *Mustang*. The much cheaper rate for in-state tuition meant that Elings focused his recruitment efforts on California campuses, which supplied most of the program's students.

68. "Oil Company Gives Computer to UCSB," unknown date, author, and publication (although most likely the *Santa Barbara News-Press*). Found in UCSB Department of Physics basement archive.

69. Ibid.

70. "Two UCSB Students Develop Device to Aid Blind with Sound Meters," *Santa Barbara News-Press*, 30 March 1975.

71. "UCSB Students to Describe Undersea Monitoring Device," undated but probably 1975, unknown author and publication (but most likely the *Santa Barbara News-Press*). Found in UCSB Department of Physics basement archive.

72. UCSB Office of Public Information, "Physics Student Wins DuPont Fellowship," 12 October 1973, folder "Elings, Virgil," box 11, UArch 11, Office of Public Information Biographical Files, UCSB LSC.

73. Bob English, "It's Called a Chromophone: Invention Enables Deaf to 'See' Sounds on Color TV," UCSB Office of Public Information press release, 12 February 1973, in folder "Physics—Scientific Instrumentation, 1973–1975," box 56, UArch 12, Office of Public Information Subject Files, UCSB LSC. The underlining in the press release title is in the original, and for further emphasis, these words also appeared in a different color from the rest of the title. This press release was taken up by papers at UCSB and in Goleta and even by the *Los Angeles Times*. See George Alexander, "'OO' Is Blue: Deaf Children to 'Read' Sounds by Color on TV Set," *Los Angeles Times*, 13 February 1973.

74. V. Elings and D. Phillips, "An Interdisciplinary Graduate Curriculum

in Scientific Instrumentation," *American Journal of Physics* 41 (1973): 570–73 (quotation on 571).

75. Ibid., 573.

76. Robert English, "Scientific Instrumentation: 'Practical Scientists' Produced by New UCSB Program," 7 November 1972, folder "Physics—Scientific Instrumentation, 1973–1975," box 56, UArch 12, Office of Public Information Subject Files, UCSB LSC.

77. Arnold A. Strassenburg, "Preparing Students for Physics-Related Jobs," *Physics Today*, October 1973, 23–29.

78. Untitled, unsigned (but must be by Buchin), undated (but probably early 1974) report, probably intended to help UCSB's press office write a press release about the thermodilution cardiac-output computer. Underlining and punctuation as in original. Found in folder "Physics—Scientific Instrumentation, 1973–1975," box 56, UArch 12, Office of Public Information Subject Files, UCSB LSC. In the same folder, immediately preceding Buchin's report, is the press release it was probably written for: Bob English, "A Student Project: Simple System Tells How Much Blood Heart Pumps per Minute," 24 May 1974.

79. David Phillips to Marian Nester, director of education, American Society for Psychical Research, 10 September 1972, folder 1, box 4, SBPC.

80. John Placer, Robert L. Morris, and David T. Phillips, "MCTS: A Modular Communications Testing System," in *Research in Parapsychology 1976*, ed. J. D. Morris, W. G. Roll, and R. L. Morris (Metuchen, NJ: Scarecrow Press, 1977), 38–40.

81. Notebook "UCSB Parapsychology 1974 Plethysmograph Agi Ban / D. T. Phillips," folder 1, box 2, SBPC.

82. Elings and Phillips, "Interdisciplinary Graduate Curriculum," 572.

83. David F. Nicoli, Paul H. Barrett, and Virgil B. Elings, "Masters in Instrumentation," *Physics Today* 31, no. 9 (1978): 9.

84. Christophe Lécuyer, *Making Silicon Valley: Innovation and the Growth of High Tech, 1930–1970* (Cambridge, MA: MIT Press, 2005).

85. Kristen Haring, *Ham Radio's Technical Culture* (Cambridge, MA: MIT Press, 2006).

86. Fred Turner, *From Counterculture to Cyberculture: Stewart Brand, the Whole Earth Network, and the Rise of Digital Utopianism* (Chicago: University of Chicago Press, 2006); Walter Isaacson, *Steve Jobs* (New York: Simon and Schuster, 2011); and John Markoff, *What the Dormouse Said: How the Sixties Counterculture Shaped the Personal Computer Industry* (New York: Viking, 2005).

87. I've interviewed several scientists who have commented on this. Probably the clearest in describing his joy at finding that he could order op amps from a catalog (and the leap in experimental capability that resulted) was Clayton Teague (interview by the author, 28 June 2002, Gaithersburg, MD).

88. Nicoli, Barrett, and Elings, "Masters."

89. Ibid.

90. Virgil Elings, review of *The Art of Electronics* by Horowitz and Hill, *Physics Today* (March, 1981): 69, 71.

91. Virgil Elings, "'Invent or Die' Is the Key to Success in Science," *R&D Magazine*, March 1995, 21.

92. Kelly Moore, *Disrupting Science: Social Movements, American Scientists, and the Politics of the Military, 1945–1975* (Princeton, NJ: Princeton University Press, 2008); and Henry Trim, "The New Alchemy Institute: A Countercultural Alternative to Big Science, 1969–1980," paper presented at History of Science Society annual meeting, 4 November 2011, Cleveland, OH.

93. David Turner Phillips, "Intensity Correlation Spectroscopy" (PhD diss., University of California, Berkeley, 1966); and Virgil Elings, biographical data form for application for UCSB position, folder "Elings, Virgil," box 11, UArch 11, Office of Public Information Biographical Files, UCSB LSC.

94. Elings, biographical data form.

95. Letter from Advertising Department of Rainbow Enterprises to Virgil Elings, 9 April 1973, re: display advertising in *Probe the Unknown* magazine; followed by draft ad, "Electronic Instrument Detects Centers of Psychic Energy"; both in folder 3, box 1, SBPC.

96. "SPOT: The Simple Speech Display," advertisement in folder "Elings, Virgil," box 11, UArch 11, Office of Public Information Biographical Files, UCSB LSC; and note summarizing call from Jay Alan Barker, teacher at Utah School for Deaf and Blind, to Phillips residence, 14 June 1981, folder 2, box 2, SBPC. I have also found a mostly blank notebook (that must have belonged to Phillips) entitled "Santa Barbara Technology Model 1700 Parts/Order/Production Planning" (folder 7, box 2, SBPC), implying that Phillips was working on SBT's product line.

97. I've heard versions of this story from Elings's later employees but not from Elings himself. It's possible those employees may have heard it not from Elings but from web and newspaper accounts sourced to Michael Levin, the founder of Optigone. See "Mirage® in the Media," Opti-Gone International website, accessed 24 January 2012, http://www.optigone.com/in_the_media .htm. That site quotes at length Tom Carter, "Mirage Relies on Light to Create Illusion," *Lexington Herald-Leader*, 24 November 1992.

98. Virgil B. Elings and Caliste J. Landry, "Optical Display Device," US Patent 3,647,284, granted 7 March 1972.

99. Virgil B. Elings and David T. Phillips, "Apparatus and Method for Measuring Cardiac Output," US Patent 4,015,593, granted 5 April 1977.

100. Glenn Schiferl, interview by the author, 13 October 2011, Santa Barbara, CA.

101. Barrett, departmental history, 28–29.

102. Vettel, *Biotech*; Peter Galison, *Image and Logic: A Material Culture of Microphysics* (Chicago: University of Chicago Press, 1997).

103. V. B. Elings, G. E. Jahn, and J. H. Vogel, "A Theoretical Model of Regionally Ischemic Myocardium," *Circulation Research* 41 (1977): 722–29.

104. Mike Buchin, interview by the author, 28 March 2012, Palo Alto, CA.

105. Fatt, Hahn, and Jackson, "Evaluation," 16.

106. "David Turner Phillips," obituary posted by cantorjoeocho, accessed 23 January 2012, http://boards.ancestry.com/topics.obits/66031/mb.ashx?pnt=1.

107. Philip Wyatt, interview by the author, 13 October 2011, Santa Barbara, CA.

108. Philip J. Wyatt, "Days of Wine and Lasers," *Science* 212, no. 4500 (12 June 1981): 1212–14.

109. Philip Wyatt, interview by the author, 13 October 2011, Santa Barbara, CA; Philip J. Wyatt, "The Taste of Things to Come," *Applied Optics* 21, no. 14 (15 July 1982): 2471–72; and "Oenological Laser," *Popular Mechanics* 156, no. 4 (October 1981): 194.

110. Phillips's notebook, July 1984–November 1985, folder 4, box 2, SBPC.

111. Cyrus C. M. Mody, *Instrumental Community: Probe Microscopy and the Path to Nanotechnology* (Cambridge, MA: MIT Press, 2011).

112. Neal Bascomb, *The New Cool: A Visionary Teacher, His FIRST Robotics Team, and the Ultimate Battle of Smarts* (New York: Crown Publishing, 2011).

113. For a statement of faith in the transformative economic and social consequences of such start-ups, see, e.g., Richard L. Florida, *Rise of the Creative Class: And How It's Transforming Work, Leisure, and Everyday Life* (New York: Basic Books, 2004).

114. For a case study of some of the difficulties in reaping that dividend, see Andrew Schrank, "Green Capitalists in a Purple State: Sandia National Laboratories and the Renewable Energy Industry in New Mexico," in *State of Innovation: The US Government's Role in Technology Development*, ed. Fred Block and Matthew R. Keller (Boulder: Paradigm, 2011), 96–108.

115. David Kaiser, *How the Hippies Saved Physics: Science, Counterculture, and the Quantum Revival* (New York: W. W. Norton, 2012).

116. Philip Mirowski, *Science-Mart: Privatizing American Science* (Cambridge, MA: Harvard University Press, 2011); and Paul Forman, "The Primacy of Science in Modernity, of Technology in Postmodernity, and of Ideology in the History of Technology," *History and Technology* 23 (2007): 1–152.

117. David C. Mowery, Richard R. Nelson, Bhaven Sampat, and Arvids A. Ziedonis, *Ivory Tower and Industrial Innovation: University-Industry Technology Transfer before and after the Bayh-Dole Act* (Stanford: Stanford University Press, 2004); Elizabeth Popp Berman, "Why Did Universities Start Patenting? Institution-Building and the Road to the Bayh-Dole Act," *Social Studies of Science* 38, no. 6 (2008): 835–71; Mark Peter Jones, "Biotech's Perfect Climate: The Hybritech Story" (PhD diss., University of California, San Diego, 2005); Sally Smith Hughes, *Genentech: The Beginnings of Biotech* (Chicago: University of Chicago Press, 2011); Doogab Yi, "The Recombinant University: Genetic Engineering and the Emergence of Biotechnology at Stanford, 1959–1980" (PhD diss., Princeton University, 2008); and Jeannette Colyvas, "From Divergent Meanings to Common Practices: Institutionalization Processes and the Commercialization of University Research" (PhD diss., Stanford University, 2007).

118. Though I sense that narrative is starting to be reconsidered. See, e.g., Matthew Wisnioski, "Why MIT Institutionalized the Avant-Garde: Negotiating Aesthetic Virtue in the Postwar Defense Institute," *Configurations* 21, no. 1 (Winter 2013): 85–116. To a greater extent than Wisnioski's earlier work, this article highlights the present-day survival (victory, even) of some late-1960s technocultural experiments.

119. Turner, *From Counterculture to Cyberculture*; and Vettel, *Biotech*.

Part Two: Seeking

4 Between the Counterculture and the Corporation: Abraham Maslow and Humanistic Psychology in the 1960s

Nadine Weidman

In 1968 Abraham Maslow (1908–70) was probably the best-known psychologist in America.[1] One of the founders of humanistic psychology and author of the best-selling *Toward a Psychology of Being*, he was president of the American Psychological Association and head of the Brandeis University Department of Psychology. Within that year, he would resign his Brandeis position to take up a lucrative fellowship funded by Saga Administrative Corporation, a food service company in Menlo Park, California, whose board chairman was impressed by Maslow's ideas about human needs and personal growth. Meanwhile, Maslow was also a regular visitor and lecturer at Esalen, a hippie retreat down the California coast at Big Sur, where his concepts of "self-actualization" and "peak experience" became watchwords. What accounts for the breadth of Maslow's appeal in the late 1960s? More specifically, how did this Establishment figure become celebrated by and interact with the anti-Establishment counterculture?

A central figure on the psychological scene in the 1950s and 1960s, Maslow believed that his new approach to human nature, the "third force" of humanistic psychology, would soon take its place beside the two other dominant psychological paradigms, behaviorism and psychoanalysis. Though he challenged the views of humanity presented by B. F. Skinner and Freud, Maslow never dismissed them as wrong, only as limited. Humanistic psychology, Maslow hoped, would subsume and transcend these other two schools of thought; he was explicit about the psychological revolution he foresaw, and he was rewarded for his originality. Over the course of his career, Maslow cemented his reputation as a theorist and idea man. He rarely did experiments and hardly ever strayed into the clinic, but left his ideas for others to test.

Maslow, however, never resided comfortably within the Establishment. He was critical of academic psychology and professional science, arguing that they had lost the amateur's sense of wonder at the world; indeed, his move to the West Coast toward the end of his life was an attempt to escape academia and find a more authentic existence. Even more, the content of his ideas gave his youthful followers a license to rebel. As Maslow said, each person had within him- or herself an "instinctoid" inner core, an intrinsic nature that pressed for fulfillment, even if that meant breaking the conventions of society. For certain members of the youth movements of the 1960s, this was a message of hope and liberation. Release of one's inner nature offered "a very close parallel to the emergence of sexual feelings at puberty. But this time," Maslow assured his listeners, "Daddy says it's all right."[2] Abbie Hoffman, the flamboyant antiwar activist, political radical, and self-ordained "Yippie," a Maslow student at Brandeis in the 1950s, recalled that "more than anyone, I loved Abe Maslow."[3]

Yet, just as he was never easy within the Establishment, Maslow never made a perfect fit with the counterculture either, regarding his youthful followers with a mixture of interest, admiration, and disgust, much like a bemused father would his boisterous and misbehaved children. He dismissed Hoffman as a clown.[4] Although Maslow approved of the hippies' values and goals, encouraging their self-growth through encounter groups, Eastern religion, and drugs, he hated their demand for instant fulfillment, for easy gratification, for "Nirvana now!" Self-actualization, Maslow stressed, took disciplined work and effort. And even though he led seminars at Esalen, enjoyed massages, and soaked in its hot springs up to the end of his life, he also filled hundreds of pages in his journals struggling to explain why the youth had turned out so differently from what his theory of human nature predicted.[5]

In the last several years of his life, Maslow was celebrated everywhere but not comfortably situated anywhere. In this chapter, I will explore how Maslow's explicitly revolutionary humanistic psychology presented a new view of human nature (as grounded in biology) as well as a new vision of science (as much more thoroughly integrated with values and as therefore allied with religion). That his ideas caught on so readily at Esalen shows that a certain kind of explicitly unconventional science held enormous appeal for the counterculture. Some of its radicals were, perhaps ironically, seeking the kind of scientific validity and academic legitimacy that Maslow provided. And the apparent ease with which Maslow traveled between corporate boardroom and hippie retreat indicates a broad crossover or exchange of people, practices, and ideas between the Establishment and the counterculture, as the precepts of humanistic psychology pervaded both. The ideal society in which Maslow imagined that self-growth was possible drew on his experiences in countercultural contexts. Even as he grew increasingly puzzled and irritated by the antics of his countercultural followers, his journal entries document his numerous attempts to deal with the challenge he believed that their behavior posed to his theory of human motivation. That he shaped and reformulated his ideas about psychological needs, growth, and self-fulfillment in response to the unrest he witnessed in the 1960s points to a reciprocal influence between this major psychological theorist and his anti-Establishment contexts.

Maslow and a "Hierarchy of Needs"

Abraham Maslow was born in 1908 in New York City. His parents were Eastern European Jews and recent immigrants; his father worked with other relatives in the family's cooperage business. The eldest of six children, young Abe grew up believing he was worthless and ugly, an attitude inculcated mainly by his mother, whom he despised, and by the anti-Semitic gangs in his Brooklyn neighborhood. Maslow rejected his parents' narrow and rigid Judaism and his father's plans to send him to law school. He attended college at City College of New York, Cornell, and the University of Wisconsin at Madison, from which he graduated in 1930 with a major in psychology. Halfway through his college career he married his first cousin, Bertha Goodman, with whom he had been very much in love since they were both nineteen.[6]

   Maslow was thoroughly trained in the psychological traditions he would ultimately come to reject. Directly upon finishing his bachelor of arts degree, he enrolled as a doctoral student in psychology at Madison, studying with the comparative psychologist Harry Harlow. Maslow wrote a dissertation on the relationships between sexual dominance, power, and social hierarchy in monkeys, both running experiments and making naturalistic observations at the local zoo. He received his doctorate in 1934 and went to work for Edward L. Thorndike at the Institute for Educational Research at Columbia University Teachers' College. Thorndike's main contribution to Maslow's career was to reveal to the younger man his "astounding IQ of 195."[7] The revelation enormously improved Maslow's self-confidence. He made a study of sexuality and dominance in women (the human counterpart to his primate studies) and took his first full-time teaching job at Brooklyn College in 1937.

During Maslow's time in New York, which lasted from 1935 until 1951, when he moved to Brandeis, the young psychologist became acquainted with a range of important social scientists: the neo-Freudian psychoanalysts Alfred Adler, Erich Fromm, and Karen Horney; the psychiatrists Abram Kardiner and David Levy; the Gestalt psychologists Max Wertheimer and Kurt Koffka; and the neuropsychiatrist Kurt Goldstein, whose term "self-actualization" Maslow later adopted and turned toward his own purposes. He also interacted with the Columbia anthropology department, especially Ruth Benedict, who encouraged Maslow to undertake ethnographic research. This he did, supported by a grant from the Social Sciences Research Council in the summer of 1938, among the northern Blackfoot Indians in Alberta, Canada. The result of his fieldwork, however, was to drive him away from cultural relativism: "It now seems more important," he wrote, "to know what kind of personality a man has than to know he is a Blackfoot Indian."[8] Back at Brooklyn College, Maslow combined his teaching with informal psychotherapy sessions for his students. In 1943 he published "A Theory of Human Motivation" in *Psychological Review*, the article that became the basis of his theorizing about human nature for the rest of his career.[9] In the fall of 1951, he moved to Brandeis, where he remained until his retirement eighteen years later (fig. 4.1).[10]

Maslow's 1943 paper presented the image of a hierarchy of needs at the core of each person. At the bottom of the hierarchy were the most basic physiological needs, for food, sleep, elimination, and sex. When these were largely satisfied, the human organism became dominated by needs at the next step up the hierarchy, for safety and security. Even the partial gratification of these needs brought new, higher ones, now for "love, affection, and belongingness," for "man is a perpetually wanting animal."[11] The love needs were surmounted in the hierarchy by the esteem needs: for "achievement, for adequacy, for confidence in the face of the world, and for independence and freedom," as well as for reputation and prestige.[12] Finally, when one had the lower needs largely

FIGURE 4.1 Abraham Maslow, as a professor of psychology at Brandeis University in the mid-1950s. Reprinted by permission of the Robert D. Farber University Archives, Brandeis University.

fulfilled, the needs at the very top of the hierarchy emerged. In the basically satisfied person, Maslow wrote, "new discontent and restlessness will soon develop, unless the individual is doing what he is fitted for. A musician must make music, an artist must paint, a poet must write, if he is to be ultimately happy. What a man *can* be, he *must* be. This need we may call self-actualization."[13] Maslow defined self-actualization as "the desire for self-fulfillment," the drive to become *actually* what one is *potentially*. "This tendency might be phrased as the desire to become more and more what one is, to become everything that one is capable of becoming."[14]

Maslow emphasized three important features of his hierarchical model. First, he considered the needs he identified to be basic to the human organism and to transcend cultural differences, which he regarded as "superficial."[15] People in every culture felt these needs and were impelled toward their fulfillment; in that sense they were common human characteristics. Maslow's discovery of the hierarchy reinforced his rejection of cultural relativism. Second, Maslow envisioned the needs as biological, not only the lowest physiological needs but those for love and esteem as well; to be deprived of their fulfillment was to be unhealthy, just as a man deprived of essential vitamins was unhealthy. "Since we know the pathogenic effects of love starvation, who is to say that we are invoking value questions in an unscientific or illegitimate way, any more than the physician does who treats pellagra or scurvy?"[16] Conscious desires, whether expressed in behavior or only verbally, were for Maslow a transparent membrane through which he could peer into the human's deepest (because biological) essence. Finally, Maslow believed that the right social conditions were vital to the achievement of self-actualization. The healthy society, he wrote, "permitted man's highest purposes to emerge by satisfying all his prepotent basic needs"; the sick society thwarted human needs and kept people at the most basic level of wanting.[17] Thus, for Maslow the exploration of human motivation and personality was also at the same time a search for the good society.

These three features of Maslow's model—its universalism and antirelativism, its biological essentialism, and its explicit connection between the healthy person and the healthy society—became elaborated and developed in all his future work. He also used these emphases to set his work apart from Freudian psychoanalysis and behaviorism. In contrast to Freud's view, in which civilization repressed basic instincts—which remained unsatisfied, might burst out or otherwise manifest themselves, and which we felt guilt for harboring—Maslow argued that (in a healthy person) a need once satisfied ceased to be a motivator, and the person moved on to higher needs.[18] And in contrast to behaviorism, Maslow believed that altering environmental conditions simply could not mold a human being's behavior in any way desired. Social conditions could not alter a human's basic essence: the best conditions could only permit it to be fulfilled, while the worst would thwart it.

Human nature was not, for Maslow, a blank slate. Like a number of other important social scientists in the 1950s and 1960s (Pitirim Sorokin, Ashley Montagu, Carl Rogers, Floyd Matson, and others whose influence was as great at that time as it has since then been forgotten or obscured), Maslow believed that human nature had biological, inbuilt tendencies, that these tendencies were for the good, and that they could be discovered by studying the best specimens that humanity had to offer. Like his colleague Sorokin, the Harvard sociologist who studied saints and altruists, Maslow studied the most psychologically healthy people he could find—those whose lower needs had been largely fulfilled and were now at the top of the hierarchy—whether among Brandeis students, Maslow's own teachers and acquaintances, or (his preferred group of subjects) historical or contemporary thinkers, philosophers, writers, artists, scientists.[19] He stressed, however, that being psychologically healthy did not mean being a "genius"; an ordinary person could also be self-fulfilling in this healthy way. Self-actualizing people were marked by an openness to experience and perception; a trust in their own judgment, even if it meant departure from societal norms; an autonomy and independence from other people's approval or opinions; a capacity for love, enjoyment, humor, and sympathy; and an ability to "sacralize" whatever it was they did, whether this was writing a poem or driving a bus, making a scientific discovery or caring for a child.[20] Self-actualizers saw and appreciated the beauty of life, cultivating an almost-childlike wonder at the world; supported by a good culture and good circumstances, they allowed their inner selves to flourish. Maslow believed that though self-actualizers were all highly individualistic, they showed what the human species could be at its best. A self-actualizing Blackfoot Indian and a self-actualizing Brooklyn Jew would share a common core of humanness despite their cultural and personal differences, and it was this common set of values that Maslow attempted to identify and define.

Many self-actualizers, but not all, reported experiences of a mystical sort that Maslow termed "peak experiences" and that he described as "feelings of limitless horizons opening up to the vision."[21] Such ecstatic moments were not to be explained by divine intervention; they were in fact, according to Maslow, entirely natural and normal parts of human experience. Any self-forgetful, intensely sensuous experience could bring them on, whether of music or art, of landscape, sex, or natural childbirth, or simply of gazing at a beloved's face. Even the ordinary (when suddenly deeply appreciated) could trigger a peak experience. A young mother watching her husband and children happily chatting at the breakfast table one morning recalled the moment as a peak.[22] Peak experiences ranged from intense to mild, and as Maslow emphasized, usually everyone was capable of them and could remember one when questioned. All it took to experience a peak was—along with fulfillment of lower needs—being open to the wonder of life. When experienced, the peak could give even the most neurotic and anxious a glimpse of what life was like for the self-actualizer.

In 1954 Maslow began compiling a mailing list of psychologists and other scholars who agreed with him that healthy people needed to become the subjects of a new human science, and who shared his dismay with mainstream academic psychology. He called his list of approximately 175 names the "Creativeness, Autonomy, Self-Actualization, Love, Self, Being, Growth, and Organismic People" and used the list to exchange reprints and mimeographed materials.[23] Maslow was, according to Roy DeCarvalho, "personally involved in the recruitment of many of these discontents," and they formed the first group of subscribers to the *Journal of Humanistic Psychology*, which Maslow founded with the help of the psychotherapist Anthony Sutich and which published its first issue in spring 1961.[24] About his 1961 version of the list, Maslow wrote that it was "made up to encourage intercommunication among people in different fields who should know each other's work."[25] Though its members were mainly psychologists, the list also included biologists, engineers, economists, artists, theologians, and cultural critics.[26] Conceived as an alternative to conventional academic channels, Maslow's informal exchange network helped circulate the ideas of humanistic psychology among a broad and diverse group of intellectuals.[27]

By the mid-1950s, Maslow was defining the basic needs, including the need for self-actualization, as actually instinct-like. In line with other biologically influenced social scientists in the middle decades of the twentieth century, Maslow rejected cultural relativism and behavior modification through environmental control, to argue instead for the instinctual grounding of human behavior. These social scientists did not necessarily agree on what the human instincts were, but all were convinced that they existed and that the blank slate was a dangerous myth. Maslow called the basic needs "instinctoid," to stress that they were weak, subtle, and delicate and were easily deformed, though never entirely expunged, by a bad culture.[28] Unlike animal instincts, human instinctoid needs were not strong, overpowering, inexorable, or irresistible; it was not the case that civilization had to hold them in check. On the contrary, a good civilization must nourish and foster them so that— the layers of deforming cultural influence peeled back—the Real Self could grow up and flourish. If the culture was bad, people were right to resist it—hence the autonomy and independence Maslow found in his self-actualizers. The conditions had to be right, and people had to learn to listen to and trust their inmost nature, the instinctoid voice of the species.

Maslow wrote that the purpose of psychotherapy was to release a person's intrinsic essence: "I think of uncovering, insight, depth therapies . . . to be, from one point of view, an uncovering, a recovering, a strengthening of our weakened and lost instinctoid tendencies and instinct-remnants, our painted over animal selves, our subjective biology." Personal-growth workshops—including the encounter groups ubiquitous in the counterculture—were dedicated to this same aim of instinctoid release, though the effort was "expensive, painful, long drawn out . . . , ultimately taking a whole lifetime of patience, struggle, and fortitude." But, Maslow continued, "how many cats or dogs or birds need help to discover how to be a cat or dog or bird? Their impulse voices are loud, clear, and unmistakable, where ours are weak, confused, and easily overlooked so that we need help to hear them." "This explains why animal naturalness is seen most clearly in self-actualizing people, least clearly in neurotic or 'normally sick' people. I might go so far as to say that sickness often consists of just exactly the loss of one's animal nature."[29]

Maslow was aware that he lacked the evidence needed to definitively prove his instinctoid claim ("the hypothesis cannot be directly proved today since the direct genetic or neurological techniques that are needed do not yet exist," he conceded).[30] Nonetheless, he used increasingly strong biological terms to describe his vision of human nature. In a 1960 radio interview, for example, Maslow called the yearning for goodness and beauty in self-actualizing people "innate instincts," asserting that his theory refuted malleable-personality "blank-slate" theories. "Most psychological theorists have given that up. Some version of biological theory, or instinct theory, or basic need theory . . . is absolutely necessary for such a conception as I've outlined."[31] Because the frustration of these needs so evidently caused psychological ills, Maslow was convinced that the basic needs were good ("we just don't have any intrinsic instincts for evil. . . . [I]nstincts, at least at the outset, are all 'good'"), and that "mass techniques" were needed to help people discover "this precious human nature deep within" themselves.[32] Defending his view against Viktor Frankl's existentialist critique that Maslow's theory denied people the ability to choose their own course in life, Maslow again emphasized the biological: "The creating [of one's personality and destiny] is not arbitrary. I cannot make myself into a woman, except in a very inefficient and unsatisfactory way. The real job there is, in a certain sense, discovering what you are . . . , discovering your own trends, your own bent, your own tendencies, your own intentions, and then sort of bringing them to pass."[33] Similarly, in his journal entries for February 1968, Maslow wrote: "All the basic-need gratifications, which help growth toward SA [self-actualization] & toward peaks, are also intrinsically biologically valuable, Darwinian value! . . . My whole psychology is fundamentally biological in a way that nobody else is. . . . Everything I've done could be said to deal with the (obscure) biological destiny and nature of man."[34]

Maslow's emphasis on human instincts was never rigidly deterministic, foreordaining people to their fate; on the contrary, his biological essentialism was intended to be freeing, to help people fight back against the strictures of an oppressive culture, to stress their innate, natural resilience, and to give them in effect a license to rebel against bad conditions. Maslow argued that his theory of instinctoid human nature was actually liberating; he tied biological essentialism to liberal, democratic politics, and behaviorism and cultural relativism to oppressive totalitarian and authoritarian regimes that demanded blind conformity. The appeal of Maslow's essentialism to the youthful rebels of the 1960s was obvious: use your inner nature as a guide rather than what the culture might be telling you. In the 1960 radio interview, Maslow said: "the cultural pressures are really not as pressing as people sometimes make them out to be, at least in America. It is very frequently possible, especially for the young person who doesn't yet have many commitments, to simply get off the merry-go-round—to say 'nuts' to cultural pressures."[35] In his 1962 book *Toward a Psychology of Being*, Maslow put his license to rebel in even stronger terms. Conformity and adjustment, when the conditions were bad, interfered with self-growth and were themselves signs of sickness: "Perhaps it is better for a youngster to be *unpopular* with the neighboring snobs or with the country club set. Adjusted to what? To a bad culture? To a dominating parent? What shall we think of the well-adjusted slave? A well-adjusted prisoner?"[36] Calling for a new attitude toward juvenile delinquency—one that acknowledged that behavior problems might sometimes exist for good reasons—Maslow declaimed:

> Clearly what will be called personality problems depends on who is doing the calling. The slave owner? The dictator? The patriarchal father? The husband who wants his wife to remain a child? It seems quite clear that personality problems may sometimes be loud protests against the crushing of one's psychological bones, of one's true inner nature. What is sick then is *not* to protest while this crime is being committed.[37]

For Maslow, as for the thousands of his college-age readers, a biological, instinctoid inner nature offered hope that rebellion through self-expression was healthy, possible, and in the long run likely to succeed: "Even though denied, [this inner nature] persists underground forever pressing for actualization."[38] The instinctoid human nature also confirmed once and for all the weakness of cultural relativism. The principles and values by which Maslow's self-actualizers lived were "broadly human" and "cross-cultural": they stood above culture and allowed one to assess and judge any behavior or any culture. Without access to such biological standards, we would be forced into relativism, and relativism, Maslow argued, would give us "no criterion for criticizing, let us say, the well-adjusted Nazi in Nazi Germany."[39]

A New Kind of Science

Maslow's hierarchy of needs, and the concepts of self-actualization and peak experience that flowed from it, offered a new view of human nature as positively grounded in a biology of health seeking. Humanistic psychology sought not so much to overturn Freudianism and behaviorism as to integrate them both into a larger framework, to explore the healthy and self- moving aspects of the person that Maslow believed the other two schools in psychology largely ignored. He also made it clear that humanistic psychology was to offer a new view of science itself: to supplement, if not entirely replace, "technologizing," mechanistic science (the kind he associated particularly with behaviorism) with a value-laden, emotion-imbued science much better suited to plumbing human depths. Both the content and the methods of such a science were for Maslow decisively different from those of conventional science.

Maslow was aware of the unconventionality of his own scientific methods. In his chapter on self-actualizing people in *Motivation and Personality*, for example, he noted that his study began, not as serious research, but only to fulfill his own curiosity, and that "for those who insist on conventional reliability, validity, sampling, etc." it would therefore appear insufficient.[40] But given the unusual nature of his inquiry into the values by which healthy people lived, the normal scientific demands for full data, for repeatability, and for public availability of data simply could not be met. Maslow's method consisted of "the slow development of a global and holistic impression" rather than the gathering of "discrete facts," an approach that rendered quantitative presentation impossible.[41] Not only were his findings different from those of the established psychological schools, but a different kind of scientific method altogether was necessary to pursue them.

In his 1966 book *The Psychology of Science: A Reconnaissance*, Maslow made explicit his critique of conventional science. The fact/value distinction that scientists were accustomed to drawing was utterly invalid in Maslow's humanistic paradigm. The study of self-actualizing people revealed not only the values by which they lived—these were the "Being-values" (or "B-values")—but also those by which everyone should strive to live. *What is* transformed seamlessly into *what ought to be*. In investigating how people surmounted the needs of the motivational hierarchy, Maslow maintained that his science "can and does discover what human values are, what the human being needs in order to live a good and happy life, what he needs in order to avoid illness, what is good for him and what is bad for him."[42] He stressed *discovery*. These B-values were as independent of the psychologist's own wishes and desires as the astronomer's discovery of new stars: like stars, B-values existed in an objective, natural realm. The claim served to strengthen Maslow's defense against the relativist criticism that the thief (for example) was self-actualizing by stealing, or the murderer by killing. Such destructive behaviors simply had no place in those at the top of the motivational hierarchy, he asserted, and they therefore could not be normative.

In his 1966 book, Maslow argued that science needed to take a new approach to nature. Instead of always trying to predict and control, to be masterful and dominant—to "rubricize" was his term for it—scientists should cultivate a more Taoistic approach, one that was open to the wonder and awe of nature, more observant, and more accepting.[43] Receptiveness and noninterference, "perception of the suchness and realness of things," "witnessing and savoring," being intensely absorbed, spellbound, fascinated, as when "psychedelic drugs have their best effects"—these were the qualities that scientists needed, not to the exclusion of the more analytical and rational rubricizing, but as a complement to it.[44] Feeling empathic and intuitive about one's subject could foster the ability to perceive more deeply: the scientist should, in Maslow's view, come to fuse with the object of his knowledge. "Can all the sciences, all knowledge be conceptualized as a resultant of a loving or caring relationship between knower and known?" he asked.[45] Emotional involvement with one's subject was not opposed to cognition but actually assisted truth finding, just as hunches, intuitions, and dreams often led to discovery.

Taken together, these capacities for direct perception, for intuitive and loving involvement with the object of knowledge, set the Maslovian scientist apart from his conventional professional counterpart. Academic scientists too often bent on rubricizing had lost or suppressed the qualities of reverence and awe, humility and wonder that amateurs and children possessed in abundance. All people were capable of the "scientific spirit," Maslow emphasized, and trained scientists had to learn to recover this central human capacity. In this sense, there was something fundamentally democratic about good science, in Maslow's view. Rather than defining science as "a matter for the expert, something to be done by a certain kind of highly trained professional and nobody else," he saw science simply as the intensification of the general human ability to perceive, of which everyone was in principle capable.[46] It was in fact deeply unscientific—and antidemocratic, perhaps even antihuman—to trust the authority, the a priori, to look the answer up in a textbook, rather than to go outside and look with your own eyes. Such keeping in touch with reality was "almost a defining characteristic of humanness itself," Maslow wrote. "A humanistic view of science and scientists would certainly suggest such a domestication and democratization of the empirical attitude."[47]

As Maslow's pursuit of a science of values caused him to revise traditional scientific methods and attitudes, his study of self-actualizing people also accompanied his search for the healthy society that such people might create. This psychologically healthy utopia (which he called "Eupsychia") would fulfill not just lower, or physiological, needs, but it would set the conditions for and further nurture personal growth. What would the world be like, Maslow wondered, "if all people had developed to the level of maturity and wisdom of these healthy people? Or what kind of a culture would be generated by a thousand of these mature individuals if they were placed on a desert island and not confronted with outside cultural forces? What kind of values would they have?"[48] Such a Eupsychia, Maslow imagined, would probably be democratic, permissive, tolerant, and respectful of individual differences; it would have low crime, no armies, little need for laws; its inhabitants would have no patriotism or nationalism but a sense of universal brotherhood.

Once he imagined it, Maslow set out to find whether and where Eupsychia existed. Working from the mailing list he had begun compiling in the 1950s, Maslow now developed a more extensive mailing list that he called the "Eupsychian Network." Published as an appendix to *Toward a Psychology of Being* in 1962, this network included—instead of individuals as on his original mailing list—organizations, institutions, and journals that shared Maslow's "humanistic and transhumanistic outlook." Such groups helped "the individual grow toward fuller humanness, the society grow toward synergy and health, and all societies and all peoples move toward becoming one world and one species."[49] In addition to organizations for humanistic psychology and the departments of business administration at Harvard, MIT, and UCLA, the list also included many entities that could be called countercultural. The diversity testified to Maslow's crossover appeal. Number one on his list was the Esalen Institute, at Big Sur Hot Springs, whose president was Michael Murphy. Maslow also listed organic farmers, natural-childbirth advocates, LSD "grads" ("individuals who have profited and grown from psychedelic experiences"), Synanon (a drug rehabilitation clinic), "Utopian and Eupsychian nudists," student-run schools outside the conventional public school system, "back to the land and small brotherhood communities," and a number of personal-growth centers connected with Esalen on the West Coast.[50] That many of the places on Maslow's list were in California was no accident; its physical and spiritual distance from the "Freud-bound" approaches of the East Coast and Midwest made it home to an astonishing array of self-growth practices making up the "human potential movement" and inspired largely by humanistic psychology. "We are in the midst of a post-Freudian revolution," one news article on Esalen noted. "And its real innovation center is Southern California."[51] Maslow's sojourns in the California counterculture during the 1960s were part of his attempt to find Eupsychia on earth. And at the center of that counterculture was the Esalen Institute.

Esalen Encounters

The story goes that Maslow stumbled on Esalen one night in the summer of 1962 while he and his wife were driving along Highway 1 looking for a place to stay. When they signed in and the man at the front desk caught their names, he leapt up in excitement; everyone in the place was reading Maslow's latest book, *Toward a Psychology of Being*.[52] Maslow was thereafter a regular guest at Esalen; during the last four years of his life, he led seminars there and became good friends with the institute's cofounder Michael Murphy. Murphy (b. 1930) was from a wealthy California family, held a BA in psychology from Stanford, and followed his interests in Eastern mysticism and philosophy by meditating in the ashram of Sri Aurobindo in Pondicherry, India, in 1956–57. Along with his Stanford classmate Richard Price, a recovering mental patient with whom Murphy had reconnected at a meditation center in San Francisco, Murphy founded Esalen in 1962 on his family's property at Big Sur (fig. 4.2). For Murphy, Esalen would allow the exploration of the spiritual and philosophical aspects of human potential, while Price sought more humane therapeutic alternatives to the ones he had endured as a psychiatric patient. For both, humanistic psychology was an essential point of reference. The two founders situated their new institution on a cliff overlooking the Pacific, amid natural hot springs and redwood forests, and Esalen quickly became a haven for people looking to bare both body and soul—through massage, nude bathing, meditation, yoga, and group therapy—and thereby experience psycho-

FIGURE 4.2 Michael Murphy and Richard (Dick) Price, cofounders of the Esalen Institute, on the Esalen Lodge deck, 1985. Murphy directed Esalen's programming toward humanistic psychology and Eastern mysticism, while Price was interested in making the place into a center for therapeutic alternatives to conventional psychiatry. Photograph © Kathleen T. Carr, www.kathleentcarr.com.

logical and spiritual growth of the very kind that Maslow was talking about.[53] "Let me tell you about Big Sur Hot Springs," Maslow said after one of his first visits there. "The operative word is *hot*. This place is hot."[54] He reportedly described Esalen as "in potential, the most important educational institution in the world."[55] When Murphy opened a San Francisco branch of Esalen in 1966, Maslow gave the inaugural lecture, "Toward a Psychology of Religious Awareness," to an overflow crowd at Grace Cathedral.[56]

What actually went on at Esalen shifted during its first decade. In its early years, which Murphy and Price dubbed its "Apollonian" period, guests sat in wooden chairs and participated in academic-style seminars on humanistic topics. The first Esalen catalog, for 1962, advertised a lecture series entitled "The Human Potentiality."[57] Murphy assigned works by Maslow, Rogers, and Erich Fromm, and humanistic psychologists led seminars on human potential and growth. Things took a more "Dionysian turn" with the arrival of the German psychiatrist and psychoanalyst Frederick (Fritz) Perls in 1964 and with the importation of encounter group therapy by the social psychologist William Schutz in 1967.[58] Visitors to Esalen were now demanding to experience the peaks and self-actualization that Maslow had been theorizing about. By the

middle of the decade, Esalen had become the premier among a burgeoning number of growth centers on the West Coast.

As a key location for the human-potential movement, Esalen's practices fell into three broad types. In Perls's Gestalt therapy, the patient sat
in the "hot seat" and in a form of psychodrama acted out her dreams
and her relationships with the other people in her life. Meanwhile, the
therapist—Perls himself—proceeded to analyze her every word, movement, and nervous tic, usually until the patient broke down in tears.
Having thus stripped his subject of all her accustomed defenses and
habits, Perls built her back up into a presumably healthy whole, newly
aware of and in touch with her feelings and sensations. The second type
of practice, encounter group therapy, was an adapted form of the group
therapy for sensitivity training (or "t-groups") developed by Kurt Lewin
at the National Training Laboratories in Bethel, Maine. T-groups garnered a substantial following among business executives in the 1950s.[59]
In their Esalen incarnation, encounter groups involved total and brutal
honesty among group members; Schutz urged participants in a marital
encounter group, for example, to tell their spouses three secrets that
would threaten their marriages. In addition to the verbal openness, encounter also commonly included physical interaction: everything from
hitting, slapping, wrestling, and dancing to hugging and kissing was
fair game. As with Gestalt therapy, the aim was emotional catharsis,
and outbursts and breakdowns were considered successful results. The
third type of Esalen practice was "body work": meditation, yoga, tai
chi, massage, and nude coed bathing in the natural, spring-fed hot tubs.
Despite the emotional and physical stresses of Gestalt therapy and encounter (or perhaps because of them), both became wildly popular at
Esalen, and Perls and Schutz vied for status as Esalen's supreme guru.
Along with rampant LSD use and uninhibited sexuality—both of which
Perls and Schutz encouraged—the experiential era was by the late 1960s
in full swing.[60]

Maslow was never a guru in the way that these two were, but rather
a constant intellectual guide and influence. Though he visited Esalen
regularly, his presence there was much less frequent than that of Perls
and Schutz, both of whom lived on the grounds. Maslow seems, moreover, to have disliked the notion of being a guru; he spoke of "withdraw-
ing" whenever anyone became too enamored of his ideas. Nonetheless,
Murphy and Price eagerly appropriated Maslow's writings, depending
on the academic and intellectual justification that Maslow provided for
the Esalen approach.[61] The very name of the countercultural ground-
swell that Esalen represented—the human-potential movement—encap-

sulated Maslow's main idea: that one possessed an inner self waiting to be actualized, and that such release was healthy and fulfilling. Jane Howard, a *Life* journalist who visited Esalen in 1968, called Maslow a "chief patron" of the human-potential movement (along with Carl Rogers).[62] Notably, even when contemporary news articles failed to mention Maslow's name, his terms pervaded their descriptions of the place. The *Los Angeles Times* explained that Esalen's techniques of heightening "body awareness" drew on "those trends in religion, philosophy, and behavioral science which emphasize the potentialities of human existence."[63] On its assortment of group therapies: "It is so new (and new *in kind*) to most people that they describe it as a 'peak experience' of their lives." The group sessions were intended to help people "regain a more child-like joyful view of the world."[64] The emphasis at Esalen was on "growth" rather than therapy.[65]

Howard emphasized, however, that the human-potential movement was never a simple application of the founders' ideas; it was, rather, "perplexingly amorphous," a "thing of many overlapping sects and synods," "awash in a glossary siphoned half from the hippies and half from the social scientists."[66] Its proponents and their followers took the spirit of humanistic psychology and recombined and reworked it in different ways, giving it a distinctive imprimatur in the process. For example, in the summer of 1967 when Schutz inaugurated his residential program to train "open encounter" group leaders, Maslow was there on the first night to give a "fatherly pep talk" to the attendees.[67] But while the qualities that Schutz hoped to foster in encounter group participants resembled those on which Maslow also focused, Schutz never mentioned "self-actualization"; rather—more tangibly and directly—he spoke about "joy."[68] Perhaps Maslow's term sounded too cerebral, while for Schutz the whole point of encounter was emotional expression and release. "The theories and methods presented here," Schutz declared, "are aimed at achieving joy. Joy is the feeling that comes from the fulfillment of one's potential."[69] Schutz regarded Maslow as a guide "to the promised land that was being vaguely sought by all the protestors and peace activists," the hippies and flower children, who were now showing up at Esalen in ever-greater numbers.[70] But Schutz never followed slavishly in Maslow's footsteps.

Maslow in turn approved—to some extent at least—of Esalen's instantiation of his license to rebel: the nudism and the psychedelic drug use, Maslow believed, along with other psychological and spiritual techniques, could provide release from conventional constraints, help set the conditions for peak experiences, and foster growth toward self-actualization. In September 1967, Maslow wrote about Esalen in his journal: "[W]ith the baths, the massage, the beauty, everybody smiling and amicable (to me) it can be a happy time for me, a real relaxation and unwinding of the guts." Besides his own enjoyment, he continued, "then obviously they're starting important things there. It's a most important, even revolutionary enterprise. And if it catches on, there may be hundreds of them shortly in the world. It may change all of education." His delight was nonetheless tempered with critique: Esalen "still hovers on the knife-edge of self-indulgence, mere experientialism, anti-science, anti-intellectualism, anti-writing."[71] Maslow's hundreds of journal entries on Esalen were often ambivalent in this way. In a January 1968 entry, for example, Maslow noted that his stress on biology and on instinct set him apart from most other professional psychologists, while it tied him to Esalen's emphasis on "the body and its joys": "I certainly enjoy nudism as at Esalen and have no trouble with it. And I certainly think sex is wonderful, even sacred. And I approve in principle of the advancement of knowledge and experimentation with anything," a reference perhaps to Maslow's oft-stated belief that psychedelic drugs (especially LSD and psilocybin) could help produce peak experiences.[72] But even here there was a criticism: "The nudism at Esalen is absolutely clean and refreshing and non-phony—but they make nothing of it."[73]

When Esalen took its experiential turn in the mid-1960s, Maslow disapproved. He hated and was humiliated by Perls, who mocked Maslow's overly "intellectual" and "academic" style at the latter's first Esalen seminar.[74] Maslow particularly objected to any approach that tried to take a shortcut to self-actualization. "Experiential," for him, became a synonym for "lazy"; discipline and work were required for the lifelong effort of personal growth, as he admonished the Esalen peak seekers. "I spoke—or rather thundered like an Old Testament prophet—of duty, responsibility, real guilt, selfishness, etc. I suppose deep down they all agree with it and will come around. But it reminds me of the real dangers of hippie-dom, of spiritual masturbation, of the far-out."[75] His disillusionment with Esalen was progressive; by 1969 he was accusing its devotees of having been "seized by a kind of experiential 'elite' whose mystique is spontaneity-impulsivity a la Fritz Perls," who valued only "enriched or enlarged or peak or novel consciousness" for its own sake, and who used other people as mere means to achieve such peaks. For Maslow, this meant nothing less than giving up the goal of producing the good person, the ultimate aim of his humanistic psychology. With the exception of Mike Murphy, Maslow wrote, Esalen types were "not compassionate or affectionate with the squares, with 'they,' with outsiders."[76] "Too many shits at Esalen," he wrote in his journal in April 1969, "too many selfish, narcissistic, noncaring types. I think I'll be detaching myself from it more and more."[77] And in early 1970, denouncing its lack of a library, Maslow characterized the people at Esalen as "anti-intellectual, anti-rational, anti-scientific, and anti-research."[78]

Though increasingly disgusted by its Esalen incarnations, Maslow appreciated how effective the encounter group could be as a hothouse for personal growth. He observed and participated in group therapies at Synanon, a drug rehabilitation clinic in Santa Monica run by former addicts, with a branch on Staten Island called Daytop Village. The "no-crap therapy" that he observed at Daytop Village in 1965, for example, "which served to clean out the defenses, the rationalizations, the evasions" and restore addicts to full human functioning, made a deep impression on Maslow.[79] "These people cure people who are considered to be incurable," he exulted in his journal in 1968, "and make me feel I've been too pessimistic about the possibilities for growth and the push toward health in practically anyone."[80] Indeed, from its beginnings as a rehab clinic, Synanon grew into a center where people of all kinds came to play the "Synanon game" of total and unflinching honesty.[81] Similarly, Maslow worked with the executives at Saga Corporation to use the concepts of a needs hierarchy and the honesty fostered in encounter groups to improve worker-management relations.[82] His hope seems to have been that places like Synanon and Saga could provide the conditions for needs gratification and self-actualization in a more disciplined way. If Esalen did not turn out to be Eupsychia, Maslow continued to seek it elsewhere.

That he did shows us that Maslow's Eupsychian network did not just exist on paper. Esalen was part of an actual network of institutions throughout which circulated the ideas, practices, and representatives of humanistic psychology. Notably, these institutions included both the countercultural and the mainstream, just as Henry Trim, in the next chapter in this volume, has noted in the case of environmentalism. As an Esalen regular, Maslow found himself also courted by businessmen and entrepreneurs in California who used his ideas about motivation and growth to inspire worker productivity. For example, during the summer of 1962, when he happened upon Esalen, Maslow was observing innovative assembly-line organization at Non-Linear Systems, a manufacturer of digital voltmeters and other electronic instruments in Del Mar. Its CEO, the engineer Andrew E. Kay, was inspired by Maslow's *Motivation and Personality*, just as the Saga executive William P. Laughlin would be a few years later. Kay in turn funded a fellowship for Maslow at the Western Behavioral Sciences Institute in La Jolla, a nonprofit human-relations research center. Maslow spent part of the same summer of 1962 at UCLA's conference center at Lake Arrowhead, observing t-groups run by the Western Training Laboratories (a California offshoot of the National Training Laboratories in Washington, DC, and Bethel, Maine).[83] The t-group became the prototype for the encounter groups that Maslow observed at Synanon and that taught him so much about the resilience of human nature. T-groups also inspired the group therapy approaches at Esalen. Practices like group therapy circulated from the mainstream to the margins and back again, just as Maslow moved between the corporation and the counterculture.

Though Esalen attracted its share of "dirty hippies," drug users, and flower children, it was never isolated from the Establishment. The experiments in expressiveness, spontaneity, and sensitivity that its encounter groups offered were never limited only to a countercultural fringe but were deliberately designed to appeal to the mainstream. Murphy and his Esalen colleague Ed Maupin emphasized that the place was for growth, not therapy or treatment, and it attracted the "worried well"— not those with serious psychological ills but those well-adjusted, solid citizens seeking an alternative to the "phoniness" of contemporary society. "White collar hippies," as one news report explained, used Esalen-type approaches to get in touch with their feelings and to open up to life.[84] For such otherwise "normal" spiritual seekers, the self was a vast wilderness, full of potential and waiting to be explored. Business executives attended seminars at Esalen and then brought the insights back to their companies. Even while Esalen overturned the conventions and assumptions of "mainstream" society, then, its leaders sought to reach out to that mainstream.

Diagnosing the Counterculture

No matter how disappointed and disgusted Maslow grew with Esalen, he never washed his hands of it and walked away. Instead, he spent hundreds of pages in the journals he kept in the last decade of his life characterizing the countercultural hippies at Esalen and elsewhere, among whom he included his own beloved younger daughter Ellen, and analyzing his differences from them in both psychology and politics.[85] He wondered what was wrong with them that they did not seem to be progressing toward self-actualization, or making any effort toward it. Their lower needs for food and safety were fulfilled—indeed, they had hardly suffered a day in their lives, Maslow believed—yet they got stuck in their progress up the hierarchy, never getting past the love needs or mistaking sex for love. All they seemed to want was easy self-gratification— "Nirvana Now!" as a popular slogan put it—whether through drugs or sex or, more disturbingly, through violence and dropping out of society.

In these reflections, Maslow struck a politically conservative note.[86] During the 1960s, he withdrew from the American Civil Liberties Union (ACLU) and from the National Committee for a Sane Nuclear Policy (SANE). In Maslow's view, these organizations had forgotten that society required a balance between autonomy and community: that one had to give up some rights for government protection and security. "Since society is then seen primarily as a satisfier of personal needs and is judged primarily by how successful it is in *this* job of personal fulfillment for all," he wrote, "then we can go on from there to deal with the inevitable losses of freedom and autonomy that we have to pay to society in return for the far greater blessings it gives us."[87] But the ACLU favored individual rights, Maslow believed, even when doing so was not in the best interests of all. "I don't see why I have to give the criminal a head start," he noted about his ACLU withdrawal.[88] It was precisely this balance between autonomy, community, and responsibility to others that the young rebels, these members of the "Spit-on Daddy club," seemed unwilling to strike.[89]

Maslow also resented Vietnam War protesters for being insufficiently patriotic. If any culture was the one to make self-actualization possible, he asserted, it was the United States, "the best ruling society there has ever been."[90] The United States came closer than any other society to embodying his ideal of the "Taoistic, pluralistic, idiosyncratic, humanness-fulfilling democracy."[91] He faulted the antiwar protesters for focusing on the failings of the American government and military and not enough on those of the North Vietnamese and the Soviets, which he thought far worse. Refusing to condemn US involvement in Vietnam, Maslow attacked the protesters for their hypocrisy: they made not one word of criticism against the "official" North Vietnamese policy of "authorizing assassination, murder, and torture."[92] It was not that Maslow believed that the United States could never be criticized or that war was good—on the contrary, he saw his scientific contribution as "a psychology for the peace-table."[93] But in his view the protesters were too one-sided in their opposition. He could never understand the radicals' bitter hatred for the United States, which he concluded must be a perverse form of fear of the powerful and successful.[94] "I feel grateful and privileged to be an American," Maslow told an audience at Saga in

November 1969, "and I suggest that you do too."[95] Unable or unwilling to understand the protesters' political reasons for opposition to the Vietnam War, Maslow gave their stance a psychological interpretation instead, attributing their critique to their own emotional immaturity.

Though his words and concepts—even his presence—inspired elements of the counterculture, its members appropriated and used his humanistic ideals for ends that he could not understand or approve. Maslow's relationship to feminism was similarly ambivalent. In her 1963 *The Feminine Mystique*, the founding document of second-wave feminism, Betty Friedan relied on Maslow's ideas and quoted extensively from his writings. Her thesis—that "[American] culture does not permit women to accept or gratify their basic need to grow and fulfill their potentialities as human beings"—directly adapted Maslow's ideas about the needs hierarchy and self-fulfillment. Friedan argued that women had the same need, just as men did, to realize their own identity as individuals. Women were not content, and could not be content, "living through" their husbands and children and confining their interests to the sphere of the home. Despite the fact that modern American housewives lived amid unprecedented material abundance, their higher needs were left unmet. Friedan even reiterated Maslow's belief that these higher needs "for knowledge, for self-realization, are as instinctive, in a human sense, as the needs shared with other animals of food, sex, survival."[96]

Maslow, however, could never support Friedan's feminist agenda. Consistent with his biological leanings, he argued that men and women were essentially different, suited by nature to different kinds of work. His examples were often gendered: in an article published in 1962, Maslow asserted that women achieved peak experiences when they hosted a successful dinner party or glanced around their sparkling-clean kitchens, while men's peaks came through creative or intellectual effort. He believed the difference might reflect something fundamental and was worthy of further investigation.[97] Friedan, by contrast, argued that the supposed difference was the noxious by-product of a culture obsessed with keeping women confined to the home. While Maslow wondered whether it was possible "for one person to grow through another," Friedan accused both him and other male psychologists of "evad[ing] the question of self-realization for a woman."[98] Though Maslow told an interviewer in 1968 that he was "excited about the new woman we're developing—the woman who can fulfill herself," in the next breath he emphasized that such fulfillment could never be equivalent to a man's: "Man's duty is to the three books he has to write before he dies," he noted. "A woman's commitment is to her man, and to her cubs."[99] Friedan excoriated the male theorists of the self who held such essentialist and demeaning beliefs. "Bemused themselves by the feminine mystique," she wrote, these psychologists "assume there must be some strange 'difference' which permits a woman to find self-realization by living through her husband and children, while men must grow to theirs. It is still very difficult, even for the most advanced psychological theorist, to see woman as a separate self, . . . no different in her need to grow than a man."[100]

Friedan's critique accentuates the irony of Maslow's position. Even as his humanistic psychology inspired the leaders of radical political movements in the 1960s, his own political views diverged sharply from theirs. Like Immanuel Velikovsky, as Michael Gordin shows in chapter 7 of this volume, Maslow was a charismatic leader who "neither liked nor trusted" his countercultural followers. Those followers, in turn, transformed his ideas in ways that he could neither sanction nor control.

Maslow's conservatism even merged into elitism. Toward the end of his life, he began to toy with the notion that self-actualizers were actually biologically or genetically superior to everyone else, an idea that sat uneasily with his long-standing democratic claim that all people should be free to fulfill their inner selves. Now he wondered whether the "fully evolved" should be considered a "biological elite" or a "biologically privileged class," setting the standard that the rest of humanity should have to reach.[101] Perhaps a "board or commission of sages" should help to decide what values people should live by, Maslow mused, and perhaps because that elite took those values as their own reward, they would require little else in the way of material privileges or luxuries.[102] Perhaps different social or governmental structures should apply to the different classes of people (authoritarian for those at the lowest levels of the motivational hierarchy, freedom-maximizing anarchy for those at the highest).[103] Even more radically, Maslow allowed himself to imagine the possibility of biological selection, a kind of eugenics, that could eliminate the "surplus" population and those who (in his view) contributed nothing useful to the world. "I haven't dared even to myself to accept" some of the consequences of this line of thinking, Maslow conceded in his journal. But perhaps it would have been best, after all, if the drug addicts he saw one day in Washington Square Park did voluntarily kill themselves in "a kind of biologically unselfish act," or if countries that refused to practice birth control were denied international aid. He even had "the lurking thought that wars and famines are after all doing nature's work."[104] Strange notions indeed for a man who was proudly, even defiantly Jewish (though never conventionally religious), who argued for an antiauthoritarian, grassroots, from-below democracy, and whose humanistic psychology celebrated and valued each individual human being.[105]

We should not whitewash these complexities and even contradictions in Maslow's thought. At the same time, however, we should note that Maslow's divergence from the political radicals was more on the matter of means than of ends. He believed that their fundamental values were good, as he often said; where he departed from them was on their ways of achieving those values, especially their impatience, their desire to see all goals reached in an instant. Like the hippies, Maslow opposed conformity and adjustment to social norms, but he believed rebellion and social change had to come about by working from within society, not by trying to destroy it from without. A quiet, gradual, almost "unnoticed" revolution was what he wanted, not a violent overthrow. "Many youngsters, students especially, espouse the highest and most intrinsic values and reject the destroyers and vulgarizers of those values," Maslow wrote in January 1968. But, he continued, the problem is that they "are not wise enough and experienced enough (and not tolerant and democratic enough . . .) to try to persuade and educate, e.g. the racists, but must fight them instead, or lose all hope because a racist doesn't become a saint in 24 hours."[106] Maslow never endorsed the racists. Rather, he objected to the political radicalism of the counterculture for the same reason that he criticized the hippies at Esalen: both were unwilling or unable to be patient about achieving their goals, whether political or psychological. "All the kids are looking to become gods in a weekend in one big bang," he lamented.[107] The generational divide between "Daddy" and his "kids" could hardly have been starker.

As much as Maslow's ideas filtered into and shaped practices and beliefs at Esalen, his experiences there shaped him in turn, causing him to rethink and to some extent modify his theory of human nature. Again and again Maslow attempted in his journal to explain the hippies' behavior and develop a framework that could accommodate it, and he worried about what was wrong with his theory of psychological growth that it failed to predict and explain the hippies' behavior. "Actually struggling with these 'radical' kids has been very good at pushing me to think through my revolution theory, my grumbles theory, politics 3, etc.," he noted in December 1968. "Maybe I should endure the irritation for the sake of my counterthinking."[108]

In the last year and a half of his life, Maslow advanced several psychological explanations for the counterculture phenomenon. In March 1969 he deemed a "pathology" the inability to recognize and be grateful for what one has. How else to explain "the fortunate ones who have everything and yet reject society, lose compassion, get nasty, mean and violent and destructive"?[109] How else to explain the liberals and radicals who condemned the United States? Perhaps, Maslow theorized, his hierarchy of needs had to be adjusted so that both the lower needs and the higher could be fulfilled in tandem—in order that the lower gratifications were not instantly forgotten and discounted and that people could remain grateful for them.

A few months later, Maslow made another attempt at explanation. The hippies, he observed, suffered from "the peak experience interpretation of life. . . . Human nature is at its best," in their view, "in the turn-on of enthusiasm, of mob excitement and happiness, when anything seems possible . . . in the exhilarated, ecstatic moment."[110] This belief they maintained even though Maslow had emphasized that peak experiences were not a way of life and could not be regularly achieved via LSD or weekend workshops at Esalen. All of his self-actualizing subjects were older people who held a beloved job. The question was why the hippies persisted in their stubborn desire for "big bang revolutions" despite Maslow's accumulated evidence to the contrary. He did not stop at criticizing them but tried to account for their behavior, concluding that they lacked a "theory of evil": "The hippies and commune-ists get disillusioned because they don't fully realize the shortcomings of human nature over the long haul, when the problems start coming, when the baby has colic, . . . when the mate is not screwing, . . . or when there's repetitive dull work to do." Then the hippie peak seeker was likely to get disappointed, his illusions of quick-and-easy Nirvana shattered. "Use the hippie illusions as examples," Maslow urged himself, of how *not* to approach self-actualization. "Say it: how not to be disappointed? . . . How to get rid of illusions? *Which* are the illusory expectations?"[111] The hippies became for Maslow a psychological case study by which he could test the value of his theory.

Maslow ultimately decided that the young political and cultural rebels did not have the awareness of death that older people had, and that this prevented them from having empathy for others, especially their parents, and from being grateful for their blessings. "We say of the kids, and I remember about myself, that they are really selfish about their parents and elders," he wrote, "not fully aware of their parents as ends, only as means and as suppliers; they don't count their blessings and aren't even aware of them often."[112] Just a few months before his death, Maslow recalled a poignant moment when he had stood on a beach and been forcibly struck by the contrast between its permanence and his mortality. That contrast made the moment all the more precious to him. "I still wonder that maybe what I've called SA [self-actualization] has reconciliation with death as a sine qua non. . . . [M]aybe death helps to create the feeling of sacralization." Applying the insight to his hippie test case, he continued: "Kids can't do this; they are still positivists perceiving only what their (external?) senses can bring them." And then, modifying his theory to account for them: "Does death awareness produce the transcendent, transpersonal, transhuman?" Finally, proposing a therapy: "Could exercises in deprivation educate us faster about all our blessings?"[113]

Conclusions

The fact that Maslow was so concerned, up to the very end of his life, to explain the hippies' behavior, develop a theoretical framework that could accommodate it, and even propose therapeutic interventions to alter it shows the extent of his involvement with the counterculture. I draw several conclusions from my examination of that involvement. First, contrary to Theodore Roszak's claim that the counterculture flat out rejected science,[114] Maslow's example shows that there clearly was such a thing as "groovy science"—specifically, in this case, groovy psychology. Maslow's psychology held particular appeal to members of the counterculture, and it took root and flourished in quintessentially countercultural contexts like Esalen.[115] Michael Murphy and other Esalen leaders were seeking intellectual justification for their beliefs and practices, not just from Beat poets and Eastern mystics, but from scientists too. That Maslow's psychology explicitly rejected a technocratic, value-free, conventional model of science and embraced individuality, spirituality, and the higher reaches of human nature served to increase its appeal.[116]

Second, Maslow's case shows that the counterculture was never an isolated fringe; there was a continuous circulation and exchange of humanistic psychological ideas, practices, and practitioners between the counterculture and what we might call the "Establishment." Maslow was a regular presence not only at Esalen but also at corporations like Non-Linear Systems and Saga, as well as at hotbeds of humanistic activity like the Western Behavioral Sciences Institute and the Western Training Laboratories and at therapeutic enterprises like Synanon. That most of these places were located in California hints at a regional subculture that facilitated and encouraged the exchange. By circulating in this way, Maslow, Carl Rogers, Timothy Leary, Fritz Perls, Will Schutz, and others became the vehicles by which countercultural ideals reached a broad and mainstream audience.

Finally, Maslow's ideas did not simply "diffuse" into a popular counterculture; rather, certain individuals with their own agendas appropriated his terms and theories and brought them to Esalen—and their agendas did not always coincide with his. This was no top-down filtering of elite professional science into a popular realm that watered down and misused the "real" science. Rather, the denizens of Esalen chose, recombined, and reinvented the precepts and elements of humanistic psychology to suit their own purposes. And the influence ran very much in both directions. As Maslow's thinking informed (but did not exclusively determine) that of the young rebels, Maslow was in turn changed by his contact with the counterculture, as his journals demonstrate. As the hippies became for him a case study by which his theory of human nature could be put to the test, Maslow shaped his humanistic psychology in response to the contexts in which it took root.

Acknowledgments

My thanks to David Kaiser and Patrick McCray for their helpful feedback and encouragement, and to Lizette Royer Barton and the staff at the Archives of the History of American Psychology for their knowledgeable assistance with the Maslow Papers. I benefited from comments on an earlier draft from Lara Friedenfelds, Joy Harvey, Susan Lanzoni, Rachael Rosner, Kara Swanson, and Conevery Valencius. Tim Leonard provided expert research assistance and much insightful discussion.

Notes

1. Mary Harrington Hall called Maslow "by far the most popular psychologist in the country today among his peers—and among those who will never be peers." Mary Harrington Hall, "The Psychology of Universality: A Conversation with the President of the American Psychological Association, Abraham H. Maslow," *Psychology Today* 2 (July 1968): 35.

2. Abraham H. Maslow, "Lessons from the Peak-Experiences," *Journal of Humanistic Psychology* 2, no. 1 (Spring 1962): 15.

3. Abbie Hoffman, *Soon to Be a Major Motion Picture* (New York: Putnam, 1980), 26. "Yippie" stands for Youth International Party, a term invented by Hoffman and Jerry Rubin and synonymous with their theatrical political and antiwar activism. Donald Moss discusses the relationship of Maslow's humanistic psychology to the counterculture in "Abraham Maslow and the Emergence of Humanistic Psychology," in *Humanistic and Transper-*

sonal Psychology: A Historical and Biographical Sourcebook, ed. Donald Moss (Westport, CT: Greenwood Press, 1999), 24–37.

4. See, e.g., Maslow's journal entry for 5 December 1968, in *The Journals of A. H. Maslow*, ed. Richard Lowry, 2 vols. (Monterey, CA: Brooks/Cole, 1979), 2:1090 (hereafter cited as Maslow, *Journals*).

5. Howard Brick has urged historians not to reify the counterculture, noting that it comprised a complex mixture of "defiantly non-conformist attitudes, uninhibited behavior, and generalized dissent." Howard Brick, *Age of Contradiction: American Thought and Culture in the 1960s* (New York: Twayne, 1998), 113–14. For Maslow, the counterculture was located largely in California and was marked as much by radical political activism, "dropping out," and rebellion from convention as by a search for higher realms of conscious experience (especially via Eastern mysticism and spirituality) and uninhibited attitudes toward the body, sex, and drugs. Maslow also often used the term "hippie" to refer to student activists and advocates of communal living and drug use, and though it has a broader meaning than this, I will use the term as he used it.

6. The details of Maslow's life are drawn from Edward Hoffman, *The Right to Be Human: A Biography of Abraham Maslow* (Los Angeles: Jeremy P. Tarcher, 1988). On the development of Maslow's theorizing, see also Richard Lowry, *A. H. Maslow: An Intellectual Portrait* (Monterey, CA: Brooks/Cole, 1973); Roy Jose DeCarvalho, *The Growth Hypothesis in Psychology* (San Francisco: EM Text, 1991); Roy Jose DeCarvalho, *The Founders of Humanistic Psychology* (New York: Praeger, 1991); and Frank Goble, *Third Force: The Psychology of Abraham Maslow* (New York: Grossman, 1970).

7. Hoffman, *Right to Be Human*, 174.

8. Maslow quoted in ibid., 137.

9. Abraham Maslow, "A Theory of Human Motivation," *Psychological Review* 50 (1943): 370–96. Reprinted as chapter 4 in Abraham Maslow, *Motivation and Personality*, 2nd ed. (New York: Harper and Row, 1970), 35–58.

10. Brandeis University was founded in 1948 in Waltham, Massachusetts, by a group of "uneducated immigrant Jews" who had set out to found "the Harvard of the Jews." Maslow was among its first faculty members. In the 1960s, Brandeis became a center of student activism. See Ethan Bronner, "Brandeis at 50 Is Still Searching, Still Jewish, and Still Not Harvard," *New York Times*, 17 October 1998.

11. Maslow, "Human Motivation," 380.

12. Ibid., 381.

13. Ibid., 382.

14. Ibid.

15. Ibid., 389.

16. Ibid., 394.

17. Ibid., 394n.

18. Ibid., 393.

19. Maslow's list of self-actualizing subjects, including such public figures as Abraham Lincoln, Albert Einstein, Eleanor Roosevelt, Jane Addams, Aldous

Huxley, and William James, appears in "Self-Actualizing People: A Study of Psychological Health," in *Motivation and Personality*, 152.

20. On the characteristics of self-actualizers, see ibid., 153–74.

21. Ibid., 164.

22. Maslow, "Lessons from the Peak-Experiences," 11.

23. "'Creativeness, Autonomy, Self-Actualization, Love, Self, Being, Growth, and Organismic' People Mailing List," July 1961, folder 3 "Miscellaneous (2)," box M4484, Abraham Maslow Papers, Archives of the History of American Psychology, Center for the History of Psychology, University of Akron, Akron, OH.

24. Roy J. DeCarvalho, "The Institutionalization of Humanistic Psychology," in *The Humanistic Movement: Recovering the Person in Psychology*, ed. Frederick J. Wertz (Lake Worth, FL: Gardner Press, 1994), 14.

25. Quoted in ibid.

26. Jessica Grogan, *Encountering America: Humanistic Psychology, Sixties Culture, and the Shaping of the Modern Self* (New York: Perennial, 2013), 20.

27. Maslow's mailing list, and the even more diverse one that succeeded it in 1962, foreshadowed similar informal correspondence networks that grew up in the counterculture in later decades and that provided alternatives to academic publication channels. For example, David Kaiser explains how in the 1970s Ira Einhorn, a self-taught physicist, community organizer, and conduit between Philadelphia's hippies and its business elite, organized and maintained a mailing list that circulated unconventional scientific ideas. See David Kaiser, *How the Hippies Saved Physics: Science, Counterculture, and the Quantum Revival* (New York: Norton, 2011), 131–34. W. Patrick McCray describes how in the 1970s the organizers of the L5 Society, a group of space-colony enthusiasts, relied on an extensive mailing list to circulate copies of their self-published newsletter. See W. Patrick McCray, *The Visioneers: How a Group of Elite Scientists Pursued Space Colonies, Nanotechnologies, and a Limitless Future* (Princeton, NJ: Princeton University Press, 2013), 88–97. Although Maslow did not distribute a regular newsletter, he used his "huge" list to send out papers, both his own and those by others, to keep its members "informed of new theory and research that contributed something to the advancement of the cause"; Walter Truett Anderson, *The Upstart Spring: Esalen and the Human Potential Movement, the First Twenty Years* (1983; Lincoln, NE: iUniverse, 2004), 184. See also Henryk Misiak and Virginia Staudt Sexton, *Phenomenological, Existential, and Humanistic Psychologies: A Historical Survey* (New York: Grune and Stratton, 1973), 111. Like Einhorn's network, Maslow's was notable for the diversity of its membership.

28. Abraham Maslow, "The Instinctoid Nature of Basic Needs," in *Motivation and Personality*, 77–95. Maslow defines "instinctoid" on p. 82. Social scientists who agreed that human nature was instinctive include Sorokin, Montagu, Gordon Allport, and Robert Ardrey.

29. Ibid.

30. Ibid., 88. See also Maslow, "Toward a Humanistic Biology," *American Psychologist* 24, no. 8 (August 1969): 724–35; and Abraham Maslow, "Theory

of Meta-motivation: The Biological Rooting of the Value-Life," *Journal of Humanistic Psychology* 7, no. 2 (Fall 1967): 93–127.

31. Abraham Maslow, "Eupsychia: The Good Society," *Journal of Humanistic Psychology* 1, no. 2 (Fall 1961): 6.

32. Ibid., 7–8.

33. Abraham Maslow, interview by Willard B. Frick, 23 November 1968, in Willard B. Frick, *Humanistic Psychology: Interviews with Maslow, Murphy, and Rogers* (Columbus, OH: Charles E. Merrill, 1971), 24.

34. Maslow, *Journals*, 2:885 (15 February 1968).

35. Maslow, "Eupsychia," 9.

36. Abraham Maslow, *Toward a Psychology of Being*, 2nd ed. (New York: Van Nostrand, 1968), 7–8.

37. Ibid., 8.

38. Ibid., 4.

39. Maslow, "Eupsychia," 5.

40. Maslow, "Self-Actualizing People," 149.

41. Ibid., 152, 153.

42. Abraham Maslow, *The Psychology of Science: A Reconnaissance* (New York: Harper and Row, 1966), 125.

43. For Maslow's definition of "rubricizing," see ibid., 81–83. Ian Nicholson argues that Maslow's revision of scientific norms caused him to worry that his science was losing its "masculine" character; see Ian A. M. Nicholson, "'Giving Up Maleness': Abraham Maslow, Masculinity, and the Boundaries of Psychology," *History of Psychology* 4, no. 1 (2001): 79–91.

44. Maslow, *Psychology of Science*, 96, 100, 101.

45. Ibid., 110.

46. Ibid., 136.

47. Ibid., 136–37.

48. Maslow, "Eupsychia," 10.

49. Maslow, *Toward a Psychology of Being*, 237.

50. Ibid., 238–40.

51. Frederick Stoller quoted in Eleanor Links Hoover, "The Great 'Group' Binge," *Los Angeles Times*, 8 January 1967.

52. The story is told by Anderson, *Upstart Spring*, 67. It is retold by Hoffman, *Right to Be Human*, 272; by Jeffrey Kripal, *Esalen: America and the Religion of No Religion* (Chicago: University of Chicago Press, 2007), 136–37; and by Grogan, *Encountering America*, 158.

53. On Murphy and Esalen, see Anderson, *Upstart Spring*; Kripal, *Esalen*; Grogan, *Encountering America*; Jeffrey Kripal and Glenn Shuck, eds., *On the Edge of the Future: Esalen and the Evolution of American Culture* (Bloomington: Indiana University Press, 2005); and the website (accessed 25 June 2013) of the Esalen Institute, www.esalen.org/page/esalen-founders.

54. Maslow quoted in Anderson, *Upstart Spring*, 112.

55. Maslow quoted in "Coast Group Spearheads a Movement Seeking Clue to Human Feelings," *New York Times*, 8 October 1967, 55.

56. Hoffman, *Right to Be Human*, 288.

57. Grogan, *Encountering America*, 162.

58. Ibid., 163.

59. On the connections between business corporations and the counterculture, see Thomas Frank, *Conquest of Cool: Business Culture, Counterculture, and the Rise of Hip Consumerism* (Chicago: University of Chicago Press, 1997).

60. On the types of therapy offered by Esalen, see Grogan, *Encountering America*, chaps. 8–9.

61. On the overlap between humanistic psychology and Esalen's aims, see Jessica Lynn Grogan, "A Cultural History of the Humanistic Psychology Movement" (PhD diss., University of Texas at Austin, 2008), esp. chap. 6; and Grogan, *Encountering America*, 159–63. Grogan writes: "Esalen became, over the course of the 1960s, a cultural beacon of humanistic psychology, and in this sense served, for many Americans, as a proxy for any direct orientation with the founders or their theory. Here the ideas that Maslow and others generated for understanding human psychology were grafted onto lived experience, where they mingled with other approaches and morphed into novel practices" (*Encountering America*, 159).

62. Jane Howard, "Inhibitions Thrown to the Gentle Winds: A New Movement to Unlock the Potential of What People Could Be—but Aren't," *Life* 65, no. 2 (12 July 1968): 48–65. Howard wrote her own memoir of her time at Esalen, called *Please Touch: A Guided Tour of the Human Potential Movement* (New York: McGraw Hill, 1970).

63. Jack Smith, "Body Awareness Key to Cult's Doctrine: Devotees Meet at Big Sur," *Los Angeles Times*, 24 April 1966.

64. Hoover, "Great 'Group' Binge."

65. Bill Dicke, "Amid Coastal Beauty: Esalen Institute Seeks Self-Awareness," *Daily Trojan*, 26 October 1967.

66. Howard, "Inhibitions Thrown to the Gentle Winds," 56–57.

67. Anderson, *Upstart Spring*, 159.

68. Ibid., 158.

69. William Schutz, *Joy: Expanding Human Awareness* (New York: Grove Press, 1967), 15.

70. Anderson, *Upstart Spring*, 160.

71. Maslow, *Journals*, 2:827 (19 September 1967).

72. Ibid., 883 (31 January 1968). On Maslow's beliefs about psychedelic drugs, see, e.g., his *Religions, Values, and Peak Experiences* (New York: Penguin, 1964), 27. Maslow was interested in and supportive of Timothy Leary's psychedelic experimentation, at least at first, and traveled to Washington, DC, in July 1965 to testify on behalf of Richard Alpert, Leary's colleague, when the American Psychological Association charged him with ethics violations. On Leary, see Maslow, *Journals*, 1:242, 269, 440; on Alpert, see 1:527–29 (16 July 1965).

73. Maslow, *Journals*, 2:884 (31 January 1968).

74. Hoffman, *Right to Be Human*, 289–93.

75. Maslow, *Journals*, 2:828 (19 September 1967).

76. Ibid., 968 (25 May 1969); see also p. 1117.

77. Ibid., 953 (6 April 1969).

78. Abraham Maslow, "Beyond Spontaneity: A Critique of the Esalen

Institute," in *Future Visions: The Unpublished Papers of Abraham Maslow*, ed. Edward Hoffman (Thousand Oaks, CA: Sage Publications, 1996), 130.

79. Abraham Maslow, "Synanon and Eupsychia," *Journal of Humanistic Psychology* 7, no. 1 (Spring 1967): 28.

80. Maslow, *Journals*, 2:882 (27 January 1968).

81. "Coast Group Spearheads a Movement"; Hoover, "Great 'Group' Binge"; and Maslow, "Synanon and Eupsychia."

82. Abraham Maslow, "On Eupsychian Management," in *The Farther Reaches of Human Nature* (New York: Viking, 1971), 237–38; and Abraham Maslow, *Eupsychian Management: A Journal* (Homewood, IL: R. D. Irwin, 1965).

83. On Maslow's activities during the summer of 1962, see Hoffman, *Right to Be Human*, 267–73.

84. Dicke, "Amid Coastal Beauty."

85. Ellen Maslow (1940–2009) was an antiwar and civil-rights activist in the 1960s; she worked for Timothy Leary as a research assistant and was close to Abbie Hoffman (see Hoffman, *Right to Be Human*, 265, 294). She later became a clinical psychologist. Maslow writes about his difficulty understanding her in his *Journals*; see entries for 3, 5, and 23 June 1966 (2:733–36).

86. On Maslow as politically conservative, see Ellen Herman, "Being and Doing: Humanistic Psychology and the Spirit of the Sixties," in *Sights on the Sixties*, ed. Barbara L. Tischler (New Brunswick, NJ: Rutgers University Press, 1992), 87–101, esp. 94; and Ellen Herman, *The Romance of American Psychology: Political Culture in the Age of Experts* (Berkeley: University of California Press, 1995), 269–75.

87. Maslow, *Journals*, 2:1248 (22 March 1970).

88. Ibid., 824 (30 August 1967).

89. Ibid., 877 (27 December 1967).

90. Ibid., 902 (20 January 1968).

91. Ibid., 1249 (22 March 1970).

92. Abraham Maslow, "See No Evil, Hear No Evil: When Liberalism Fails (Jan 1967)," in Hoffman, *Future Visions*, 163.

93. Maslow, *Journals*, 2:895 (20 August 1967).

94. Ibid., 903–4 (14 and 24 February 1968).

95. Abraham Maslow, "The Dynamics of American Management: Remarks at Saga Corporation (Nov 1969)," in Hoffman, *Future Visions*, 185.

96. Betty Friedan, *The Feminine Mystique* (1963; New York: W. W. Norton, 2013), 77, 378.

97. Maslow, "Lessons from the Peak-Experiences," 11.

98. Friedan, *Feminine Mystique*, 393.

99. Maslow quoted in Hall, "Psychology of Universality," 56.

100. Friedan, *Feminine Mystique*, 393. On Maslow's influence on Friedan, and on the relationship between humanistic psychology and feminism, see Herman, *Romance of American Psychology*, 276–303, esp. 290–92.

101. Abraham Maslow, "Humanistic Biology: Elitist Implications of 'Full Humanness' (March 28, 1968)," in Hoffman, *Future Visions*, 71.

102. Ibid., 71–72.

103. Maslow, *Journals*, 1:632 (25 May 1966).

104. Ibid., 2:1230 (31 January 1970).

105. Ibid., 1250 (22 March 1970).

106. Ibid., 902 (20 January 1968).

107. Ibid., 980 (30 July 1969).

108. Ibid., 1090 (5 December 1968).

109. Ibid., 951–52 (31 March 1969).

110. Ibid., 1180 (20 September 1969).

111. Ibid., 1180–81 (20 September 1969).

112. Ibid., 1258 (28 March 1970).

113. Ibid., 1260 (28 March 1970).

114. Theodore Roszak, *The Making of a Counter Culture: Reflections on the Technocratic Society and Its Youthful Opposition* (Garden City, NY: Anchor / Doubleday, 1969).

115. Maslow himself recognized his influence over the hippies in his interview with Willard Frick, acknowledging that his widely read *Toward a Psychology of Being* was helping to produce "idealistic goals" in young people "of truth and honesty and pure justice and excellence and the renunciation of hypocrisy and phoniness" (Frick, *Humanistic Psychology*, 36). On the trends toward nude therapy and sensitivity groups popular in the counterculture, Maslow said, "I've had something to do with all of [these trends], kind of setting them in motion" (48).

116. Similarly, Stephen J. Whitfield writes that Abbie Hoffman, Allen Ginsberg, Stanley Kubrick, and other countercultural icons sought to "rattl[e] the bars of the iron cage of bureaucracy, technology, rationality." See Stephen J. Whitfield, "The Stunt Man: Abbie Hoffman (1936–1989)," in Tischler, *Sights on the Sixties*, 112.

5

A Quest for Permanence: The Ecological Visioneering of John Todd and the New Alchemy Institute

Henry Trim

"Eco-catastrophe!" screamed the title of Paul Ehrlich's jeremiad in the September 1969 issue of *Ramparts*. In the scenario Ehrlich imagined, pesticides killed the oceans as population growth, food shortages, and pollution unleashed calamity across the globe.[1] Among the millions frightened by Ehrlich's grim prediction were ethologist John Todd, his wife, Nancy Jack Todd, and his friend and colleague ichthyologist William McLarney.[2] Fed up with the "doom watch" biology he practiced at San Diego State University, which allowed him only to catalog the biosphere's ills and mainstream society's increasing destructiveness, Todd decided they needed to act if they wanted to save the world from ecocatastrophe.[3]

Determined to help Americans and Canadians live within the world's ecological limits, the Todds and McLarney founded the New Alchemy Institute (NAI) and incorporated it as a nonprofit scientific institution in 1970.[4] The NAI's articles of incorporation stated it would "engage in scientific research in the public interest on ecologically and behaviourally planned agriculture systems [and] . . . on methods to reduce environmental contamination and to restore natural waters and landscapes."[5] Led by Todd, the countercultural scientists of the NAI imagined an environmentally sustainable future and, going a step further than most, designed and built the technological systems they believed would make such a future possible. By end of the 1970s, Todd and the New Alchemists had pioneered "living machines," highly efficient systems of greenhouse aquaponics, and constructed the iconic Ark Bioshelters.[6] In the 2000s, Todd's living machines were adapted for biofiltration and urban farming, but perhaps more importantly his work helped to define and launch the ecological design movement.[7]

The New Alchemists' embrace of a sort of technoecological sublime challenges the dominant understanding of both environmentalism and the counterculture. Charles Reich's and Theodore Roszak's description of the counterculture as a "subversion of the scientific world view" has defined it as suspicious of science, particularly anything produced by the "Big Science" of the Cold War military-industrial complex.[8] Similarly, Roszak and other scholars have suggested that an almost-religious concern for wilderness and an antipathy for technology bordering on antimodernism defined environmentalism.[9] As a result, countercultural environmentalists are presented in two primary ways: as indolent apolitical hippies communing with nature on isolated communes or as angry radicals out to tear down North America's technoscientific systems and declare the idea of human progress misguided.[10] While there is a degree of truth to these depictions, they tend to perpetuate an oversimplified understanding of both. Most problematically, such representations assume that countercultural environmentalists always conceptually divided nature from culture and attempted to protect some iteration of pure nature from the invariably corrupting influences of humans and technology.[11]

Todd and the New Alchemists rejected the idea of pure nature and employed science and technology in an attempt to construct a sustainable fusion of human culture and the nonhuman environment. Complementing Andrew Kirk's broad discussion of this "ecopragmatism" in chapter 10 of this volume, my essay examines one influential ecopragmatist's merger of technology, nature, and humanity to construct an alternative to declensionist environmentalism and countercultural rejections of science. Todd's combination of scientific knowledge and ecological vision defies easy analysis, as it straddled the lines between futurology, scientific optimism, and catastrophic environmentalism. He embraced environmentalist concerns about ecological catastrophe and actively criticized mainstream applications of Big Science as the cause of social and environmental problems. Yet he also grounded his work in painstaking research, developed new technologies, and tirelessly promoted his vision of technologically mediated sustainability.

W. Patrick McCray's concept of "visioneering" offers a particularly useful lens through which to approach Todd's seemingly contradictory mixing of science, environmentalism, and imagination. McCray devised the concept to analyze Gerard O'Neill's and K. Eric Drexler's optimistic attempts to design, produce, and promote space colonies, nanotechnology, and a limitless future. McCray describes visioneering as "developing a broad and comprehensive vision for how the future might be radically changed by technology, doing research and engineering to advance this vision, promoting one's ideas to the public and policy makers in the hopes of generating attention and perhaps even realization."[12] Todd and his New Alchemists did exactly this as they attempted to define and engineer a sustainable future. Todd and space-colony visioneer O'Neill both belonged to the community of technological environmentalists connected by the networks surrounding Stewart Brand's *Whole Earth Catalog*.[13] Although Todd's ultimate goals differed, he shared O'Neill's optimistic belief that new "technologies could shape future societies, upend traditional economic models, and radically transform the human condition."[14] Todd also shared O'Neill's charisma and willingness to tirelessly promote an alternative vision of the future.

The lens of visioneering allows me to broaden the history of the counterculture by highlighting the significant influence that science, particularly NASA's Big Science research, had within the counterculture. This approach also highlights the importance some environmentalists attached to the concept of progress and illustrates how such groups adapted ideas of progress, commonly associated with endless expansion, to a world characterized by environmental limits.[15] Finally, conceptualizing Todd as a visioneer helps my analysis take seriously the financing of research and underlines the close connections between the counterculture and mainstream funding organizations and governments. Much like O'Neill and Drexler, Todd was a groovy scientist on the margins with radical ideas for how to save the world. To legitimize his ideas and make his vision concrete, he founded institutions and built networks of intellectual and financial support connecting the counterculture to the mainstream. These links illustrate how the relationships of cooperation between countercultural environmentalism and governments emerged from the efforts of Todd and others to use scientific and technical expertise to make a place for their ideas within broader discussions of environmentalism and development during the 1970s.

Building the New Alchemy Institute

Immediately after founding the NAI in 1970, John Todd, Nancy Jack Todd, and William McLarney left San Diego for Woods Hole, Massachusetts. Todd and McLarney joined the prestigious Woods Hole Oceanographic Institution on Cape Cod to undertake intensive research away from the demands of teaching and administration.[16] Todd, however, did not give up on his nascent institute or his desire to save the world. After settling on Cape Cod, the small group quickly found a rundown eleven-acre farm near Woods Hole where their institute could at last take on a physical existence.[17] An assortment of similarly concerned scientists and graduate students, as well as a cross section of organic-agriculture devotees, countercultural tinkerers, and architects joined the Todds and McLarney to staff the NAI over the course of the 1970s. Their research focused on renewable energy (principally solar and wind power), small-scale organic farming (primarily greenhouse vegetables), and aquaculture, which the group gradually combined with hydroponics to produce fish and leafy greens as efficiently as possible.

Starting in 1971 Todd, McLarney, their friends and colleagues, and an assortment of back-to-the-land enthusiasts began building the physical structures of the NAI. The institute began with a dilapidated barn, overgrown fields, and scavenged refrigerators serving as aquaculture tanks (fig. 5.1).

In the first year, the group assembled a geodesic dome greenhouse for year-round aquaculture and agriculture experiments.[18] True to the countercultural style of the NAI, the construction of the first large geodesic dome involved a barn raising in which communards and hippies, including one experienced dome builder called Multi-Facet, gathered to assemble the conspicuous structure. Soon the New Alchemists and their friends had built multiple greenhouses and geodesic domes, laid out fields, dug fishponds for their tilapia and trout, and even designed and built a series of water-pumping and electricity-generating windmills.[19] Voicing the ethos of the group, Bill McLarney emphasized the measured optimism of the NAI: "I don't think anyone should be fool enough to think we can save the world. But if each of us were to look at some of the directions we'd liked to see it go in, then put your own little bit of force behind them—and have a hell of a good time while we're doing it—then that's what we should do."[20]

Around these structures grew a staff of researchers determined to

FIGURE 5.1 A view of the New Alchemy Institute grounds from the top of the group's water-pumping windmill. Reprinted from *Journal of the New Alchemists* 2 (1974).

"restore the lands, protect the seas, and inform the earth's stewards."[21] Officially rejecting the hierarchical organization of academic science, the NAI adopted an egalitarian organizational structure based on the participatory models of the New Left and the communes of the counter-culture.[22] This official equality attracted a number of communards, students, and young scholars who shared Todd's environmental and social concerns and his optimistic view of science and technology. According to Todd, the interconnected nature of ecological systems and the holistic philosophy of Taoism inspired this integrated view of science and society. The group even decorated their first newsletter with a Taoist yin-yang symbol.[23] Similarly, the name "New Alchemy" referred to medieval and early-modern practices of science, which Todd and McLarney believed approached knowledge construction from an interdisciplinary perspective better suited to solving the world's environmental problems.[24] In effect, the NAI imagined itself as an egalitarian and holistic commune bringing together scientists, engineers, and philosophers to save the world.

The combination of optimistic scientific and technical research with a lighthearted countercultural vibe almost immediately drew visitors. Soon after the first geodesic dome took shape and the New Alchemists began their "biotechnic" research, a menagerie of hippies, back-to-the-land groups, and curious middle-class families began to appear at the NAI. As the number of visitors continued to grow in 1973–74, the New Alchemists abandoned this informal practice and organized official tours every Saturday. This schedule allowed the New Alchemists to focus on their research during the week and then provide a carefully structured tour capable of showing off their recent projects and providing brief talks on the wider value of holistic science and the importance of sustainability.[25] Even as the tours became structured and educational, the countercultural feel remained in the form of post-tour potluck lunches and informal conversation with guests and New Alchemists.

For the New Alchemists, however, the reality diverged substantially from the egalitarian ideal. Contrary to the NAI's rhetorical commitment to egalitarianism, Todd dominated the institute. His ideas defined it, he led its research, and he was its public face. One could accurately think of Todd as a sort of groovy-science guru blending scientific expertise and countercultural concern while leading a band of friends and followers in a quest to save the world.[26] Todd actively embraced the persona of the countercultural guru as he experimented with ideas and tried to generate support for the NAI's work. With shoulder-length blond hair, an intense manner, and a penchant for hyperbole, he certainly looked and sounded the part. Allusions to Taoism and extensive discussions of society's potential for improvement through a holistic approach and the right technology further accentuated his sage like qualities.[27]

Todd, however, was far more than a countercultural guru able only to turn people on to the holistic reality of the world (fig. 5.2). He had considerable successes in academia. Todd, who grew up near Hamilton, Ontario, had been intrigued by sustainable farming from a young age, and he studied agriculture as an undergraduate at McGill University and pursued biology at the University of Michigan under the guidance of Marston Bates.[28] Todd's doctoral research, which examined how DDT pollution damaged the communal behavior of fish, was well received and was published in *Science*.[29] At San Diego State, he became

FIGURE 5.2 Dr. John Todd speaking to visitors at the New Alchemy Institute on Cape Cod. Reprinted from *Journal of the New Alchemists* 2 (1974).

the dean of biology while still in his thirties. Todd remained active as a scientist throughout the 1970s and beyond. He worked at the prestigious Woods Hole Oceanographic Institution early in the decade, participated in most NAI experiments, and oversaw further research into his living machines in the 1980s and 1990s. Todd's charisma and scientific expertise were a potent combination. Andy Wells, an adviser to the government of Prince Edward Island who worked closely with Todd in Canada, remembers him as an absorbing speaker and compelling thinker who seemed to possess the technical knowledge to support his big ideas.[30] This willingness to think big and undertake highly experimental research coupled with a talent for self-promotion would be the foundation of Todd's success.

Todd's Vision

Todd's visioneering began in earnest with the publication of his paper entitled "A Modest Proposal." As Nancy Jack later remembered, this article, published in 1971 in the thick, sepia-toned pages of *New Alchemy Institute Bulletin*, generated national interest in the group for the first time and defined the institute's goals and approach.[31] The self-published *New Alchemy Institute Bulletin, New Alchemy Newsletter,* and *Journal of the New Alchemists* functioned as the primary means through which the NAI presented its research and Todd's ideas to supporters.[32] A membership fee designed to help generate revenue for the NAI's research supported publication. The journal was advertised by the iconic *Whole Earth Catalog.*[33] Specific articles, such as "A Modest Proposal," circulated widely, but it is difficult to say if other articles received substantial exposure within the counterculture.[34] Instead of advertising the NAI, these publications functioned primarily as a repository of its research and ideas that the group could send to prospective supporters and charitable organizations as examples of their work and future plans.

In "A Modest Proposal," Todd accomplished two things. He embraced science and technology as a means of creating positive environmental change and he outlined a vision of a sustainable society. This vision would guide his work and gain him support throughout the 1970s and beyond.[35] Viewing earth through the eyes of an "ecologist-philosopher" from another planet, Todd began his article as a classic environmental narrative of decline.[36] Borrowing from systems ecology and the language of the space age, he identified the loss of "the required amount of biological variability in our life-support bases" as a serious threat to environmental stability and human society.[37]

According to Todd, this loss of diversity resulted from human ignorance and disregard for the fundamental laws of ecology. Pointing to agribusiness, Todd argued that large-scale industrial science had carefully removed diversity from ecological systems through its narrow search for productivity and control.[38] In his view, this disregard for ecology's basic law—diversity produces stability and uniformity produces instability—ensured that the distorted systems North Americans relied upon for their sustenance would collapse.[39] Riffing on the counterculture's critique of the large-scale technology and the concentrated bureaucratic power of the military-industrial complex, Todd added an ecological twist by likening them to the monocultures of agribusiness and connecting the cultural alienation and instability that resulted from social conformity to the ecological instability that resulted from uniformity.[40] Ecology, in Todd's view, provided a means of addressing both social and environmental problems.

After describing the world's ecological problems Todd shifted gears to provide readers with an optimistic model for creating the sustainable communities of the future. Extolling a technoscientific solution, Todd imagined a technological system that combined renewable energy and

intensive organic agriculture and aquaculture to provide energy and food
to a community without damaging the environment. Carefully modeled
on leading ecologists Eugene Odum's and Howard Odum's studies of
natural ecosystems, Todd's proposed "biotechnology" included plants,
animals, and fish to create a technologically mediated stable climax ecosystem.[41] Human users also had a central place in Todd's ecotechnology.
Moving nutrients throughout the system, they would employ technology to ensure each feedback loop functioned effectively. Managing such
hybrid systems would, in Todd's view, allow operators to gain ecological
knowledge and become both users of nature's resources and agents of
the biosphere's overall stability.[42]

Todd's vision of sustainability drew heavily on perhaps the most
iconic product of Cold War Big Science in the 1960s: the NASA space
program. As part of the Apollo program, NASA adapted earlier experiments with enclosed environments on submarines and in bomb shelters
to the challenges of keeping astronauts alive for long periods by experimenting with self-sustaining space capsules.[43] These efforts to understand and minutely control human/machine/environment interaction to
create a stable feedback system of "cabin ecology" provided the basis
for Todd's "biotechnical" systems.[44] Imagined as something similar to
a space capsule, Todd's "biotechnology" was supposed to function as
a complete system capable of both providing sustenance for its users
and recycling their waste.[45] In effect, Todd's vision took Kenneth Boulding's idea of "Spaceship Earth" literally and imagined community-sized
systems, which merged technology and biology, as the foundation of a
future sustainable society.[46]

To Todd these biotechnological systems would be a means of transforming North America through a recolonization of the rural landscape. Todd thought of North America as a space doomed to ecological
and social collapse by resource depletion and pollution. Echoing and
inverting contemporary ideas of outer space as a new frontier where
Americans could revitalize their society, he redefined the troubled North
American landscape as a frontier ripe for the spread of new technologies and ecological science.[47] In his view, this new terrestrial frontier offered possibilities for revitalization similar to those of outer space, since
it would unleash the energy of twenty-first-century pioneers who would
use his "cabin ecology" to create a truly sustainable society and ensure
humanity a permanent place on Spaceship Earth.[48]

Todd's embrace of the visioneering principle that the "future might
be radically changed by technology" and the New Alchemists' merger
of machine and nature seem to contradict the foundations of environ-

mentalism. This approach, however, had a substantial following within countercultural environmentalism, as Todd and the NAI figured prominently in the *Whole Earth Catalog* and back-to-the-land publications.[49] As Kirk argues in chapter 10 of this volume, a significant minority of the counterculture saw technology as an effective means of preserving the environment and achieving social change by mediating human interactions with nature. Thus, rather than representing an apolitical disengagement from society, technological experimentation was the ecopragmatists' favored method for transforming American society. Defining technology as the key to saving the environment, Todd asserted that his integrated "biotechnic" systems provided the most effective means of "re-establishing [a] much needed link with the organic world" and creating a sustainable future.[50]

Combining Big Science research, environmentalism, and technological optimism, Todd's vision internalized several fundamental contradictions. For instance, Todd criticized Big Science when used by the military or agribusiness, while he adopted NASA research and the Odums' ecosystems ecology, which originated from their efforts to trace radioactive material through the environment for the Atomic Energy Commission.[51] The Odums themselves further enhanced the technocratic potential of their ecological theories as they advocated using them to manage both human society and the natural environment. Indeed, a fundamental conclusion of their work held that the feedback between human and environment necessitated careful scientific management.[52]

Within the counterculture and environmentalism, some avidly employed the knowledge produced by Big Science projects even as others attacked those projects. While such contradictions should not be ignored, neither should they be used to dismiss or marginalize Todd and other pragmatic countercultural environmentalists simply because they defy easy categorization. Their willingness to borrow and combine contradictory ideas was the basis of their creativity and the pragmatic solutions they offered in the 1970s.[53] Instead of rejecting or attempting to escape the influence of Big Science, Todd and others attempted to repurpose ideas and technologies to construct a workable alternative and paid little attention to intellectual provenance. In their view, the possibility of creating a workable alternative outweighed concerns about its underlying politics.

Building a Support Network

Indispensable for any visioneer, as McCray argues, is the ability to imagine and shape the future.[54] With his vision of an ecotechnical future outlined, Todd set about promoting this work and building a network of support. As in the world of academic science, mobilizing funds played an instrumental role in success. Building networks, seeking media attention, and generating financial support were all central to the work of ecopragmatists. Todd proved highly skilled at all these activities, as he successfully created a support network that embraced the counterculture, the environmental movement, elite charitable foundations, and government policy makers. Todd's vision, hard work, and personal charisma, along with the NAI's scientific research and publications, ultimately enabled the group to construct the technologies Todd hoped would make his "biotechnological" vision a reality.

Todd's vision and his focus on the practical application of science and technology quickly gained him a following among the countercultural environmentalist community. Soon after publishing "A Modest Proposal" Todd convinced the Rodale Foundation, a pioneer in the organic-farming community, to fund the NAI's research into organic agriculture.[55] Writing in the Rodale Press's *Organic Gardening and Farming*, Todd argued that an alternative to ecologically dangerous factory farming could be created only through intensive scientific research.[56] Forming a small program of citizen science, Todd recruited gardeners and farmers across America to help him develop techniques of intensive aquaculture and organic pest control that he believed would enable the construction of self-sustaining enclosed ecosystems.[57]

Another source of support for Todd's research materialized when Stewart Brand visited the NAI in 1973. Impressed by what he saw, as well as Todd's plans to expand the NAI's research to include solar and wind energy, Brand promised the NAI $16,000 in funding from the Point Foundation.[58] After 1973 the NAI would continue to have a close relationship with Brand's *Whole Earth Catalog* and the network of countercultural environmentalists and ecopragmatists it supported. Todd, for instance, took part in the community's debates over space colonies.[59] His work also provided Brand with a useful example of the potential of the holistic merger of technology and ecology Brand advocated in the *Whole Earth Catalog*.[60] In the late 1970s the two groups grew even closer as J. Baldwin, the editor of *CoEvolution Quarterly*, joined the NAI hoping to help commercialize the technologies it had developed.[61]

The NAI's research also received attention outside the counterculture. In the early 1970s international concern over environmental limits and ecological collapse had exploded with the publication of the Club of Rome's *Limits to Growth* and the Stockholm Conference on

the Human Environment and continued to grow with the oil shocks
of 1973. Reporters began arriving at the NAI asking what sort of solutions these groovy scientists and their solar-heated fishponds might have
for the world's problems. In 1973 the *New York Times* reported that
"a group of oceanographers" had managed to "create a decentralized,
nature-loving, but still comfortable way of life."[62] *Time* and *Science*
published similarly flattering reviews praising the group's commitment
to ecological design.[63] Todd did everything he could to expand this interest. According to David Bergmark, an architect who helped design
and build the NAI's Ark Bioshelters, Todd succeeded. The crowds that
came to Cape Cod to see the NAI's experiments with enclosed ecosystems included astronauts curious about space capsules and representatives of DuPont who thought Todd might be interested in new transparent and durable plastics under development.[64]

The allure of Todd's work spread to Canada as well. An article in the
Canadian Magazine entitled "The World That Feeds Itself" lauded his
work and argued that Todd's ecosystem approach might hold the best
promise of future survival if energy and resource shortages actually occurred.[65] In 1974 Robert Durie, the director of Environment Canada's
Advanced Concepts Branch, traveled to Cape Cod to see the New Alchemists' work firsthand and assess its relevance to Canada. He left convinced that Todd's vision had great potential and the group's research
could help Canadians adapt to a future dominated by environmental
limits.[66] Durie's support would help the NAI impress other Canadian
policy makers, most notably the government of Prince Edward Island
(PEI), and eventually receive the hundreds of thousands of dollars in
federal funds necessary to build the PEI Ark Bioshelter.

Anxious to impress visitors and potential supporters and ensure that
this positive publicity continued, the NAI started self-publishing the
Journal of the New Alchemists in 1973. It amalgamated the different
styles of do-it-yourself (DIY) magazines, countercultural newspapers,
and scientific journals.[67] Aesthetically, the *Journal of the New Alchemists* resembled classic countercultural magazines such as the *San Francisco Oracle* and the *Great Speckled Bird*. It emulated their visual, even
psychedelic, style: unicorns, phoenixes, and fish decorated its covers,
and Celtic-inspired designs and artistic illustrations adorned its articles
(fig. 5.3). The journal also lent physicality to the NAI with numerous
black-and-white photographs. Printed in a large, twelve by eight and a
half inch format on thick paper and running close to a hundred and fifty
pages on average, these journals represented a substantial investment on
the part of the NAI.

FIGURE 5.3 The rising-phoenix cover of the second volume of *Journal of the New Alchemists*. Subsequent volumes pictured other mythical creatures and continued to feature elaborate cover art. Reprinted from *Journal of the New Alchemists* 2 (1974).

The journal's articles covered a range of topics. The majority of articles were academic in tone and content and presented detailed reports of Todd's, McLarney's, and other NAI scientists' research. For instance, in the second volume of the *Journal of the New Alchemists* McLarney presented a short but highly technical and extensively researched article describing his research on midge larvae.[68] The Woods Hole Oceanographic Institution, where McLarney and Todd worked in the early 1970s, actually funded this research, as their experiments at the NAI and their professional scientific careers overlapped. Along with these academic articles, the NAI also occasionally used the journal to give instructions for DIY projects, including windmills, integrated agriculture and aquaculture systems, and solar collectors.[69] Nearly every volume also included essays in which Todd expounded upon his vision of a sustainable future and fleshed out his theories of how humans, technology, and nature could be merged to achieve ecological permanence.[70]

The NAI's investment in the journal, helped by Todd's determination and salesmanship, paid off in the mid-1970s. With the beautiful and informative journal to present and a growing list of research projects to laud, Todd impressed foundations and gained much-needed financial support.[71] By 1974 the NAI had reached a two-year agreement worth $50,000 with the Rockefeller Brothers Fund and had obtained pledges of support from other less prestigious organizations.[72] Grants from the Canadian government and the National Science Foundation followed in 1975 and 1978.[73] These funds enabled a dramatic expansion of the NAI's facilities and gave Todd the financial stability to leave Woods Hole and dedicate himself full-time to his vision.

The majority of this time and money went into research on enclosed ecosystems. Determined to better understand these systems and create technologies capable of integrating humans into nature's feedback systems, Todd turned to NASA's experiments with self-sustaining ecosystems. Applying ecology to the problem of space travel, scientists attempted to design self-contained ecosystems capable of sustaining life over the vast distances of space.[74] Envisioned as "a little piece of this biosphere," space cabins were planned as stable climax ecosystems capable of supporting a small number of astronauts.[75] Seeking to replicate NASA's systems, Todd and the New Alchemists experimented with terrestrial space cabins. This resulted in "living machines," large plastic cylinders of algae and herbivorous fish that employed hydroponic gardens to filter the nutrient-rich water and return it to the fish.[76] This integrated system would be the heart of the NAI's bioshelters, recycling wastes and providing food for the human "passengers" of these terrestrial space cabins.

As the NAI's resources and research expanded, Todd began crisscrossing North America presenting his research and elaborating on his vision. Throughout the 1970s, Todd attended conferences ranging from a symposium on political ecology organized by anarchist Murray Bookchin in Vermont to Denis Meadow's more prestigious Limits to Growth '75 conference in Texas.[77] Todd's writings, research, and speeches, along with his circumambulations, established him as an important figure within the small community of ecological designers and pragmatic environmentalists. Peder Anker, who has written extensively about ecological design, notes that Todd's growing corpus of work with cabin ecology brought together ecological designers' interest in cybernetics and ecology and placed him at the leading edge of the movement in the mid-1970s.[78]

Todd also used the NAI's growing research and his increasingly detailed vision of the future to gain the support of policy makers. In Canada, Todd leveraged his scientific credentials, the NAI's support network, and post–oil shock concerns over energy shortages to make his vision attractive to federal and provincial policy makers. Anxious to impress, Todd highlighted the connections between his ecological designs and NASA's cabin ecology. He identified his work as a "spaceship" approach and emphasized the high-tech character of his designs, which employed "micro-computers" as "control elements."[79] According to Todd, his advanced designs could act as a sort of space-age ark preserving entire communities if the oil supplies started to run out. Impressed by Todd's work, which he encountered at a conference in Ottawa, and well aware of Environment Canada's interest, Andy Wells brought it to the attention of PEI's premier, Alexander Campbell, who agreed that Todd's research might be a possible solution to PEI's dependence on expensive imported oil.[80]

Todd's designs offered the provincial government more than an insurance policy: he promised that they could reshape local economies by meeting the "food, shelter and power needs of urban and rural families" and provide the technology to found local wind energy and sustainable housing industries.[81] This vision of ecological and economic transformation coming from a successful scientist with his own institute and the support of the ecological design community as well as the Rockefeller Foundation convinced both the provincial and federal governments. In 1975 the Canadian government provided the NAI with land and promised all the financial support necessary to construct Todd's proposed space-age ark on PEI.[82]

Todd's ability to draw substantial support from diverse sources made the NAI's research possible. The New Alchemist's Ark on PEI could not have existed without Todd's skill at promotion and network building. The importance of mainstream funding sources for Todd's visioneering challenges the often-assumed distance between mainstream science and the counterculture. Todd's ability to generate mainstream interest and financial commitments suggests that translation occurred in the area of ecological design as scientists moved between academic biology and the counterculture with the support of elite funding organizations.[83] Nor were the projects of ecopragmatists the purview of only grassroots groups, although they provided instrumental assistance in the early years of the NAI. The Canadian state had a central role in one of the most important ecological design projects of the decade. Rather than walling themselves off, groovy scientists engaged in constant dialogue as they both criticized and worked with mainstream institutions in an attempt to transform North America.

Constructing Terrestrial Space Cabins

Todd's efforts to transform North American society reached their apogee with the PEI Ark Bioshelter, which stands as a milestone in ecological design. Its construction highlights the popularity of NASA research among the counterculture as well as the ways in which environmentalists adapted the concept of progress to a limited world. A central focus of the ecological design movement in the 1970s was to re-create the house along ecological lines.[84] Ecological designers, including Todd, believed that by refashioning this ubiquitous technology, they could transform the foundations of American society by reconnecting people with nature and providing the personal means to live in a sustainable manner. To reimagine the home, these countercultural and environmentalist designers drew heavily on the Spaceship Earth concept popular in the 1970s.

The connection between space and ecology has several foundations. One emerged from Stewart Brand's own campaign for a photograph of the earth from space, which he hoped would transform humans' understanding of their fragile planet and usher in an age of ecology.[85] The work of Buckminster Fuller, the design guru for much of the counterculture, provided another connection to space. He popularized the idea of Spaceship Earth and even argued that the planet should be managed according to the principles of NASA's cabin ecology in his popular book *Operating Manual for Spaceship Earth*.[86] The Odum brothers, who helped found ecosystems ecology, gave this claim a more scientific foundation when they applied systems ecology to human societies and asserted that American society should be managed in a manner analogous to a stable climax ecosystem on a space capsule.[87] In Howard Odum's view, "the biosphere [was] really an overgrown space capsule" amenable to the same methods of scientific control.[88] Effectively, this "spaceship" approach to environmental problems assumed that living in harmony with nature was possible if one adopted the technology and lifestyle of astronauts and transitioned to the steady-state economy of a climax ecosystem.[89]

Todd and the New Alchemists used this spaceship approach to environmental problems to build a number of unique bioshelters. The most advanced of these "life support systems" was the PEI Ark.[90] The physical embodiment of the merger of human, machine, and nature that Todd had envisioned years earlier in "A Modest Proposal," the PEI Ark attempted to provide an almost completely self-sufficient system capable of providing twenty-first century pioneers with abundant food, comfortable shelter, and inexhaustible energy while restoring the environment.

"Weaving together the sun, wind, biology, and architecture," the NAI's space-cabin design perfectly accorded with contemporary concerns about limits, as it promised to efficiently produce fish, vegetables, and energy with almost no fossil fuel by recycling nutrients.[91] To optimize its energy efficiency the NAI carefully employed Howard Odum's theories of energy flow to design the Ark along ecological lines.[92] When constructed, the Ark's feedback system was a living machine, which integrated a series of solar fishponds, a small greenhouse, a solar-heating system, and water-pumping and electricity-generating windmills (fig. 5.4). A barnlike plastic and wood structure sunk into the ground to maximize thermal efficiency enclosed these complex components. The Ark also included a three-bedroom house, a workshop/garage, and a laboratory for "21st century pioneers."[93] In effect, the Ark merged a space cabin with a suburban home to come as close as possible to being

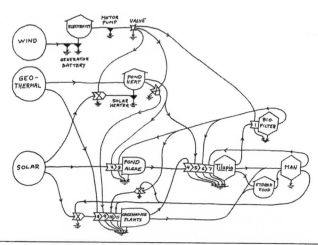

ENERGY FLOW DIAGRAM FOR SOLAR-HEATED GREENHOUSE AND AQUACULTURE COMPLEX

1) Circulation of fish pond water through biological filter. 2) Increased growing temperature for algae.
3) Nutrient cycling from fish wastes. 4) Terrestrial plants to Tilapia. 5) Increased temperature for Tilapia.
6) Removal of growth inhibitors. 7) Fish husbandry. 8) Earth heat warming the pit-greenhouse. 9) Night
warming of greenhouse. 10) Irrigation by nutrient/rich water. 11) Vegetable gardening.

FIGURE 5.4 Ark Bioshelter energy flow diagram. In this diagram, the New Alchemists used Howard Odum's cybernetic approach to ecology pioneered by NASA to conceptualize how their Ark might function as a stable ecosystem. Reprinted from *Journal of the New Alchemists* 2 (1974).

its own ecosystem able to support its inhabitants and preserve Spaceship Earth.

Excited by the building of a structure that incorporated every aspect of his vision, Todd wrote a triumphant article entitled "Tomorrow Is Our Permanent Address" describing how his Ark could transform North America. According to Todd, the Ark was an "adaptive structure" that "optimized personal lives" through ecological design.[94] Embodying his attacks on destructive large-scale technologies and the alienating and emasculating influence of industrial society, the Ark functioned on the small scale and enhanced local and personal self-sufficiency by providing food and energy. In his view, it also counteracted social alienation by directly involving its users in what he believed to be the most meaningful activities of all: providing for oneself and restoring the environment.[95] Todd also carefully pointed out that all this could be accomplished because the NAI did not reject science but rather "redefined it and redirected it" to achieve a "new synthesis" of human, technology, and landscape.[96]

With this "new synthesis," Todd reconceptualized progress to fit within the limits of Spaceship Earth. He replaced the physical expansion normally associated with progress with temporal expansion and suggested that ensuring human existence through sustainability was as great an achievement as exploring new worlds. Similarly, he argued that a society rebuilt around his sustainable technology represented both scientific progress, particularly the extension of biology and ecology, and an improvement in the human condition, because Arks, or similar technology, helped people to overcome their destructive artificial separation from nature and to recognize their place in the greater holistic unity of life.

Unfortunately for Todd, only some of the technologies within the Ark performed as well as he had promised. In its first year of operation, the central component of the Ark, its living machines, worked well, producing fish, vegetables, and seedlings for reforestation projects. The systems with which the NAI had less experience, however, did not operate quite so effectively. The Ark's solar-heating system malfunctioned when piping connecting the solar collectors in the living area leaked and caused a cascade of problems that undermined the rest of its heating systems.[97] In fact, the structure's complexity meant that managing it effectively required a considerable degree of training and knowledge, an ironic result, considering Todd's desire to produce technologies that individuals and communities could operate with little formal training.

As disappointing as this malfunction was it did not compare with the failure of the Ark's wind turbines. Developed by the NAI, the turbines employed a novel system using hydraulics to control the blades and generate electricity.[98] Despite their emphasis on experimentation and research, the NAI did not test them in their rush to complete the Ark. Pierre Trudeau, the prime minister of Canada, wanted to ceremoniously open the Ark in early September 1976, and not wishing to disappoint the prime minister, the New Alchemists rushed construction. Soon after the grand opening ceremony, at which both Prime Minister Trudeau and Todd lauded the bioshelter's potential, PEI's high winds overwhelmed the turbines and their hydraulics seized up, forcing the Ark to draw electricity from PEI's grid—an ignoble result for a structure promoted as self-sufficient.[99] Compounding the seriousness of this failure was the fact that the provincial government presented the turbines as the potential foundation for a local industry.[100]

Locals viewed the breakdown of the wind turbines as a betrayal, a harsh judgment for an experimental project but reasonable considering Todd's and the provincial government's promises of space-age technology and new industries.[101] The structure's $354,000 price tag and the tens of thousands of dollars in annual costs also rankled, and locals began to see the Ark as an expensive boondoggle.[102] Frightened by collapsing local support for the Ark, the federal and provincial governments attempted to take a more active role in the project. To this end, they instituted strict controls on the New Alchemists' budget.[103] Intolerant of interference and well aware of the changing mood on the island, Todd and the rest of the New Alchemists left PEI abruptly in 1978.[104] Taking a parting shot at the government that had recently lavished support on him, Todd told the local paper that the parochial province would regret abandoning the opportunity to pioneer sustainability that he had provided.[105] The provincial Institute of Man and Resources took over the Ark and operated it as a research and public-demonstration facility until 1981, when it closed permanently.[106] Due to the project's acrimonious end, locals remember Todd primarily for his aggressive "salesmanship."[107] Nationally the Ark received more latitude and remained an intriguing, albeit disappointing, attempt to live sustainably.[108] Similarly, the ecological design community saw the Ark as a successful experiment and interpreted both its successes and its failures as a demonstration of ecological design's growing potential to combine humanity, nature, and technology for the benefit of all.[109]

Ecological Design and Living Machines

Undeterred and largely unscathed by the failures of the PEI Ark, Todd refused to give up on his vision. Pursuing the transformative possibilities he had identified in the Ark, Todd continued to promote his sustainable designs. Inspired by Margaret Mead's interest in the Ark, the New Alchemists decided they needed to apply Todd's designs on a larger scale, starting with small communities and eventually moving to large cities and entire bioregions.[110] To that end the NAI hosted a large conference in 1979 to discuss ecological design and its potential to solve North American social and environmental problems. Organizing the event around the concept of "the village as solar ecology," Todd managed to involve influential members of the counterculture, the environmental movement, and the ecological design community, including solar energy advocate Amory Lovins, anthropologist Mary Catherine Bateson, and designer J. Baldwin.[111]

Building on Todd's visioneering, the group extended his systems to encompass entire communities. Baldwin, whose ecopragmatist designs are discussed elsewhere in this volume by Andrew Kirk, extolled the possibilities of community-sized geodesic dome bioshelters. Imagining a twenty-first-century village built on the periphery of a 1.5-acre bioshelter, Baldwin argued that such a design could re-create many of the advantages of Todd's Arks on a large scale. In a re-creation of the PEI Ark, houses connected to the dome's edge would be heated by the bioshelter as their inhabitants used it to produce the community's food. Echoing Todd's arguments, Baldwin claimed that the design's transformative properties would help create self-sufficient and ecologically sustainable communities while cutting fossil fuel consumption and undermining mainstream consumer culture.[112] The 1979 "village as solar ecology" conference, other ecopragmatist meetings, and the PEI Ark laid the foundations for ecological design.

As concerns over limits faded in the 1980s, the NAI began to struggle financially, and Todd realized that the substantial levels of funding required for projects, such as the PEI Ark, were unlikely to materialize in the future.[113] Shifting his focus to defining and championing ecological design and experimenting with his "living machines" on a smaller scale, Todd and his wife left the NAI to found Ocean Arks International in 1982.[114] Anxious to keep his research on sustainable systems relevant and uphold his vision of a sustainable future, John and Nancy Todd began to write extensively about green architecture and published their first book dedicated to the subject in 1984.

The Todds' book, *Bioshelters, Ocean Arks, City Farming*, distilled John Todd's vision and the NAI's research into a theory of ecological design. Refashioning Todd as an architectural visionary rather than a groovy-science guru, the Todds outlined nine "precepts" for sustainable construction, including "the living world is the matrix for all design," "design should be sustainable through the integration of living systems," and "design should follow a sacred ecology." Readers were urged to incorporate John Todd's merger of humanity, environment, and technology into their designs.[115] Readers listened. This book, along with Todd's extensive work in the field, helped him shape the ecological design movement.[116]

In the 1990s the Todds continued to contribute to the ecological design movement as it professionalized and became more popular. They wrote a second book outlining Todd's approach to ecological design, *From Eco-cities to Living Machines: Principles of Ecological Design*, which once again advocated Todd's technologically mediated approach to sustainability.[117] The journal *Ecological Engineering* republished his work, which helped maintain interest in his ideas. Drawing on his continued research with "living machines" Todd and his assistant, Beth Josephson, identified and elaborated upon twelve "principles required for the design of task-oriented mesocosms."[118] These articles brought mainstream attention to the Todds and their institute, Ocean Arks International, which enjoyed a brief explosion of interest in the 1990s. As a result, Todd was offered a position in the University of Vermont's environmental studies program, which he accepted.[119]

While Todd contributed to ecological design in the 1980s and 1990s, he continued to develop the integrated aquaculture and agriculture systems that had been the foundation of his work since the early 1970s. In the mid-1980s, Todd shifted his focus from constructing terrestrial space cabins, designed to carry their users through ecological catastrophe, to the more mundane, but widely needed, work of waste treatment. Employing the living machines he had pioneered at NAI, Todd once again re-created tropical estuaries, this time for their ability to purify. By 1989 he had managed to produce a system capable of treating the sewage of a small community.[120] In the early 1990s he redesigned this system to fit on a small, floating platform and enable it to biofilter industrial wastewater lagoons. Tyson Foods put this system to the test in a lagoon attached to a poultry-processing plant in 2001.[121] After successfully bringing that facility's wastewater within EPA standards, Todd's design began to spread, and it has been copied by former employees and others interested in a biological approach to waste treatment.[122]

Besides being commercialized as biofiltration systems, Todd's living machines were copied, albeit largely unknowingly, by urban farmers using "aquaponics" in the 2000s.[123] This newly recognized branch of aquaculture has re-created or copied the Arks' combination of hydroponics and aquaculture to culture fish and grow vegetables almost exactly as the NAI did in the 1970s. In 2008 Will Allen, an urban farmer from Milwaukee, received a $500,000 genius fellowship from the MacArthur Foundation for using aquaponics to supply the city's food deserts (communities with few grocery stores and little access to fresh food) with fish and vegetables.[124] This "new" approach to farming adopts Todd's original, "spaceship" approach: it recycles nutrients to ensure high productivity of solar greenhouses adapted to the small spaces and marginal resources of urban or personal farms. Today, kits for such systems can be found on the Internet, an achievement ecopragmatists would surely find fitting.[125]

Despite this success, we are still a long way from becoming a society dedicated to using cabin ecology to live sustainably. Nonetheless, Todd's visioneering remains important. Predicated on the merger of human, machine, and nature, Todd offered a pragmatic vision of sustainability. Occupying a middle ground between the limitless growth imagined by space enthusiasts and the declensionist views of many environmentalists, Todd provided a path toward modern living within earthly limits. Todd's pragmatic vision captured the attention of environmentalists, hippies, and politicians, and it became central to an influential community of ecopragmatists seeking solutions to the world's environmental problems. Garnering support and substantial funds, Todd was able to experiment with his ideas and even create the system he hoped would give North American society a sustainable future. Not all the components worked. Much to the disappointment of Canadians, the complexity of self-sufficient components defeated him and his fellow designers on PEI. However, other technologies and ideas worked, contributing to ecological design, industrial biopurification, and urban farming. In fact, Todd's visioneering can be seen as an important contribution to what some environmental historians have characterized as the "light-green society."[126] This hybrid of environmentalism and industrial modernity created through the blurring of the boundaries between culture and nature resulted from efforts like Todd's and could have few better exemplars that his living machines.

Todd's visioneering reveals an often-overlooked side of the counter-culture and environmentalism. Above all, it illustrates the importance of science and technology to countercultural environmentalism. His efforts pinpoint the fundamentally problematic nature of any analysis that attempts to understand the counterculture or the environmental-ism of the 1970s as antiscience. While condemning some applications of science and technology and even openly discussing the collapse of American civilization, Todd, and many other members of the counter-culture and environmental movements, never abandoned the concept of progress or the conviction that solutions to the world's problems could be found through the application of science.

Beyond illustrating the importance of science and technology to countercultural environmentalism, Todd's visioneering highlights the surprising degree of cooperation and translation that existed between the counterculture and the mainstream in the 1970s. Parallels exist be-tween Todd and better-known countercultural figures, such as physicist Fritjof Capra, who drew out and expanded upon the connections be-tween the counterculture and quantum physics.[127] Working in the me-dium of ecology rather than physics, Todd used his scientific training and talent for promotion to enlist the support of governments, counter-cultural groups, and elite institutions for his technologically mediated vision of sustainability. The ease with which Todd and his ideas moved between these communities and the fundamental importance of their financial contributions underline the presence of cooperative relation-ships between governments, elite institutions, and countercultural envi-ronmentalists and their impact on environmental politics. Without this support, the PEI Ark, one of the most iconic efforts to live sustainably and a foundational project in ecological design, could not have been built. Cooperative and mutually beneficial relationships existed and even thrived alongside the better-known conflicts between governments and countercultural environmentalists.

Notes

1. Paul Ehrlich, "Eco-catastrophe!," *Ramparts* 8, no. 3 (September 1969): 24–28.

2. Nancy Jack Todd, "New Alchemy: Creation Myth and Ongoing Saga," *Journal of the New Alchemists* 6 (1979): 10.

3. Nancy Jack Todd and John Todd, *Bioshelters, Ocean Arks, City Farming: Ecology as the Basis of Design* (San Francisco: Sierra Club Books, 1984), 3.

4. New Alchemy Institute, Articles of Incorporation, San Diego, CA, 1970, folder 11, box 1, New Alchemy Institute Records, Special Collections Depart-

ment, Iowa State University, Ames (hereafter cited as NAI Records). Todd, being a Canadian, hoped his work would influence all of North America.

5. Ibid.

6. For a description of these living machines and how Todd conceived of them, see Nancy Jack Todd, *A Safe and Sustainable World: The Promise of Ecological Design* (Washington, DC: Island Press, 2006), 167–68.

7. Peder Anker, *From Bauhaus to Eco-house: A History of Ecological Design* (Baton Rouge: Louisiana State University Press, 2010), 108–9.

8. John Greene, *America in the Sixties* (Syracuse, NY: Syracuse University Press, 2010); Charles Reich, *The Greening of America: How the Youth Revolution Is Trying to Make America Livable* (New York: Random House, 1970); and Theodore Roszak, *The Making of a Counter Culture: Reflections on the Technocratic Society and Its Youthful Opposition* (Garden City, NY: Anchor / Doubleday, 1969), 50 (quotation).

9. Frank Zelko, "Making Greenpeace: The Development of Direct Action Environmentalism in British Columbia," *BC Studies* 142/143 (Summer/Autumn 2004): 197–239; and Thomas Dunlap, *Faith in Nature: Environmentalism as Religious Quest* (Seattle: University of Washington Press, 2004).

10. Timothy Miller, *The 60s Communes: Hippies and Beyond* (Syracuse, NY: Syracuse University Press, 1999); and Christopher Manes, *Green Rage: Radical Environmentalism and the Unmaking of Civilization* (Boston: Little, Brown, 1990).

11. The examination of the tangled existence of nature and culture has been a major theme in environmental history since the 1990s. See William Cronon, "The Trouble with Wilderness; or, Getting Back to the Wrong Nature," in *Uncommon Ground: Rethinking the Human Place in Nature*, ed. William Cronon (New York: W. W. Norton, 1995), 69–90.

12. W. Patrick McCray, *The Visioneers: How an Elite Group of Scientists Pursued Space Colonies, Nanotechnologies, and a Limitless Future* (Princeton, NJ: Princeton University Press, 2013), 13.

13. Andrew Kirk, *Counterculture Green: The "Whole Earth Catalog" and American Environmentalism* (Lawrence: University Press of Kansas, 2007), 165, 170.

14. McCray, *Visioneers*, 10.

15. Richard White, "Frederick Jackson Turner and Buffalo Bill," in *The Frontier in American Culture*, ed. James Grossman (Berkeley: University of California Press, 1994), 25; and Leah Ceccarelli, "At the Frontiers of Science: An American Rhetoric of Exploration and Exploitation," paper presented at University of British Columbia Colloquium Series, 17 January 2013, Vancouver, BC.

16. John Todd, "The New Alchemists," in *Ecological Design: Inventing the Future*, ed. Chris Zelov and Phil Cousineau (Easton, PA: Knossus, 1997), 172.

17. Nancy Jack Todd, "The New Alchemy Institute—East: Cape Cod," *New Alchemy Newsletter* 1 (Spring 1972): 4.

18. Nancy Jack Todd, "Readers' Research Program," *New Alchemy Newsletter* 1 (Spring 1972): 11.

19. William McLarney and John Todd, "Walton Two: A Complete Guide to Backyard Fish Farming," *Journal of the New Alchemists* 2 (1974): 79–117; and Earle Barnhart, "An Advanced Sail-Wing for Water-Pumping Windmills," *Journal of the New Alchemists* 3 (1976): 25–27.

20. N. J. Todd, *Safe and Sustainable World*, 191.

21. This was the NAI's motto throughout the 1970s. John Todd, "Introduction," *New Alchemy Institute Bulletin* 1 (Fall 1970): 1.

22. John Todd, "Realities from Ideas, Dreams and a Small New Alchemy Community," *New Alchemy Newsletter* 2 (1973): 5–6. For a discussion of the participatory models of the New Left and the counterculture, see Francesca Polletta, *Freedom Is an Endless Meeting: Democracy in American Social Movements* (Chicago: University of Chicago Press, 2002), 122–30; and Miller, *60s Communes*, 67–69, 89–90.

23. See *New Alchemy Institute Bulletin* 1 (Fall 1970). Todd often pointed to Taoism as a guide for his efforts to design holistic systems integrating human, machine, and nature. See John Todd, "Pioneering for the 21st Century: A New Alchemist's Perspective," paper presented at Limits to Growth '75, 19–21 October 1975, The Woodlands, TX.

24. William McLarney to New Alchemists, "What NAI Means," folder 12, box 38, NAI Records.

25. Nancy Jack Todd, "Visitors," *Journal of the New Alchemists* 1 (1973): 5–6.

26. As Miller (*60s Communes*, 166–69) notes, a dominant charismatic leader was common among countercultural groups even though it contradicted the commitment to participatory democracy.

27. J. Todd, "Pioneering for the 21st Century."

28. John Todd, "Marston Bates," *Journal of the New Alchemists* 2 (1974), front matter. See N. J. Todd, *Safe and Sustainable World*.

29. See John Todd, Jelle Atema, and John Bardach, "Chemical Communication in Social Behavior of a Fish, the Yellow Bullhead," *Science* 158, no. 3801 (3 November 1967): 672–73.

30. Andy Wells, interview by Alan MacEachern, 27 July 1999, series 11 Alan MacEachern interviews, Institute of Man and Resources Fonds, Prince Edward Island Public Archives and Records Office (hereafter cited as IMR Fonds).

31. N. J. Todd, "New Alchemy: Creation Myth and Ongoing Saga," 11.

32. Nancy Jack Todd continues to write about sustainability today in *Annals of the Earth*, which is published by Ocean Arks International, a nonprofit organization that she and John Todd founded in 1982 to disseminate the ideas and practices of ecological sustainability.

33. Stewart Brand et al., *The Last Whole Earth Catalog: Access to Tools* (Menlo Park, CA: Portola Institute, 1971), 180.

34. The NAI's records provide very little information about their publication practices or the circulation of their journals and newsletters and those who subscribed to them. Some documents do, however, mention that the journals were used to assist with fundraising efforts.

35. Kirk, *Counterculture Green*, 158. Visioneers also used the trope of the

frontier to depict space as a place possible to colonize and the crucible of a new American society; see McCray, *Visioneers*, 68–72.

36. John Todd, "A Modest Proposal," *New Alchemy Institute Bulletin* 2 (Spring 1971): 3.

37. Ibid., 1; John Todd, "The Dilemma beyond Tomorrow," *Journal of the New Alchemists* 2 (1974): 123.

38. J. Todd, "Modest Proposal," 4.

39. Systems ecology of the 1970s emphasized progression and the stability produced by the diversity of climax ecosystems. See Sharon Kingsland, *The Evolution of American Ecology, 1890–2000* (Baltimore: Johns Hopkins University Press, 2005), 210–11.

40. J. Todd, "Modest Proposal," 7.

41. Ibid., 9.

42. Ibid., 12.

43. See Allen Brown, "Regenerative Systems," in *Human Ecology in Space Flight*, ed. Doris Calloway (New York: New York Academy of Sciences, 1963), 82–119. Peder Anker discusses the politics and broader influence of this research in "The Ecological Colonization of Space," *Environmental History* 10, no. 2 (April 2005): 239–68.

44. Anker, "Ecological Colonization of Space," 240.

45. J. Todd, "Modest Proposal," 12. Todd would elaborate on this theme in later writing. See John Todd, "The World in Miniature," *Journal of the New Alchemists* 3 (1976): 54–79.

46. Kenneth E. Boulding, "The Economics of the Coming Spaceship Earth," in *Environmental Quality in a Growing Economy: Essays from the Sixth Resources for the Future Forum*, ed. Henry Jerrett (Baltimore: Johns Hopkins University Press, 1966), 3–14.

47. McCray, *Visioneers*, 48.

48. J. Todd, "Pioneering for the 21st Century."

49. David Lees, "Aboard the Good Ship Ark," *Harrowsmith* 9 (September/October 1977): 49–53, 81–82, 97–98.

50. J. Todd, "Modest Proposal," 8.

51. Emory Jessee, "Radiation Ecologies: Bombs, Bodies, and Environment during the Atmospheric Nuclear Weapons Testing Period, 1942–1965" (PhD diss., Montana State University, 2013). Also see Stephen Bocking, *Ecologists and Environmental Politics: A History of Contemporary Ecology* (New Haven, CT: Yale University Press, 1997), 73–75.

52. Eugene Odum, "The Strategy for Ecosystem Development," *Science* 164, no. 3877 (18 April 1969): 262–70; and Howard Odum, *Environment, Power, and Society* (New York: Wiley-Interscience, 1971).

53. Kirk, *Counterculture Green*, 16.

54. McCray, *Visioneers*, 12.

55. N. J. Todd, "Readers' Research Program," 11.

56. John Todd, "Shaping an Organic America," *Organic Gardening and Farming* 18, no. 9 (September 1971): 50–55; and John Todd, "The Organic Gardener and Farmer as a Scientist," *Organic Gardening and Farming* 18, no. 11 (November 1971): 63–69.

57. John Todd and William McLarney, "The Backyard Fish Farm," *Organic Farming and Gardening* 19, no. 1 (January 1972): 99–109; and John Todd and Richard Merrill, "Insect Resistance in Vegetable Crops," *Organic Farming and Gardening* 19, no. 3 (March 1972): 140–54.

58. Kirk, *Counterculture Green*, 145.

59. John Todd, "Biological Limits of Space Colonies," *CoEvolution Quarterly* 9 (Spring 1976): 20–21; and John Todd, "Comment," in *Space Colonies*, ed. Stewart Brand (San Francisco: Whole Earth Catalog, 1977), 48–49.

60. Kirk, *Counterculture Green*, 165.

61. J. Baldwin to Gary Hershberg, 14 December 1981, folder 12, box 1, NAI Records. See also J. Baldwin, "A Dome Bioshelter as a Village Component," *Journal of the New Alchemists* 7 (1981): 151–53.

62. John Hess, "Farm-Grown Fish: A Triumph for the Ecologist and the Sensualist," *New York Times*, 6 September 1973.

63. "The New Alchemists," *Time* 105, no. 11 (17 March 1975): 100; and Nicholas Wade, "New Alchemy Institute: Search for an Alternative Agriculture," *Science* 187, no. 4178 (28 February 1975): 727–29.

64. David Bergmark, interview by Alan MacEachern, 29 July 1999, series 11, IMR Fonds.

65. Barry Conn Hughes, "The World That Feeds Itself," *Canadian Magazine*, 9 February 1974.

66. Robert Durie to John Todd, "Re: Visit to the New Alchemy Institute on Cape Cod, October 28, 1974," Background 1974–1981 File, series 5, subseries 2, IMR Fonds.

67. For a useful history of the underground press in the 1960s, see John McMillian, *Smoking Typewriters: The Sixties Underground Press and the Rise of Alternative Media in America* (New York: Oxford University Press, 2011).

68. William McLarney, "An Improved Method for Culture of Midge Larvae for Use as Fish Food," *Journal of the New Alchemists* 2 (1974): 118–19.

69. Marcus Sherman, "A Water Pumping Windmill That Works," *Journal of the New Alchemists* 2 (1974): 21–27; J. Todd and McLarney, "Walton Two"; and Earle Barnhart, "Solar Collector for Heating Water," *Journal of the New Alchemists* 3 (1976): 30–32.

70. J. Todd, "Dilemma beyond Tomorrow"; J. Todd, "World in Miniature"; and John Todd, "Tomorrow Is Our Permanent Address," *Journal of the New Alchemists* 4 (1977): 85–106.

71. John Todd to Michaela Walsh (Rockefeller Brothers Fund), 4 October 1974, folder 21, box 26, NAI Records. Todd included a copy of the *Journal of the New Alchemists* with his letter asking for support.

72. Rockefeller Brothers Fund, "Proposed Grant to New Alchemy Institute," 6 December 1974, folder 21, box 26, NAI Records.

73. Nancy Jack Todd, "Finances," *New Alchemy Newsletter*, Summer 1978, 3; and N. J. Todd, *Safe and Sustainable World*, 45, 98.

74. Brown, "Regenerative Systems," esp. Odum's comments on p. 85.

75. Anker, "Ecological Colonization of Space," 242.

76. Robert Angevine, Earle Barnhart, and John Todd, "New Alchemy's Ark: A Proposed Solar Heated and Wind Power Greenhouse and Aquaculture

Complex Adapted to Northern Climates," *Journal of the New Alchemists* 2 (1974): 36.

77. J. Todd, "Dilemma beyond Tomorrow"; and J. Todd, "Pioneering for the 21st Century."

78. Anker, *From Bauhaus to Eco-house*, 109.

79. J. Todd, "Pioneering for the 21st Century"; and John Todd, "An Ark for Prince Edward Island: A Family Sized Food, Energy and Housing Complex Including Integrated Solar, Windmill, Greenhouse, Fish Culture and Living Components," New Alchemy Institute Vertical File, PEI Collection, Library of University of Prince Edward Island Special Collections, Charlottetown (hereafter cited as PEI Collection).

80. Alan MacEachern, *The Institute of Man and Resources: An Environmental Fable* (Charlottetown, PEI: Island Studies Press, 2003), 23.

81. J. Todd, "An Ark for Prince Edward Island"; and Andrew Wells, "Discussion of the New Alchemy Institute's ARK in PEI and the Institute of Man and Resources," 19 November 1975, IMR Vertical File, PEI Collection.

82. MacEachern, *Institute of Man and Resources*, 25; and Robert Durie, "Technical Review Meeting: Ark for Prince Edward Island," New Alchemy Institute Vertical File, PEI Collection.

83. David Kaiser's work demonstrates that translation also occurred between elite scientific institutions and New Age groups to the benefit of both. See David Kaiser, *How the Hippies Saved Physics: Science, Counterculture, and the Quantum Revival* (New York: W. W. Norton, 2011).

84. Anker, *From Bauhaus to Eco-house*, 116.

85. Neil Maher, "Neil Maher on Shooting the Moon," *Environmental History* 9, no. 3 (July 2004): 526–31.

86. R. Buckminster Fuller, *Operating Manual for Spaceship Earth* (Carbondale: Southern Illinois University Press, 1969). For a critical discussion of Buckminster Fuller and his ideas, see Peder Anker, "Buckminster Fuller as Captain of Spaceship Earth," *Minerva* 45, no. 4 (December 2007): 417–34.

87. H. Odum, *Environment, Power, and Society*; and E. Odum, "Strategy for Ecosystem Development."

88. H. Odum, *Environment, Power, and Society*, 125.

89. Anker, "Ecological Colonization of Space," 246.

90. J. Todd, "An Ark for Prince Edward Island."

91. New Alchemy Institute, *An Ark for Prince Edward Island: A Report to the Federal Government of Canada* (Souris, PEI, 30 December 1976).

92. Angevine, Barnhart, and J. Todd, "New Alchemy's Ark," 42.

93. New Alchemy Institute, *An Ark for Prince Edward Island*.

94. J. Todd, "Tomorrow Is Our Permanent Address," 89.

95. Ibid., 102.

96. Ibid., 90.

97. Durie, "Technical Review Meeting"; and Bergmark, interview by MacEachern, 29 July 1999, series 11, IMR Fonds.

98. Joe Seale, "New Alchemy Hydrowind Development Program," *Journal of the New Alchemists* 5 (1979): 44–52; and J. Todd, "An Ark for Prince Edward Island."

99. Lees, "Aboard the Good Ship Ark"; and Durie, "Technical Review Meeting."

100. Wells, "Discussion of the New Alchemy Institute's ARK in PEI and the Institute of Man and Resources."

101. MacEachern, *Institute of Man and Resources*, 53.

102. Durie, "Technical Review Meeting."

103. MacEachern, *Institute of Man and Resources*, 57.

104. John Todd to Andy Wells, "Re: Taking Over the Ark," February 1978, series 5, subseries 2, IMR Fonds.

105. MacEachern, *Institute of Man and Resources*, 58.

106. Rob Dykstra, "The Ark Sinks," *Atlantic Insight* 3 (August/September 1981): 30–32; and Silver Donald Cameron, "Floundering of the Ark," *Maclean's* 93 (1 June 1981): 13–14.

107. MacEachern, *Institute of Man and Resources*, 121.

108. Constance Mungall, "Space Age Ark: A Brave New Home," *Chatelaine* 50 (November 1977): 52–53, 103–9.

109. See J. Baldwin and Stewart Brand, eds., *Soft-Tech* (San Francisco: Point, 1978), 166–67.

110. N. J. Todd and J. Todd, *Bioshelters, Ocean Arks, City Farming*, 12.

111. Chris Zelov and Phil Cousineau include all three as important contributors to ecological design. See Zelov and Cousineau, *Ecological Design: Inventing the Future*.

112. Baldwin, "Dome Bioshelter as a Village Component," 153.

113. Memo from John Todd to New Alchemists, "Another Damn Modest Proposal," 20 November 1980, folder 12, box 38, NAI Records; Board Meeting re: Finances, 8 October 1982, folder 1, box 2, NAI Records; and J. Baldwin to Gary Hershberg, 14 December 1981, folder 12, box 38, NAI Records.

114. John Todd to the NAI Board, re: the Presidency, 4 December 1981, folder 1, box 2, NAI Records; and N. J. Todd and J. Todd, *Bioshelters, Ocean Arks, City Farming*, 12.

115. See N. J. Todd and J. Todd, *Bioshelters, Ocean Arks, City Farming*, where John Todd's design precepts are discussed in chapter 3.

116. Anker, *From Bauhaus to Eco-house*, 111–12.

117. Nancy Jack Todd and John Todd, *From Eco-cities to Living Machines: Principles of Ecological Design* (Berkeley: North Atlantic Books, 1994).

118. John Todd and Beth Josephson, "The Design of Living Technologies for Waste Treatment," *Ecological Engineering* 6, nos. 1–3 (May 1996): 109–36.

119. N. J. Todd, *Safe and Sustainable World*, 164.

120. Steve Lerner, *Eco-pioneers: Practical Visionaries Solving Today's Environmental Problems* (Cambridge, MA: MIT Press, 1997), 50.

121. N. J. Todd, *Safe and Sustainable World*, 169.

122. Ibid., 175.

123. James McWilliams, *Just Food: Where Locavores Get It Wrong and How We Can Truly Eat Responsibly* (New York: Little, Brown, 2009), 180–83; and Cynthia Wagner, "Vertical Farming: An Idea Whose Time Has Come Back," *Futurist* 44, no. 2 (March/April 2010): 68–69.

124. Barbara Miner, "An Urban Farmer Is Rewarded for His Dream," *New York Times*, 25 September 2008.

125. Michael Tortorello, "The Spotless Garden," *New York Times*, 17 February 2010.

126. Michael Bess, *The Light-Green Society: Ecology and Technological Modernity in France, 1960–2000* (Chicago: University of Chicago Press, 2003), 8.

127. Kaiser, *How the Hippies Saved Physics*, xxiii.

6

The Little Manual That Started a Revolution: How Hippie Midwifery Became Mainstream

Wendy Kline

It's the most incredible thing I'd ever seen. It let me see that if every man could see his kid being born it would be a much more pleasant culture or world to live in. **Paul Fjerstad, reflecting on his son's home birth**[1]

The response was enormous. . . . It wasn't just a few hippies that were interested in better birth—it was all kinds of people.
Ina May Gaskin, explaining the popularity of *Spiritual Midwifery*[2]

On 5 October 1973 Mary and Paul Fjerstad welcomed their son Ernie into the world. Unlike most American births in the early 1970s, he was born at home, delivered by midwives, with his father in attendance. Mary's water had broken the night before, and she giggled with excitement as she went to the phone to alert her midwives and her husband of the impending arrival. Though she had endured two miscarriages previously, she felt confident about her labor. She had attended many home births since her last miscarriage and believed that birth on her own terms could be a joyful, rather than a fearful, event. Midwives Pamela Hunt and Cara Gillette (Mary's twin sister) soon arrived at the modest D-frame house Mary and Paul shared with another couple on Second Road in Summertown, Tennessee. Paul showed up shortly after.

Together, her three coaches massaged her during her contractions, held her hand, shared stories and encouragement. "I really loved everyone a bunch," Mary wrote later. "We felt like old buddies, lifetime friends enjoying the occasion."[3]

Ernie was one of 125 babies born that year on "The Farm," an "intentional community" formed by hippies in 1971.[4] By the time of Ernie's birth two years later, seven hundred people lived in buses, tents, and a few houses on their thousand acres of property, purchased for only $70 an acre and paid for primarily by individual members' inheritances. Though geographically isolated, The Farm was already well known, and remarkably, fifteen thousand visitors traipsed through the property that year alone, some to gawk, others seeking food, shelter, or a place to give birth. The Farm's spiritual leader, Stephen Gaskin, insisted on an "open gate" policy, refusing to turn away anyone seeking refuge, a policy that would lead to the collapse of the communal structure in the 1980s (though it remains today as a smaller collective).

Though Ernie's birth was very much a private affair, witnessed only by Mary's midwives and husband, it was one of over seventy birth stories that introduced ordinary readers to the concept of family-centered births in The Farm's popular book *Spiritual Midwifery*. From the very first edition of *Spiritual Midwifery* in 1975 to its most recent printing in 2004, Ernie remains an infant, chubby and radiant, smiling at the camera. His mother, Mary, who would join her twin sister as a Farm midwife shortly after Ernie's birth, clasps her son in the photograph, gazing directly into the lens. She appears both content and confident, as if she knows she has made the right life choices. "Cara and I were raised by parents who believed that we could do anything—gender was no barrier. My father and mother had complete and utter confidence in me," Mary reflected later.[5] At a time when almost no middle-class, educated women were choosing a home birth, Mary and her twin opted not only to have their own babies at home but also to help others do the same. Their birth stories, captured in *Spiritual Midwifery*, have inspired many readers to rethink how and where their own children should be born.

The story of Ernie's birth (and the production of *Spiritual Midwifery* more generally) raises a number of questions for historians seeking to assess the impact of countercultural practice on mainstream society. Where should we place these young women who demanded the right to birth at home and whose published experiences helped to resuscitate the practice of midwifery in the United States? As historian Gretchen Lemke-Santangelo argues, "hippie women . . . have long been ignored and marginalized, relegated to the sidelines of both the counterculture and the women's movement."[6] This is certainly the case with The Farm midwives, whose history is well known among birth activists but seldom cited within the literature on countercultural or women's history.

What accounts for the absence of historical analysis on The Farm midwives? Perhaps it is due to the fact that they defy simple categorization. Were they part of a cutting-edge movement, spiritually and environmentally ahead of their time? Or did they represent a return to traditional values, including a less evolved notion of women's roles in the community? One can find evidence to support either claim. At the same time that newly formed feminist groups were fighting to end gender-based job discrimination and to legalize abortion, The Farm advertised work opportunities by gender and urged women to adopt rather than abort.[7] Yet as Mary and Cara's background suggests, these were young women who had been raised to believe in themselves, and their confidence and ability resulted in a birthing movement that gained worldwide attention, as well as birth outcomes that continue to be cited as evidence that home births offer a safe, more humane, and effective alternative to the hospital.

This chapter analyzes the role of The Farm midwives in discovering and popularizing alternatives to hospital birth in American society. Countercultural advocates of home birth articulated a blend of ancient and modern, philosophical and practical in their writings and teachings on birth. Not everyone was pleased with the resulting "rising tide of demand for home delivery," which Warren Pearse, executive director of the American College of Obstetricians and Gynecologists (ACOG), characterized as an "anti-intellectual—anti-science revolt."[8] Pearse's assessment has obscured the actual contributions of The Farm midwives and other home birth advocates of the 1970s. They did not reject technology or science so much as they demanded that it be integrated with experience, mysticism, feminism, and faith. And though they believed that midwifery was a woman-centered practice (one that deserved a central place in modern American society), they viewed the experience of childbirth as a family-centered event, one in which the father's role in supporting the laboring mother was also key. "We believe in returning the experience of birth to the family," they emphasized.[9] They have practiced this woman-directed, family-centered model on The Farm for over forty years and promoted it further through their popular text *Spiritual Midwifery*, which has sold more than half a million copies and has been translated into six foreign languages.

How The Farm Midwives Came to Be

In 1967 Joanne Santana, a self-proclaimed hippie living in San Jose, California, discovered she was pregnant at the age of twenty-one. "I really knew nothing about having a baby," she reflected later. Her only point of reference came from the television series *Wagon Train*, ranked number one by the Nielsen ratings in the early 1960s. The popular western featured a wagon train traveling from Missouri to California and focused almost entirely on the masculine pursuits of the predominantly male cast. Every so often, however, as Santana recalls, "the wagons would pull over, you'd hear a woman scream, a baby cry, and the wagons would roll on. So I thought you screamed, and the baby came out. That was my childbirth education."[10]

Needless to say, it did little to prepare her for her first birth, which took place in O'Connor Hospital, the oldest hospital in Santa Clara County. It was a quick labor. Already seven centimeters dilated by the time she checked in, a labor nurse assured her she would make it to the requisite ten centimeters soon enough. But Santana found the pain to be nearly unendurable and begged for medication. "I thought the whole purpose of going to the hospital was that they give you something for pain," she remarked later. Instead, the nurse encouraged her to think about her breathing, which helped temporarily. Suddenly she felt she couldn't take the pain anymore, and the contractions stopped, triggering self-reflection. "And I'm sitting there going well, what are my options here? And I thought, well, somehow my body has gotten pregnant, grown this child, well on the way to having this baby, and I did not understand intellectually how to do this, but my body evidently is doing it, so I've gotta trust my body that it's going to get me out of this." The contractions began again and she felt the urge to push. Because the doctor had not yet arrived, the nurse instructed her not to push. Hospital staff wheeled her into a delivery room and strapped down her arms and legs. As Santana remembered it, this happened right after she had made a deal with herself to listen to her body. "So this is not good, but I'm trying to be a good girl trying not to push and stuff. And when [the doctor] finally gets there he cuts a huge mediolateral episiotomy for a six-pound, fourteen-ounce little girl I should have had fifteen minutes ago with the nurses. I could not sit down for a month without a pillow."[11] She left with her new, healthy baby girl, Jordana, feeling dazed by her birth experience and anything but empowered.

One year later, Santana picked up the underground newspaper *San Francisco Express Times* and learned about a class that was offered through the Experimental College at San Francisco State and was taught by Stephen Gaskin. "Monday Night Class" was a series of lectures covering everything from meditation to discussions about politics, religion, and psychedelics. "When we first got the class together we were like a research instrument, and we read everything we could on religion, magic, superstition, ecology, extrasensory perception, fairy tales, collective unconscious, folkways, and math and physics," Stephen recalled. "And we began finding things out as we went along about the nature of the mind."[12] By 1969, this class had grown to several thousand people, many of whom began to see Gaskin as their spiritual teacher.

At first glance, Stephen Gaskin does not appear to be hippie guru material. Born in 1935, he served in the Marine Corps in Korea as a rifleman in the early 1950s. He moved to San Francisco, where he attended San Francisco State College under the GI bill, getting his BA in 1962 and MA in creative writing in 1964; he then taught creative writing and semantics at the college until 1966, when his contract ran out. "It wasn't that I got fired for being a hippy," Gaskin explains; "it was just that I'd gotten too weird to rehire by the time my contract expired."[13] He approached the head of San Francisco State's more alternative Experimental College, and found out that there was a teaching slot available on Monday nights. "The idea was to compare notes with other trippers about tripping and the whole psychic and psychedelic world," Gaskin writes, by then heavily caught up in the San Francisco countercultural scene. "We discussed love, sex, dope, God, gods, war, peace, enlightenment, mindcop, free will and what-have-you, all in a stoned, truthful, hippie atmosphere. We studied religions, fairy tales, legends, children's stories, the I Ching, Zen koans—and tripping." Everything was fair game in his multidisciplinary approach. "It was easy to tell when we were onto something hot—I could see the expressions move across those thousand faces like the wind across a wheat field. It was like being inside a computer with a thousand parallel processors."[14] His computer metaphor is telling; Gaskin saw clear connections between technology, science, and spirituality that would become central to his philosophy.

By the time Santana discovered Stephen Gaskin, he was well established in the local counterculture, offering not only his regular Monday Night Class but also Sunday morning services in Sutro Park. She became a regular attendant in 1969 and, when she found she was pregnant again later that year, sought recommendations from other attendants regarding alternative birth practitioners, determined to have a different experience the second time around. Santana learned that there was a man named Bob who was calling himself a midwife and living at the Good Earth commune on Pine Street in the Haight-Ashbury district of San Francisco. Six weeks before her due date, on Mother's Day of 1970, she went to meet him. "It was quite a scene," she recalled.[15] By this time, Haight-Ashbury had been overrun by heroin and speed junkies, and many of the Haight's founding hippies had moved on. Founded in 1968 by two ex-convicts, the Good Earth commune was part of the second wave of the Haight's hippie settlements. Though membership was ever changing, over seven hundred people flowed through the commune's half dozen houses in the neighborhood.[16] In honor of Mother's Day, the men were baking bread and the women were "taking long baths and lounging around." As scholar Timothy Miller points out, the counterculture was to a large extent "male-defined," and Santana's memory of this role reversal attests to the exceptionality of the occasion.[17] It was perhaps fitting that the "midwife" Santana went to meet was male, for that, too, was nearly unheard of.

Bob had long, stringy, gray hair and looked to be in his forties. He asked Santana about her previous birth experience and assured her that she could give birth without an episiotomy this time (one of her biggest concerns about her first birth). In the forty-odd deliveries Bob had presided over, he had never done an episiotomy, recommending a slow delivery and an ice pack to prevent tearing. Unfortunately, Bob was about to move to a commune in Oregon and could not assist with her birth, but he sent her to New Age Natural Foods to purchase a midwifery manual he found useful.

Fred Rohe, originally a chemicals salesman, had opened his shop back in 1965 in the Haight-Ashbury district. A self-declared Buddhist interested in "providing nutritional and spiritual guidance," he provided everything from herbal tea to books, record albums, and a meditation area. By 1969 his sales had increased tenfold, indicative of the growing marketability of countercultural items.[18] On the shelf of his store sat a small manual, *Pregnancy, Childbirth and the Newborn: A Manual for Rural Midwives*, written by a Stanford thoracic surgeon, Dr. Leo Eloesser.

Santana leafed through the manual before taking it home. "It was just our speed," Santana later recalled. "It's like 'put warm brick in baby's bed. Remove brick before putting baby in bed.' It told you how to do, you know, a birthing pack, how to sterilize your instruments, how to deliver a breech birth, how to stop a hemorrhage, things like that."[19] Santana did not realize it at the time of her purchase, but this little manual would become crucial to her and her community of Gaskin followers in the near future.

How did a manual for rural midwives end up in an urban natural-foods store? *Pregnancy, Childbirth and the Newborn* began as a series of teaching papers to be used in midwifery training courses in Communist-controlled areas of North China, where doctors were in extremely short supply. The context could not have been more different. While Santana sought to escape from the clutches of modern obstetrics, Dr. Leo Eloesser, residing less than sixteen miles away at one of the most prestigious medical schools in the country, carried his expertise to war-torn rural North China in an effort to improve infant and maternal mortality rates. According to Eloesser, in 1948 there was one doctor for every 80,000–150,000 inhabitants, resulting in dire conditions.[20] As part of a UNICEF mission, Eloesser agreed to create a medical training program in North China to improve the situation.[21]

Eloesser faced an enormous task. The $500,000 allocated for this mission would still not produce the kind of facilities or personnel available in the West in a timely fashion. Approximately fifty medical schools generated somewhere between one and two hundred "acceptably trained" and a thousand "ill-trained" graduates per year, and at that rate, Eloesser argued, "graduates would die off faster than they could be replaced."[22] Some alternative body of health workers was needed. He proposed the creation of a new medical program in North China that would recruit young men and women with no more than a grade-school education. Six-month training courses in sanitation, communicable diseases, and midwifery would produce competent health care practitioners in a practical and efficient manner. Three months of "simple" classwork followed by supervised fieldwork in each of these areas, he claimed, would dramatically improve health outcomes.[23] As UNICEF was quick to point out in a later press release touting the program's success, it "reversed the usual procedure of selecting only a few for top-level training, frequently given outside of their country."[24] After training, these "average-level" students would in turn train others in nearby villages, thereby spreading basic training models further out into the countryside. While this process would become common after Mao's 1965 directive for the widespread establishment of "barefoot doctors" in China's rural areas, the medical training program created by Eloesser seventeen years earlier set a clear precedent.[25]

But how would he go about teaching these new courses? Eloesser himself had no background in midwifery, nor was he likely to be exposed to it back in the United States. Ironically, the increasing embrace of Western medicine in the early twentieth century led to the rapid professionalization of midwifery in China but not in the United States, where midwifery was "nearly eliminated" during the first half of the twentieth century.[26] Rural midwives in China had neither the resources nor access to the Western-funded modern training centers such as the Central Midwifery Schools in Nanjing or Beijing.[27] These rural midwives were referred to by public health officials as "Granny Midwives" and, like the aging African American midwives in the rural American South, were frequently perceived as part of the public health problem rather than part of the solution.[28] Instead, Eloesser sought the "hitherto untapped reservoir of woman power" of the student population to practice what he called "sensible midwifery"—to, for example, "sit by at a normal delivery and refrain from putting cow dung or earth on a newborn's navel," a derogatory reference to the practices of local midwives that he deemed backward and unsanitary.[29]

For teachers, Eloesser contacted the American Board Mission in Tientsin, a Christian missionary organization that had established several centers in North China by the late nineteenth century. This particular mission had two nurse midwives, Isabel Hemingway and Edith J. Galt, who were available to teach for Eloesser. Both Hemingway (a first cousin of Ernest Hemingway) and Galt had grown up in China with missionary parents, spoke the language fluently, and had studied nursing and midwifery at the Maternity Center in New York City, where they had "obstetrical training in modern New York hospitals."[30] They thus provided, in Eloesser's view, the best of both worlds—training in Western medicine and experience with the local culture.

Eloesser, Hemingway, and Galt put together a series of teaching papers to present to their students (twenty young women in 1949, and another nineteen in 1950). Sixty-six illustrations accompanied a basic text that covered anatomy, fetal development, labor and delivery, postpartum care, and hygiene. Hemingway later commented privately to Eloesser on how effective this teaching style had been: "I know that students learned the steps for delivery and did the things they would have to do several times in class so that when they really saw a delivery they were already knowing what to expect. It was a great help to them and to us."[31]

Eloesser's UNICEF mission was short-lived; he supervised only two series of medical training courses lasting just over a year. Galt moved on to Korea, Hemingway to Turkey, and Eloesser to Mexico. But perhaps this dispersion helped further disseminate the principles of the manual, which continued to be revised, expanded, and translated over the next twenty-six years. Evidence suggests that the book's impact was felt not just in developing countries but also in the United States. "In regard to your manual for rural midwives, I purchased quite a number of these when they came out and have distributed quite a few," noted Dr. Nicholas Eastman, chief of obstetrics at Johns Hopkins in 1956.[32] Nearly two decades later, Dr. Victor Richards of San Francisco Children's Hospital also expressed his enthusiasm to Eloesser: "You would be surprised, but I have heard over the years many favorable comments about your manual for rural midwives. I am delighted to know you will be putting out a new edition."[33]

Taught in China by American missionaries, published in Spanish, English, Korean, and Portuguese, this manual reached a global market. It also, strangely enough, came full circle, landing in a health food store just a few miles away from its creator's home institution. More than a decade before the first edition of *Our Bodies, Ourselves* would offer lay readers an accessible manual on women's health that later triggered a women's health movement, *Pregnancy, Childbirth and the Newborn* became a do-it-yourself tool that would provide the necessary foundation for a burgeoning home birth movement. The difference, of course, is that unlike the women's health activists who wrote *Our Bodies, Ourselves* in the 1970s, Dr. Leo Eloesser had no intention of providing urban hippies with an alternative to modern obstetrical care.

Armed with her new manual (fig. 6.1), Santana continued her search for a birth attendant. Another Gaskin follower told her about a labor and delivery nurse, Diane Mehler, who had recently delivered a baby at home. "And so I went to see her," Santana recalled. Mehler lived in a house near the top of Mount Tamalpais (along with the psychedelic rock band Sopwith Camel), about twenty-five miles north of San Francisco. Newly pregnant herself, Mehler agreed to help the couple but explained that she did not want to have to do it in San Jose because of the long drive and Santana's short first labor. So the couple moved into an old school bus to allow for greater portability. On July Fourth 1970, two weeks after her due date, Santana and her family were headed to Sutro Park in San Francisco to hear Gaskin speak when she had several strong contractions. They turned the bus around and headed to a friend's house in San Francisco to call Diane. When they couldn't reach her, they called the Good Earth commune and spoke to a woman whose baby had been delivered by Bob, and she agreed to come over and help. Upon her arrival, she filled the tub with warm water and helped Santana climb in. "The water was great," Santana recalled. They finally reached Diane, who came over, boiled her instruments, and helped Santana out of the tub to check on her progress. Baby Anthony was already on his way out, born in only three hours. In the distance, Fourth of July fire-

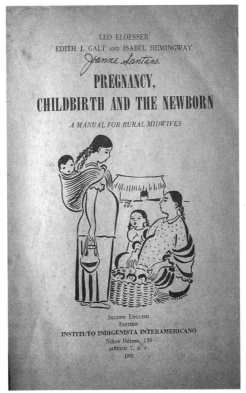

FIGURE 6.1 Joanne Santana's personal copy of *Pregnancy, Childbirth and the Newborn: A Manual for Rural Midwives*. Photograph by Wendy Kline.

works seemed to announce his arrival. In exchange for Diane's services, Santana gave her their fifty-two-inch, round, Mexican oak kitchen table and baby cradle.

Santana remembers this birthing experience as a major turning point in her life. "It was the most empowering thing that ever happened to me," she explained. "The difference between those two births was just night and day," she said of Jordana's and Anthony's births. "I just felt kind of ripped off after Jordana. I mean it was wonderful having her, and she was a great nurser and all that stuff, but I didn't feel like empowered where they took her away right away, I wasn't allowed to have her to just nurse her when I wanted, I'd get engorged, and I'd have to wait for them to bring her." Her recovery time with Anthony's birth was dramatically shorter. "This time I could sit right up, no stitches, you know, it was just incredible. And I felt like a monkey. I felt like I could do anything." More significantly, she found her calling (fig. 6.2). "I re-

FIGURE 6.2 Joanne Santana continues to practice and teach midwifery skills and has attended over one thousand births. Santana is shown here teaching an introduction to midwifery assistant skills at the Farm clinic with obstetric model "Fanny," during The Farm Midwifery Assistant Workshop, 24–30 March 2013. Photograph by Wendy Kline.

ally wanted to devote my life to bringing that empowerment to families. Because I don't care if people want to have 'em in the hospital or a birth center or at home, but I think it should be their choice."[34] Santana was one of a growing number of women in the counterculture whose own birth experiences—either a traumatic hospital birth or an empowering home birth (or, as in Santana's case, both)—helped to revive the practice of midwifery in the United States.

The following Sunday, Santana finally made it to Sutro Park for Stephen Gaskin's regular Sunday service. Sitting in the shade of a palm tree was thirty-year-old Ina May Middleton, who would divorce her first husband to marry Stephen Gaskin six years later and go on to publish five books on midwifery and childbirth. Santana joined Ina May under the shade of the tree, proudly presenting her friend with newborn Anthony. When Ina May asked where she'd had the baby, Joanne told her that she did it at home. "And she said, 'Well, how did you do that?' and I said, 'With a midwife.' And you could just see her eyes light up and the wheels start turning and she said, 'Oh, it would be really nice to have one of those, you know.'"[35]

Like Joanne and 99 percent of all US laboring women in the 1960s,

Ina May's first birth experience had taken place in a hospital. "Fear of having a repeat of what I experienced during my first birth in a hospital was what prompted me to figure out a way to learn to be a midwife," Ina May later declared.[36] Back in Illinois while in graduate school, she had delivered her daughter in a hospital, an experience that traumatized her: "During birth at the hospital, I was left alone and treated like I had done something nasty. Then I was approached by a gang of masked attendants who came in the room and treated me like a ritual victim. They used forceps, and then I wasn't allowed to see my baby for 18 hours."[37]

Ina May was raised in Marshalltown, Iowa, where she graduated from high school in 1958. After receiving her bachelor's degree from Iowa State University, she traveled to Malaysia with her first husband (whom she married at the age of nineteen) to volunteer for the Peace Corps. They planned to head directly to San Francisco after two years in Malaysia but, when they discovered she was pregnant, decided to return to the Midwest for graduate school. "We thought, well, you don't just go out pregnant and decide to become a hippie with no way to make a living." With a master's degree, she could work as a teaching assistant for better pay. After obtaining a degree in English from Northern Illinois University in 1967, she took off with her husband and young daughter for San Francisco to "become hippies," as she remembered later.[38]

Ina May found work in Chinatown, teaching English to local residents while her husband stayed home with their daughter. "I absolutely loved that job," she recalled.[39] Like so many who flocked to San Francisco in the 1960s, her world was soon turned upside down. She and her husband began attending Stephen Gaskin's Monday Night Class, where she first encountered the man who would become her mentor and life partner. She also met her future midwifery partner Pamela Hunt, an art major at San Francisco State who had taken one of Gaskin's very first courses, "Magic, Einstein, and God," in 1966. Hunt recalls that there were only eight or nine people in the class, though within a few years the number of attendants would grow to 1,500. "We talked about telepathy and how people interact with each other on a vibrational level as well as a physical level and how important it was to take care of the energy that you are dealing with," she remembers.[40]

Though Hunt had not yet given birth, her exposure to hospital birth practices, like Santana's and Gaskin's, had a profound effect on the path she chose to take. As an art major, she studied at the University of Guadalajara for two years in the early 1960s, where she took a required anatomy course. On a field trip to a state-run hospital, Hunt witnessed three births. "When they delivered the babies, they kind of pulled the babies out, tossed them up in the air, slapped them on the back a couple of times and threw them to a nurse," she recalled.[41] Each of the mothers looked "tired and forlorn" afterward, with no one to comfort them, as husbands weren't allowed in the room. Yet there was Pamela, along with her classmates, "a group of strangers who didn't know the first thing about birth," witnessing the entire event. "Why they arranged for us to be at these births and put these poor women up as models at this most vulnerable time in their lives, I'll never know. We certainly didn't learn any anatomy, or compassion for the mother or baby, either."[42] She discovered that the cesarean section rate in Mexico was 75 percent, which she found shocking, as she "expected Mexico to be pretty natural."[43] Though she had not yet seen Santana's copy of the Mexican midwifery manual, she had imagined birth practices to resemble those described in the book rather than what she saw in urban Guadalajara.

Hunt returned to her hometown of Sacramento after two years in Mexico, intending to complete her undergraduate degree at California State University Sacramento. But in 1966 she "got mixed up with a guy" that her mother really didn't like and headed down to San Francisco to finish her coursework there. "When I first got to San Francisco, I was as square as they come. And all of a sudden I was in the middle of the art department at San Francisco State in 1966. And that changed everything." She signed up for Stephen Gaskin's class that included meditation and discussions about politics, religion, and psychedelics. "The world changed overnight," she says. "Literally."[44]

"Out to Save the World": The Caravan

In 1970, while preaching at the nondenominational Glide Church, Stephen Gaskin was invited to deliver a series of lectures at schools and churches across the country, and over two hundred of his followers decided to join him in school buses for this "Astral Continental Congress," a call for a spiritual and social revolution. The Caravan, as it came to be called, generated more and more media attention as its collection of school buses wound their way through forty-two states spreading the inchoate messages of peace and revolution to students and churchgoers. Remarkably, despite his sometimes-rambling messages and sermons, Gaskin provided a logic and rationale for his vision of peace—which was based on both spirituality and science.

At the age of thirty-five, Gaskin was older than many of his followers, yet young enough that they could relate to him. "Stephen said things I had always known but had never heard anyone articulate before," remembers Cara Gillette, Mary's twin sister, who helped deliver baby Ernie. "He talked about energy, God, attention, and compassion. I felt, inside my mind, a clear bell of truth."[45] He spoke in plain language, mixing modern-day metaphors of science and spirituality in his attempts to provide simple answers to complex questions. "Religion is the wiring diagram of the way human energy is moved to relate with the universe," he explained on Halloween, 1970, at Saint Stephen's Catholic Church in Minneapolis.[46] In Iowa City, he described the human soul as "the electrical field that surrounds your equipment," while the "brain is a field generator, it's not just a hunk of meat with wiring in it."[47] Speaking at the Washington Monument, he reflected, "I was thinking this morning that praying is like communicating telepathically."[48] His frequent use of such scientific analogies suggests that his followers were technologically savvy. Intent on showing the media and the mainstream that hippies were anything but lazy or stupid, he frequently demonstrated their potential for complexity by blurring the boundaries between religion, spirituality, and science.

Yet the heart of Gaskin's message was quite simple and practical (if only loosely based on specific scientific principles). "Spiritual isn't like misty or somewhere else. Spiritual is how do we, here and now, work it out as best as we can, because this is all we have," he explained to students at Wright State University in Dayton, Ohio, on 16 November 1970.[49] Speaking primarily to youth upset by the Vietnam War and social injustices at home, Gaskin urged them to do something. "I feel like beatniks have been spaced long enough, and that we know where it's at, and that it's time we got off welfare, gave up food stamps, and began to produce with this energy which we say we have so much of."[50]

Well aware of the media attention he was getting on the Caravan, he took the opportunity to portray the counterculture as full of positive potential for social change. Just before dawn on 27 December 1970, Gaskin maneuvered the fifty-odd buses of the Caravan through the streets of Washington, DC, to the Ellipse, the park between the White House and the Washington Monument, for a sunrise service. One Caravan member reflected that their arrival was "one of those moments, full of meaning and symbolism, pregnant with possibilities."[51] CBS News was on hand to interview Gaskin, and the *Washington Post* covered the story as well.

At a time when many mainstream Americans associated hippiedom with hedonism, Gaskin and his followers sought to convey a much more activist mind-set. As if to remind onlookers that Gaskin was doing much more than "tuning in and dropping out," the destination sign on top of his Scenicruiser read "Out to Save the World." Well before the decision to establish an intentional community in Tennessee, this group believed that they carried an inclusive message relevant to everyone—male or female, rich or poor, black or white. One member, Michael Traugot, described the sense of optimism and importance that these travelers experienced as they traveled across the country:

> Several times on the Caravan Stephen had told us that we were living at a very unique time, a crossroads, an unusual moment in history. So much change was happening, people felt so confused and so desperate, so hungry for understanding, that a window was temporarily opened into their hearts and minds. People would listen to new ideas, consider points of view they would not have considered before. We had a chance to influence our culture, the group mind of America and the world, in the direction of peace and compassion and living with nature rather than trying to dominate. That's why we were doing this Caravan. But we had better act quickly, Stephen would tell us, because the window might soon close, the culture harden up again.
>
> And it was working! Everywhere we went, the authorities and the mainstream society as a whole were very sweet to us, treated us with a great deal of respect. We had the highest spiritual and ethical standards, went out of our way to be and appear harmless, to show respect to others, to work hard and not ask for handouts. All this worked in our favor, but it wasn't just us. It was partly the times.[52]

The excitement was palpable, and the potential for change seemed enormous. Young people from all over the country, frustrated with the increasing violence and heavy drug use of the counterculture, saw this as a turning point. People were listening, and they appeared willing to consider what Gaskin had to say as a legitimate alternative to the status quo. What they did not realize at the time, however, is that the longest lasting impact of the Caravan had less to do with Stephen Gaskin than it did with some of his pregnant female followers, whose determination to stay out of the hospital led to a revolution in birth practices. When Gaskin announced his plan to speak across the country, follower (and future Farm midwife) Cara Gillette was pregnant. "I knew I had to go with Stephen and the community," she remembers. "The thought of staying behind and having my baby in some hospital, with 'sharp things coming at me,' wasn't an option for me. I wasn't so much scared of having a baby as of those 'sharp things coming at me.'"[53] Her birth story, like her twin sister Mary's, would become memorialized in *Spiritual Midwifery*, introducing hundreds of thousands of readers to the idea that medical doctors were not the only ones capable of delivering babies in modern America.

Births on the Caravan

Just like in the *Wagon Train* television series of Joanne Santana's childhood, when a woman went into labor, the caravan pulled over, waited for the birth, then rolled on (fig. 6.3). "We all parked in a sort of protective formation around the bus in which the birth would take place," wrote Gaskin, "and everyone waited for the baby's first cry."[54] The first baby was born in a parking lot at Northwestern University, while Ste-

FIGURE 6.3 The caravan on 1 April 1971, just after the birth of Phil Schweitzer's daughter Sara Jean. Photograph by Gerald Wheeler.

phen was lecturing to several hundred people in an auditorium. Franz, the father, approached Stephen to ask for his assistance during the labor (because Stephen had received first-aid training in Korea), but since he was scheduled to lecture, Ina May offered to help instead. "I was no midwife at the time," Ina May wrote later, "but I was able to help the mother [Anna] stay relaxed." She brought fellow Caravanners Pamela Hunt, Mary Louise Perkins, and Margaret Nofziger with her to help in the small bus, lit by kerosene lamps. "I remember her looking beautiful all through her labor," remembered Pamela, "kind of rosy and glowing."[55] The labor lasted only three hours, and baby Immanuel came out without any problems, caught by his father.

After the birth, Ina May felt stunned. "I was in a state of amazement for several days," she wrote later. "I had never seen a newborn baby before (my baby was almost a day old before I was allowed to see her) and I was struck with how perfect this baby looked. . . . I felt a definite calling to be a midwife, but my master's degree in English had not prepared me for anything so real life as a birth."[56] In order to gain a better sense of the mechanics of birth, she turned to the one resource available to her on the Caravan—Eloesser's *Pregnancy, Childbirth and the Newborn.*

Cara Gillette felt lucky to witness this first Caravan birth, knowing that she, too, would give birth on the road. "Anna was so lovely," she remembers of the laboring mother.[57] What she did not expect was to be next in line, as her due date was six weeks away when she felt some cramps the following week. The Caravan was stopped at Bruin Lakes State Park outside Ann Arbor, Michigan. "We were concerned because it was so early," Cara explained, "and we didn't want to admit that it was really happening."[58] She scrambled to prepare for the birth. She and her partner, Michael, had scraped together their small savings to buy a tiny bread truck to travel in the Caravan, and it had broken down several times on the trip. "We were constantly falling behind," she remembers.[59] It was too small for a birth, so Mary Louise (who had also been at the first Caravan birth) and her partner, Joseph, lent them their bus for the day.

As soon as she lay down in Mary Louise's bed, twenty-five people crowded onto the bus to witness the second Caravan birth. One of the men even had a camera. Pamela recalled that the "vibes felt strained."[60] Sensing Cara's discomfort, she and Ina May asked all the men to leave except Michael, the father of Cara's child. Though Cara's twin sister, Mary, had been traveling on the Caravan with her husband, they were stuck in Minneapolis, where Paul had to rebuild the engine of their van. (She did not find out about the premature birth until she received a letter from her mother about it.)[61] Pamela and Ina May hunkered down with the manual to prepare for the impending birth, keenly aware that it would be a premature one and studying up on emergencies. Then suddenly, one of the other women there "became nervous and superstitious when she saw what I was reading, and took the book out my hands," Ina May recalled, "afraid that if I read about something negative, I would cause it to happen." What happened next taught Ina May the importance of accepting responsibility. "Instead of taking the book back, I allowed myself to be intimidated by the other woman" (who later, ironically, would become a physician).[62] According to Pamela, "the situation felt shaky."[63]

Yet Cara labored along, finding the birthing to be "surprisingly easy. . . . It felt ecstatic. Everything that happened in my body felt really natural."[64] Shortly before baby Anne was born, Cara began to doubt herself. "I wasn't sure I could do it." Then one woman took her hand, and another massaged her feet. "The loving touch made all the difference," Cara remembers. "I pushed through."[65] Baby Anne was born after six hours of labor, weighing just five pounds, and was caught by her father. She promptly "gave a small cry and then turned blue and just lay there."[66] Panic ensued. They had not yet read what to do if the baby didn't breathe. "We all watched, frozen," recalls Cara.[67] Someone ran out and got Stephen, who rushed over and breathed into the baby's mouth. "She took a breath, cried, and turned pink; our first miracle and our first heavy lesson," recalled Pamela.[68] Cara, eternally grateful to Stephen, reflected, "it was probably a good thing we learned the high stakes of birth early on."[69] Perhaps this scare had an impact on their own midwifery manual; they listed terms that pertained to birthing emergencies in easy-to-find bold print in the index of *Spiritual Midwifery*.

By the third birth, which took place in Ripley, New York, Ina May was officially established as the midwife in charge, and the only people allowed to be present for a birth were the female attendants (Ina May, Pamela, and Margaret) and the father of the child. This time, the Caravan pulled over to the side of the road in the middle of the town to await the baby's birth, close to a church. Ina May delivered the baby. Afterward, the minister rang the church bells while the locals greeted the new mother with food and warm wishes, a reminder that the alternative practices espoused by these itinerant hippies were not always met with hostility.

Eight more births would take place during the Caravan, each one generating excitement described by Ina May as "contagious." Pamela and Ina May were among the pregnant women who learned from those before them. "Each mother who gave birth became an inspiring and encouraging example to the other women," wrote Ina May. "We came to look at birth as a sort of initiation or rite of passage—something for which you could gather up your courage with the help of your friends and contemporaries."[70]

But in order for birth to become a truly inspirational event, it had to be safe. Beginning with the Ann Arbor birth (when Cara's baby didn't initially breathe), Ina May realized that for all of the beauty and naturalness of birth, it sometimes required medical expertise. Joanne's midwifery manual was a start, but the group lacked anyone with hands-on training until the Caravan reached Rhode Island. There, a local obstetrician by the name of Louis La Pere, who had read about the Caravan and the births en route, came to visit and offered Ina May Gaskin, Pamela Hunt, and Margaret Nofziger a hands-on seminar on birthing emergencies and complications. La Pere had studied medicine at the University of Bologna in the early 1950s before returning to his home town of Westerly, Rhode Island.[71] Perhaps his interest in their endeavors was a result of his European education and the fact that the University of Bologna had long had an established midwifery program. Whatever the case, he taught the three aspiring midwives sterile technique, how to resuscitate a newborn, and what to do with an umbilical cord that became tightly wrapped around the baby's neck. He also gave them obstetrical instruments, medications, and an obstetrics textbook. "It was a little handbook," she recalled. "Benson's *Handbook of Obstetrics and Gynecology*." She shared the book with Pamela, who would read it for a couple of weeks and then pass it back to Ina May so she could study it.[72] "And so I used it kind of as a guide," Gaskin explained, viewing La Pere as one of "a bunch of doctors that really thought that the U.S. had missed the boat in not having midwives. And thought that that was going to change."[73] Like the residents of Ripley, New York, Dr. La Pere seemed to view births on the buses as a potentially revolutionary act deserving of recognition.

Thus, like Stephen Gaskin, Ina May Gaskin viewed her new calling as a mixture of science, practicality, and spiritualism. Her approach to birth reflected his philosophy. This is, of course, not surprising, given their relationship. Not only were they partners (they would officially marry in 1976), but they were also a team, and in the spring of 1974, Stephen ordained Ina May as "acting minister" so that she could conduct services in his absence.[74] In her view, birth was "telepathic," "heavy," and a "trip." One's mind-set and relationship with the universe affected the birth process. "Putting out energy from one end of your tube (in the form of truth or love)," she explained, "makes it easy to put out energy from the other end (in the form of a baby). The reverse of this is true too—if a lady squinches up her mouth so that it isn't good to look at or screams or says pissed off things, her puss will cinch up too, and she'll be more likely to tear." Quoting the founder of the Sufi movement in the West, she added, "with love, even the rocks will open."[75] Thus, spirituality also had a practical element, according to Gaskin, actually easing some of the pain of childbirth.

The Farm

At the end of the Caravan tour, Stephen Gaskin and his followers decided to purchase land in Tennessee (where land was much cheaper than in California) and create a commune on a thousand acres, based on the principles put forward by Gaskin. They were back in San Francisco for only a week before they turned around and headed back East, becoming "Okies in reverse," eventually settling on a thousand acres of blackjack oak trees, which they called The Farm (fig. 6.4).[76] The sense of community that had been developed along the route turned into something more permanent as they plowed, farmed, built, and meditated together. As scholar Fred Turner points out, this was in the midst of

FIGURE 6.4 The Farm in March 2013. Photograph by Wendy Kline.

the "largest wave of communalization in American history." Perhaps as
many as 750,000 Americans lived in some ten thousand communes in
the early 1970s, though few survived as long as The Farm. Turner refers
to these pioneer visionaries as "New Communalists," arguing that even
as they opted to live apart from mainstream society, they often "embraced the collaborative social practices, the celebration of technology,
and the cybernetic rhetoric of mainstream military-industrial-academic
research."[77] This is certainly true of The Farm, where ingenuity, creativity, and determination led to a great deal of innovation: the first hand-
held Geiger counter (called the "Nuke Buster"), rechargeable electric
golf carts, and even a Doppler fetal pulse detector.[78] As Farm resident
Phil Schweitzer recalls, "the interest in all of this stuff to me is the fact
that we did it on no budget. Not that the technology didn't exist—we
were just trying to use the scraps from society to piece together the
technology that we would need to survive out here in the woods with
no money."[79]

With more babies on the way, one of the first orders of business was
to establish a protocol for deliveries on The Farm. One thousand acres
was a lot of space, and after living in such tight quarters for the previous four months, many families scattered to the far ends of the property.
Initially, this proved challenging to Ina May and Pamela, who "had no
communications other than how loud we could shout from hill to hill
to relay a message. If we needed to make a phone call, we had to drive
out three miles to a local bar to use the phone."[80] Once phone lines
were established, midwives and pregnant women received priority in
getting a line installed. Later, each midwife was provided with a pickup
truck or four-wheel-drive vehicle equipped with a citizen's band radio
so that they could be in constant communication. "We were always very
techie," Pamela remembers.[81]

Shortly after they settled on The Farm, a public health nurse visited the premises. Eager to maintain their practice of delivering babies,
Ina May asked her how they could receive more training to practice
midwifery in Tennessee. Though plenty of home births occurred in the
region, as a large Amish community lived nearby, Tennessee had no li-
censure program for midwives. The nurse returned the following day
with two men from the Bureau of Vital Statistics. "She gave me a box of
ampules of silver nitrate [to drop in the newborns' eyes], and the men
gave me a stack of birth certificates and death certificates and wished
me good luck." This goodwill gesture gave the new midwives a boost of
confidence that they could continue with their craft. It also made them
realize that they had not obtained birth certificates for all the babies

born on the Caravan, a task that proved challenging, since many public health authorities "weren't sure how to go about certifying the birth of a baby who had arrived in a schoolbus."[82] Most newborns had been weighed on produce scales in nearby grocery stores, hardly an official indicator of a child's heft.

Since midwifery school was not an option, Ina May and the "midwife crew" turned to the local family practitioner, Dr. John Williams.[83] While relations between doctors and lay midwives in the 1970s were frequently hostile, that was not the case between Williams and The Farm midwives. Shortly after they arrived, the midwives brought one of their newborns who had been premature to Dr. Williams. "He took a liking to us," Pamela remembers, "and so he started coming out here." She describes him as "a real doctor," referring to his warmth, caring, and concern. "And he was not afraid to teach us."[84] In addition to practicing at Maury Regional Hospital, Williams had been delivering babies in the homes of the local Amish community for years and had noticed that the Amish women and their babies suffered a lower rate of infection than those he delivered in the hospital. He theorized that home birth mothers had built up resistance to organisms in their own home that made them less susceptible to infection, and he was eager to see "if his theory would be borne out by the statistics of [The Farm] births. (It was.)," declared Ina May. "We were told to call him any time, night or day, if we had questions about the pregnant or birthing women and their babies."[85] Perhaps due to the rural setting and the lack of health care providers, country doctors were more likely to view midwives (with any level of experience) as helpmeets rather than hindrances, allowing for a more collaborative approach to birth.

Thus began a decades-long fruitful relationship between Dr. Williams and The Farm midwives, who consulted with him regularly on everything from fevers to breech births. "Dr. Williams helped us a lot," recalled Pamela. "He always made you feel good." When he got a CB radio for his pickup truck so that they could reach him when he was away from a telephone, his radio nickname was "Dr. Feelgood."[86] His reputation extended beyond the boundaries of The Farm; in 1981 he was selected by *Good Housekeeping* magazine as one of the Ten Most Outstanding Family Physicians in the U.S.[87] "He was an inspiration to many people in our community and to me," wrote former patient Debbie O'Neal after his death in 2003. "I have many fond memories of him. I remember how he always cared for my grandparents and my parents. I do not know of another doctor that had as much care and concern for others as he did."[88] He himself was modest about what he had done for the community of midwives, writing to Ina May toward the end of his career, "Any help I gave to you was a part of every physician's creed—i.e., to learn, to experience, then to teach—too many of us forget."[89]

Dr. Williams played an instrumental role in educating and supporting the young midwife crew as they set up operations on The Farm. In the first two years, they established their prenatal clinic, using equipment donated by doctors (most likely including him) along with some purchased at a nearby medical supply house, delivered 129 babies, and continued their studies. "All this time we were reading obstetrics textbooks and learning everything we could about the technical part," Ina May wrote. "We learned more about the vibes part with every birthing that we attended."[90] As they developed their unique style, blending science and spirituality, they believed that their relationship with the medical community became more reciprocal.

This was particularly true when laboring mothers ended up in the hospital, most typically for a premature or breech birth, which happened four times in the first two years. In many parts of the country, hippies in labor begun at home were greeted at hospitals for emergencies with hostility—blamed for being irresponsible for attempting a home birth. But in Summertown, Tennessee, the unusual mixture of people made for strange bedfellows, colorful stories, and, sometimes, greater acceptance. Shortly after the Caravan arrived at The Farm, two babies, Naomi and Brian, were born prematurely and needed to spend time in the hospital. While there, they served as young cultural ambassadors, for it was through them that their parents and midwives interacted with the community at large. "We got to be friends with a lot of Tennesseeans who we met while visiting the babies in the hospital," stated Ina May. "The nurses called Naomi 'the littlest hippy' and were fascinated that the babies had been born at home."[91] These types of interactions provided the potential for a greater understanding and appreciation between people with two different perspectives on birth.

Other encounters must have raised a few eyebrows even as they allowed for greater understanding. When the midwives discovered that Carolyn's baby was in a breech position, they decided she should give birth at the hospital, since she had never yet experienced a birth. Because the hospital would not allow husbands in the delivery room (a common hospital policy at the time), Carolyn and her husband, Harlan, opted to stay in the waiting room during much of her labor so that he could help support her. Pamela, Stephen, and Ina May stayed with them as well. There was another group of people in the room biding their time as another woman labored away in private. One can only imagine their reaction to what happened next, as the waiting room became an impromptu hippie birthing room. "Carolyn was sitting on a couch in a half-lotus, and when she'd start to have a rush [Ina May's word for "contraction"] Stephen and I would feel it and look up from our magazines at the same time, and our backs would straighten and we'd rush with her," Ina May described. "We had a lot of fun doing that. It was obviously a telepathic scene, and the other folks who were there got curious. American ladies usually put up a lot of fuss about having babies, and nobody in the obstetrics ward gets to have much fun if some lady is on a bummer about having her kid. I felt all the hospital folks [should] be grateful that they'd got to have a good time too."[92] Whether or not it lessened the pain for Carolyn, it appears to have lightened the tension in the room, at least for the hippies.

When Carolyn was ready to deliver, she was escorted into a delivery room. Though her husband could not accompany her, Ina May was allowed in, having established a positive relationship with Dr. Williams. Carolyn remembered, "the doctor was showing Ina May just how to do it and she was picking it all up. It was far out. Everyone felt stoned there in the delivery room—it felt like one smart head stoned on the energy of the birthing." Typically, Williams would put a woman under general anesthesia for a breech birth, but Ina May insisted that he allow her to experience the birth unmedicated. "The doctor and the nurses in the obstetrics ward got off to see how Carolyn tripped through her rushes. They hadn't seen anyone do it quite like that before, and they were interested."[93] Carolyn's birth offered a teachable moment for everyone involved.

With each hospital birth, the gap between midwife and doctor appeared to narrow. When Susan delivered her baby prematurely at Maury hospital in 1973, she expressed surprise about how positive her experience was. "I had some preconceived notions about hospitals and labor rooms," she wrote. "They quickly vanished as I learned to relax and exchange energy with Ina May." Ina May was once again given permission to participate in the birth. "The delivery room wasn't the weird, scary place I had anticipated," remembered Susan. "The doctor had really good vibes, and I knew the situation was really under control. I could tell that the doctor really liked Ina May's presence there and appreciated her techniques. . . . Having Ina May there was almost like giving birth on The Farm."[94] Ina May also believed that her presence was having an effect. "I noticed that with each hospital birth we had that our doctor got gentler and more compassionate about how he handled our ladies," she wrote. "At first he had been into pulling the placenta out by the cord and scraping the uterus to be sure he had got everything, which is a heavy trip if you haven't been anesthetized, but he was telepathic enough with us to notice that we thought that this was a little heavy."[95] Ina May no longer viewed herself as merely a student of birth; she was also becoming a critic and teacher, one who was beginning to have an impact on local hospital birth practices.

Writing Spiritual Midwifery

As The Farm's population grew (up to 1,500 in 1982, with about 14,000 visitors per year), so did the number of births. Over 2,500 babies have been born on The Farm by the midwives, whose favorable statistics (including a 1.8 percent cesarean section rate) have caught the attention of consumers and birth practitioners around the world.[96] Many more births have been affected by Ina May's home-birthing philosophy than just those taking place on The Farm, however. *Spiritual Midwifery*, a guide to birthing for consumers and birth practitioners published on The Farm, has sold over half a million copies, has been translated into six languages, and is still in print. *Spiritual Midwifery* started out as a section of the Book Publishing Company's first publication, *Hey Beatnik!* (1974). The earlier book provided an in-depth report about everything happening on The Farm, including its philosophy, farming, construction, soy dairy, grain mill, Farm Band, motor pool, school, and midwifery. This first version of *Spiritual Midwifery* was only 17 pages long (it would grow to 480 pages by the revised edition in 1977), a mixture of birth stories from multiple perspectives (father, mother, and midwife), photographs, and instructions for prenatal care and delivery.

Hey Beatnik! was a popular book, lively and colorful, reflective of the energy and spirit of several hundred young hippies creating a new communal life, geographically isolated from judgmental parents or the increasingly destructive drug culture that had taken over Haight-Ashbury in San Francisco by the 1970s. Paul Mandelstein, who created the Book Publishing Company on The Farm, cobbling together his print shop from used equipment he picked up in Atlanta, remembers *Hey Beatnik!* as "an amazing project." Because The Farm was a commune, no one was paid a salary, instead working on the book "for the love of it and to do the best" job they could. Over twenty artists contributed illustrations, and the array of colors that enlivened the book's pages were from remnants of ink cartridges donated by a Seventh Day Adventist printing company in Nashville ("they were vegans and so were we," Mandelstein explained of the donation). One distinctive characteristic of both *Hey Beatnik!* and the first edition of *Spiritual Midwifery* was the typeface, which was usually purple instead of black. "We didn't like black because we felt it was the dark aura. You know how hippies were back in those days," he chuckled later. "Of course over time we realized it was harder for people to read and we gave up our very conservative ideas about that."[97] By 1977 the type was set in black.

It was not initially clear what form *Spiritual Midwifery* would take. Shortly after the publication of *Hey Beatnik!*, Ina May started receiving letters from readers asking for more information about home birth, along with more birth stories, and she realized she had material for an entire book. She wanted to write the kind of book she had never had when she was first pregnant. "I remember being given a little book on pregnancy in 1966 when I was pregnant the first time and I just threw it aside because it was like a machine manual," she reflected later. "I didn't want to be told 'this is the first stage' and 'this is the second stage' and, you know, a description that a man would write about birth—what would *he* know?"[98] Looking at the birth stories written for *Hey Beatnik!*, she was struck by how good the writing was and decided to make these stories the centerpiece of *Spiritual Midwifery*. Pamela remembers that "we'd go out and pile on Ina May's bed and . . . we . . . started reading through the [birth stories], and you know, I'd get to one and say, 'Oh, this one's great—here, read this, Ina May'—and . . . she'd read one and she'd start laughing and we'd all say, 'Yeah, we gotta put that one in.'"[99] Stories on breech births, hospital emergencies, and even stillbirths fleshed out the narrative, a sobering reminder of what could go wrong mixed in with the delight of a successful and empowering home birth.

Some of the letters Ina May received made it clear that people were using her section in *Hey Beatnik!* to deliver their own babies. "And then I knew I had to write in a lot more detail than that because, you know, there's hardly anything in there." Thus, *Spiritual Midwifery* became much more than just a collection of birth stories for the expectant mother; it also served as a midwifery manual. Ina May continued reading all that she could on the medical and technical aspects of birth, drawing on her newly established connections with nurses and doctors whom she'd encountered both while traveling on the Caravan and on The Farm. They were "very helpful in helping me write up those parts, you know, about the gauge of the needle used for this and that and sterilization." But much of the material she got out of medical textbooks. "I knew that in the medical books it was a combination of life-saving information with crazy stuff. And so I got the good stuff and left out the other and said things my own way."[100] With this material, *Spiritual Mid-*

wifery became a practical how-to manual and a spiritual guide. She divided it into three main sections: first, the birth stories (adding new ones in each edition); second, a section for the expectant parents (prenatal care, exercise, and taking care of a newborn); and finally, instructions to midwives (on anatomy and physiology, the stages of labor, complications, emergencies, and other important advice).[101]

Four editions and several more printings later, the book continues to get attention. "Vanessa" discovered a copy of the book in a health food store in 2003. "This book changed my life," she posted on *Goodreads*, a social cataloging website launched in 2007. "I had no idea what I wanted to do with my life. I was lost, I was tired of trying to find direction. . . . When I opened the pages it was like the clouds parted and a beam of light spread over me," she recalls.[102] Nearly three thousand readers have reviewed the book on *Goodreads* since 2007, despite the fact that *Spiritual Midwifery* has changed little since its original printing more than thirty years earlier. Most of the photographs, including that of baby Ernie in his mother Mary's arms, remain the same, along with a majority of the stories. Only the data and instructions to parents and midwives have been updated (including new subjects such as maternal death and postpartum depression), as well as some of the language.[103] Yet the stories, now more than a generation old, continue to speak to readers. "Don't be turned off by the outdated language and fashion in this book," wrote Astrid in 2008. "It has the potential to change your whole outlook on birth and pregnancy."[104] *Spiritual Midwifery* introduced the concept of birth stories as a legitimate authoritative source of knowledge, affirming the possibility that birth could and did take place outside the hospital.[105]

Conclusion

Astrid's appeal—"Don't be turned off by the outdated language and fashion" of *Spiritual Midwifery*—serves as a reminder that readers can pick and choose what they want to get out of a particular text. Readers like Astrid most likely did not choose to move to a commune to give birth, but they did consider their options. The birth stories presented in *Spiritual Midwifery* suggest that birth has the potential to be something other than a fearful medical event dictated by hospital protocols. The resurgence of midwives and the emphasis on family-centered birth even in hospitals (where fathers are welcomed and encouraged to serve as birth coaches) suggest that the model developed and promoted by The Farm midwives touched a nerve. As Michael Traugot had said of the

Caravan, "We had a chance to influence our culture." He was not referring specifically to the out-of-hospital births, but that was, in fact, to be
the legacy of the Caravan and The Farm.

Spiritual Midwifery hit the market at the beginning of a boom in
lifestyle literature—what sociologist Sam Binkley refers to as the "countercultural lifestyle print culture of the 1970s."[106] Narratives on diet,
exercise, sex, birth, massage, and even Volkswagen repair emerged as
an alternative to the "stale living patterns of the postwar mass consumer."[107] Small, unconventional presses with minimal production budgets, such as the Book Publishing Company on The Farm, cropped up
and stunned the New York–based book market by capturing a wide
readership. Technology aided this industry, as the availability of computer typesetters inspired authors and editors to "undertake their own
production, editing, and design."[108] Scholars have begun to recognize
that the impact of these volumes went well beyond countercultural
readers. An oft-cited and heavily analyzed example of this countercultural print revolution is the *Whole Earth Catalog*. Started by Stewart
Brand in 1968, the publication was an "ad hoc collection of product
reviews, commentaries, and ecological screeds gathered from experimental lifestylists and back-to-the-landers."[109] But as historian Andrew
Kirk points out, the vast majority of the nearly two million people
who purchased copies of the *Whole Earth Catalog* were in fact urban
dwellers who never intended to move back to the land. "Dismissing the
counterculture as the apolitical sellouts of the 1960s and 1970s misses
the rich contributions this cultural mode made to politics and culture,"
Kirk argues.[110]

The role of The Farm midwives in transforming birth practices is
equally irrefutable. Not everyone agreed with what Ina May and the
other Farm midwives chose to write about, including a number of feminists who dismissed natural childbirth and communal living as a step
backward for women's liberation. But many agreed that the American
way of birth in the 1970s was far from ideal. "The response was enormous," Ina May recently said of *Spiritual Midwifery*. . . . It wasn't just
a few hippies that were interested in better birth—it was all kinds of
people."[111] Like the readers of the *Whole Earth Catalog*, most who read
Spiritual Midwifery had no intention of giving up their earthly posses-
sions and adopting the communal lifestyle. But perhaps *this* was the
book that allowed them to question the standardized labor and delivery procedures at their local hospital, to consider that birth could be a
spiritual as well as physical experience, and to believe that a laboring
mother had the right to make informed choices about where and how to

give birth. *Spiritual Midwifery* really began with a little manual written for indigenous populations, which, in the hands of determined young women on a spiritual quest, became a self-empowering tool that helped to revive an entire profession through the production of their own midwifery manual. Midwives became, on The Farm, the community's "ultimate power figures," a reminder that these particular hippie women played a crucial role in both birth culture and counterculture.[112]

Notes

1. Paul Fjerstad, "Ernest's Birth," in Ina May Gaskin, *Spiritual Midwifery*, rev. ed. (Summertown, TN: Book Publishing Co., 1977), 51.

2. Ina May Gaskin quoted in *Birth Story: Ina May Gaskin and the Farm Midwives*, documentary film produced by Sara Lamm and Mary Wigmore (2012).

3. Mary Fjerstad, "Ernest's Birth," in I. M. Gaskin, *Spiritual Midwifery* (1977), 51.

4. "The New Farm Report," 17 January 1974, Phil Schweitzer private collection, Summertown, TN.

5. Mary Fjerstad, e-mail correspondence with the author, 21 August 2013.

6. Gretchen Lemke-Santangelo, *Daughters of Aquarius: Women of the Sixties Counterculture* (Lawrence: University Press of Kansas, 2009), 181.

7. The 19 June 1978 edition of *Notes from "Amazing Tales of Real Life,"* no. 3, includes the following announcement in the Classified section: "LADIES: if you're looking for a job, check the *help wanted* board at the store." The early editions of *Spiritual Midwifery* offered adoption as an abortion alternative. "Don't have an abortion," they wrote. "You can come to the Farm and we'll deliver your baby and take care of him, and if you ever decide you want him back, you can have him" (I. M. Gaskin, *Spiritual Midwifery* [1977], 448). "Talk about women's lib, how about unborn babies' lib?," Stephen Gaskin writes in *Hey Beatnik!* (Summertown, TN: Book Publishing Co., 1974), arguing that "abortions as a means to save your ego are immoral." (*Hey Beatnik!* is not paginated.)

8. Warren Pearse, "Home Birth Crisis," *ACOG Newsletter*, July 1977.

9. "The Farm Report 1978," Phil Schweitzer private collection, Summertown, TN.

10. Joanne Santana, lecture, The Farm Midwifery Assistant Workshop, 26 March 2013, Summertown, TN.

11. Ibid.

12. Stephen Gaskin, *The Caravan*, rev. ed. (1972; Summertown, TN: Book Publishing Co., 2007), 127.

13. Stephen Gaskin, *Monday Night Class*, rev. ed. (Summertown, TN: Book Publishing Co., 2005), 8.

14. Ibid., 9.

15. Santana, lecture, The Farm Midwifery Assistant Workshop.

16. David Talbot, *Season of the Witch: Enchantment, Terror, and Deliverance in the City of Love* (New York: Simon and Schuster, 2012), 158.

17. Timothy Miller, *The Hippies and American Values* (Knoxville: University of Tennessee Press, 1991), 16.

18. Warren James Belasco, *Appetite for Change: How the Counterculture Took on the Food Industry* (Ithaca, NY: Cornell University Press, 2006), 97.

19. Santana, lecture, The Farm Midwifery Assistant Workshop.

20. Leo Eloesser, "Assembly Line for Country Midwives," *Pacific Spectator* 7, no. 2 (Spring 1953): 2.

21. J. E. B. McPhail, "Summary of CLARA-UNICEF Medical Program in North China," 6 February 1950, MSS 20, folder 2, box 20, Leo Eloesser Papers, Lane Medical Archives, Stanford University Medical Center (hereafter cited as Eloesser Papers).

22. Eloesser, "Assembly Line for Country Midwives," 2.

23. Ibid., 2–7.

24. Press Release ICEF/201, 5 April 1950, United Nations Department of Public Information Press and Publications Bureau, folder 2, box 20, Eloesser Papers.

25. According to Victor and Ruth Sidel, by 1972 there were over one million barefoot doctors in China. See Victor Sidel and Ruth Sidel, *Serve the People: Observations on Medicine in the People's Republic of China* (Boston: Beacon Press, 1973), 197. An American translation of *A Barefoot Doctor's Manual* was first published in the United States in 1974 by the US Department of Health, Education, and Welfare, publication no. (NIH) 75-695 (Public Health Service, National Institutes of Health, Bethesda, MD; translation of *Ch'ih chiao i sheng shou ts'e*), and appealed to many American readers interested in a basic reference guide to Chinese medicine, including descriptions of acupuncture, massage, and herbal treatments. Four pages cover "New Methods for Delivery of the Newborn"—nothing in comparison to what readers could find in Eloesser's *Pregnancy, Childbirth and the Newborn*, 4th ed. (Mexico City: Instituto Indigenista Interamericano, 1976) or in Ina May Gaskin's *Spiritual Midwifery*.

26. Judith Rooks, *Midwifery and Childbirth in America* (Philadelphia: Temple University Press, 1997), 30.

27. Amanda Harris et al., "Midwives in China: '*Jie Sheng Po*' to '*Zhu Chan Shi*,'" *Midwifery* 25 (2009): 205.

28. Ibid., 203–12.

29. Eloesser, "Assembly Line for Country Midwives," 3.

30. McPhail, "Summary of CLARA-UNICEF Medical Program in North China," 2.

31. Isabel Hemingway to Leo Eloesser, 24 April 1955, folder 2, box 15, Eloesser Papers.

32. Nicholas Eastman to Leo Eloesser, 5 October 1956, folder 2, box 15, Eloesser Papers.

33. Victor Richards to Leo Eloesser, 30 January 1973, folder 2, box 15, Eloesser Papers.

34. Santana, lecture, The Farm Midwifery Assistant Workshop.

35. Ibid.

36. Ina May Gaskin, "Birth Story: A 'Pregnancy' 30 Years in the Making," 10 May 2013, accessed 29 May 2013, http://www.huffingtonpost.com/ina-may -gaskin/post_4689_b_3253016.html.

37. Ina May Gaskin quoted in Katie Allison Granju, "The Midwife of Modern Midwifery," 1 June 1999, accessed 29 May 2013, http://www.salon .com/1999/06/01/gaskin/.

38. Ina May Gaskin, conversation with the author, 26 March 2013.

39. Ibid.

40. Pamela Hunt, interview by the author, 29 March 2013.

41. Gwyn Harvey, "Interview with Pamela Hunt," *Birth Gazette* 4, no. 1 (Fall 1987): 6.

42. "Pamela's Story," in I. M. Gaskin, *Spiritual Midwifery*, 4th ed. (Summertown, TN: Book Publishing Co., 2002), 22.

43. Harvey, "Interview with Pamela Hunt," 6.

44. Hunt, interview by the author, 29 March 2013.

45. Cara Gillette, e-mail correspondence with the author, 24 August 2013.

46. Gaskin, *The Caravan*, 14.

47. Ibid., 35.

48. Ibid., 165.

49. Ibid., 83.

50. Ibid., 22.

51. Michael T. quoted in ibid., 170.

52. Michael Traugot quoted in ibid., 171.

53. Cara Gillette, e-mail correspondence with the author, 24 August 2013.

54. I. M. Gaskin, *Spiritual Midwifery* (2002), 17.

55. "Pamela's Story," in ibid., 23.

56. I. M. Gaskin, *Spiritual Midwifery* (2002), 16.

57. Cara Gillette, e-mail correspondence with the author, 24 August 2013.

58. "Anne's Birth," in I. M. Gaskin, *Spiritual Midwifery* (2002), 36.

59. Cara Gillette, e-mail correspondence with the author, 24 August 2013.

60. "Pamela's Story," in I. M. Gaskin, *Spiritual Midwifery* (2002), 24.

61. Mary Fjerstad, e-mail correspondence with the author, 21 August 2013.

62. "Anne's Birth," in I. M. Gaskin, *Spiritual Midwifery* (2002), 36.

63. "Pamela's Story," in ibid., 24.

64. "Anne's Birth," in ibid., 37.

65. Cara Gillette, e-mail correspondence with the author, 24 August 2013.

66. "Pamela's Story," in I. M. Gaskin, *Spiritual Midwifery* (2002), 24.

67. Cara Gillette, e-mail correspondence with the author, 24 August 2013.

68. "Pamela's Story," in I. M. Gaskin, *Spiritual Midwifery* (2002), 24.

69. Cara Gillette, e-mail correspondence with the author, 24 August 2013.

70. I. M. Gaskin, *Spiritual Midwifery* (2002), 17.

71. Nancy Burns-Fusaro, "Dante Society Honors Four 'Bologna Boys," *Westerly Sun*, 22 November 2008, accessed 26 June 2013, http:// www.thewesterlysun.com/news/local/article_0aa29839-daa1–59c6-a94d -732b601528d4.html.

72. Margaret Nofziger decided she did not want to continue studying or practicing midwifery after the Caravan.

73. Ida May Gaskin, conversation with the author, 26 March 2013.

74. "Farm Report January 1975," Phil Schweitzer private collection, Summertown, TN.

75. Ina May Gaskin, "Spiritual Midwifery," in *Hey Beatnik!*, n.p. Hazrat Imayat Kahn is the Sufi founder.

76. Rupert Fike, Cynthia Holzapfel, Albert Bates, and Michael Cook, *Voices from The Farm: Adventures in Community Living*, 2nd ed. (Summertown, TN: Book Publishing Co., 2012), 12.

77. Fred Turner, *From Counterculture to Cyberculture: Stewart Brand, the Whole Earth Network, and the Rise of Digital Utopianism* (Chicago: University of Chicago Press, 2006), 32–33.

78. Albert K. Bates, "Technological Innovation in a Rural Intentional Community, 1971–1987," paper presented at the National Historic Communal Societies Association annual meeting, 17 October 1987, Bishop Hill, IL.

79. Phil Schweitzer, interview by the author, 30 March 2013.

80. I. M. Gaskin, *Spiritual Midwifery* (2002), 30.

81. Hunt, interview by the author, 29 March 2013.

82. I. M. Gaskin, *Spiritual Midwifery* (2002), 20.

83. The "midwife crew" changed quite a bit over the years, and some women rotated in and out of midwifery positions (to care for their children and/or to do other types of work on The Farm). Currently, five of the seven midwives on The Farm have been there since the founding of The Farm in 1971. Joanne Santana, owner of the midwifery manual pictured in figure 6.1, was originally a teacher at The Farm school, transitioning over to the midwife crew as her children grew older and more independent. She pointed out to me that the "crew" consisted of more than the actual midwives—they also had assistants. "You know, you hear about the people in the book and stuff but there was a lot of people on the crew that did nothing except go to the births, set up equipment, pack sterile packs at the clinic, help out the clinic ladies working the pharmacy, it was a huge crew behind every one of these births. Work as EMTs, do the ambulance shifts, that kind of thing." Santana, lecture, The Farm Midwifery Assistant Workshop.

84. Pamela Hunt, lecture, The Farm Midwifery Assistant Workshop, 26 March 2013, Summertown, TN.

85. I. M. Gaskin, *Spiritual Midwifery*, 4th. ed., 29.

86. "Pamela's Story," in ibid., 26–27.

87. http://www.tributes.com/show/John-O.-Williams-88512246.

88. Posted by Debbie O'Neal, 21 August 2003, http://www.tributes.com/condolences/view_memories/88512246?p=10&start_index=1.

89. John Williams to Ina May Gaskin, 24 August 1993. Letter in Pamela Hunt's possession.

90. Ina May Gaskin, "Spiritual Midwifery," selection from *Hey Beatnik!*

91. Ibid.

92. Ibid.

93. Ibid.

94. Susan Rabideau, quoted in "Spiritual Midwifery," selection from *Hey Beatnik!*

95. Ina May Gaskin, "Spiritual Midwifery," selection from *Hey Beatnik!*

96. See http://www.thefarmmidwives.org/preliminary_statistics.html (accessed 30 May 2013).

97. Paul Mandelstein, interview by the author, 16 April 2013.

98. Ina May Gaskin, conversation with the author, 26 March 2013.

99. Hunt, interview by the author, 29 March 2013.

100. Ina May Gaskin, conversation with the author, 26 March 2013.

101. Although Ina May Gaskin is listed as the sole author of the book, many midwives contributed to it as well.

102. See http://www.goodreads.com/review/show/8591519 (accessed 26 August 2013).

103. For example, "puss" has been changed to "yoni" in the 4th ed. (2002) edition. Ina May Gaskin writes in the preface to the 4th ed., "I have decided to undertake a new experiment in this edition of *Spiritual Midwifery*. With the hope of helping women to proudly reclaim all the words that refer to their reproductive organs, I will use terms that did not appear in previous editions of the book. This is because I like dealing with language so that it works for us rather than against us" (8).

104. See http://www.goodreads.com/review/show/22894701 (accessed 26 August 2013).

105. For more on the emergence of the birth story as a new literary genre, see Mary Lay, *The Rhetoric of Midwifery: Gender, Knowledge, and Power* (New Brunswick, NJ: Rutgers University Press, 2000).

106. Sam Binkley, *Getting Loose: Lifestyle Consumption in the 1970s* (Durham, NC: Duke University Press, 2007), 5.

107. Ibid., 101.

108. Ibid., 108.

109. Ibid., 5.

110. Andrew Kirk, *Counterculture Green: The "Whole Earth Catalog" and American Environmentalism* (Lawrence: University Press of Kansas, 2007), 186.

111. I. M. Gaskin, *Birth Story*.

112. Doug Stevenson, "Up-Close Relationships," in Fike, Holzapfel, Bates, and Cook, *Voices from The Farm*, 101.

Part Three: Personae

7

The Unseasonable Grooviness of Immanuel Velikovsky

Michael D. Gordin

To everything there is a season, as the classic counter-cultural song by the Byrds (and also, of course, Ecclesiastes) has it. Certainly, Immanuel Velikovsky (1895–1979) never imagined that the late 1960s and early 1970s were going to be his. For almost two decades, Velikovsky's extraordinary claims about the recent history of the solar system and its influence on the ancient history of humanity had lain in quarantine, isolated by the Establishment scientists whose approval he so desperately sought. Mainstream science would have nothing to do with him, but in the late 1960s the counterculture offered Velikovsky acceptance, even a reverence unusual for this iconoclastic cohort. In 1971 Murray Gell-Mann, who had received the Nobel Prize in Physics two years earlier, viewed the burgeoning movement with alarm: "We are seeing among educated people a resurgence of superstition, extraordinary interest in astrology, palmistry and Velikovsky; there is a surge of rejection of rationality, going far beyond natural science and engineering."[1] Velikovsky in post-1968 America was no longer "fringe": he was one of the most popular authors read by college youth.

It was Velikovsky's moment, and a most unlikely moment at that. Born in 1895 in Vitebsk—now in Belarus but then a thriving Jewish metropolis within the Russian Empire—he came to intellectual maturity in emigration first in Berlin and then in Palestine, nourished on two major cultural trends within contemporary Judaism: Zionism and Freudianism. He was, in short, a child of interwar intellectuality, not of flowers and communes. Seen from the perspective of his conceptual formation, the 1970s were an improbable setting for his moment in the limelight.

He achieved this new status largely because of a book, itself a relic of another time—in this case postwar anxieties about anti-Communism and the status of science. In April 1950 the Macmillan Company released Velikovsky's first English-language book, entitled *Worlds in Collision*, which proceeded to rocket to the top of nonfiction best-seller lists nationwide. In this blockbuster, Velikovsky argued that ancient mythological, scriptural, and historical sources from a variety of cultures—but principally from the Near East, and especially the Hebrew Bible—contained repeated homologous descriptions of major catastrophes: rains of fire, immense earthquakes, tsunamis, dragons and snakes fighting in the heavens. These passages had long been interpreted by rationalist readers as metaphors or ecstatic visions. Not so, argued Velikovsky: when compared and synchronized, they pointed toward a massive global catastrophe (more accurately, a series of catastrophes) that actually happened. Velikovsky tracked two main cataclysms: one that occurred around 1500 BC, during the Exodus of the Hebrews from Egypt; and another in the eighth century BC, which changed the length of the year from 360 days to its current 365¼ days, stunning the prophet Isaiah and chronicled as the battle between Athena and Ares in Homer's *Iliad*.[2]

Worlds in Collision ignited a firestorm of criticism from scientists, who attacked the book in reviews across the popular press. At its core was Velikovsky's mechanism for these catastrophes. He claimed that the first was caused by a comet that had been ejected from Jupiter and almost collided with Earth, remaining trapped in gravitational and electromagnetic interaction with Earth for several decades, raining petroleum from its tail, flaring across the heavens, and tilting Earth's axis. (The last caused the sun to appear to stand still over the plains of Gibeon in Joshua's famous battle.) Eventually, the comet stabilized into the planet Venus. Thus, Earth's nearest planetary neighbor was originally a comet born in historical times, as attested by proper interpretation of the records of the collective memory of humanity. Venus's movements then displaced Mars, which threatened Earth in the second series of catastrophes. Velikovsky's arguments presupposed a reformulation of the central precepts of geology, paleontology, archeology, and celestial mechanics, not to mention ancient history.

A small number of scientists, principally astronomers, wrote to Macmillan—then the nation's most prestigious scientific publisher, earning most of its revenues from textbook sales—and threatened a boycott of Macmillan texts unless Velikovsky's book was dumped. George Brett, the director of the press, arranged for the contract to be transferred to Doubleday, a trade press that had no Achilles heel in the form of a textbook division, and Velikovsky continued to publish with the latter until his death.[3] The public exposure of this fledgling boycott generated among Velikovsky's supporters (few as they then were) a rhetoric centered on Galileo against the church, a variant of David against Goliath: Velikovsky spoke alone against the powers that be, a voice of reason against hidebound dogmatism and privilege.

This is where we find Velikovsky in the early 1960s, a man doubly displaced beyond his season. The first displacement concerns the composition of *Worlds in Collision* itself. This book, which was the *succès de scandale* of 1950, was originally conceived as a riposte to the denigration of Judaism that Velikovsky perceived in Sigmund Freud's final work, *Moses and Monotheism* (1938).[4] Leaving his home in Tel Aviv, he embarked on a research trip with his family to New York City in 1939 (it was permanently extended by the outbreak of war and then the vicissitudes of life) and began composing a manuscript entitled "Freud and His Heroes" to expose Freud's neuroses about his own Judaism in his dream theories, in his depiction of Oedipus, and finally in his scandalous argument that Moses was an Egyptian who gave the Jews the Pharaoh Akhnaton's (Akhenaten) monotheism before being murdered by his own people. A sudden insight led Velikovsky to consider the ten plagues and the miracles of the Exodus to be natural phenomena, which eventually brought him to Venus and a trade-book contract for *Worlds in Collision*.[5] The seeds were planted in the season of psychoanalytic debates and were harvested in Cold War America.

That was the second displacement. Velikovsky's book had originally been pitched by Macmillan as an intervention in long-standing worries about the conflict between science and religion. Here was a book—maybe not a serious scientific monograph, as five separate peer reviewers pointed out before blithely approving it—that used ancient sources, principally the Hebrew Bible, to make claims about the dynamics of planets. Although the advance publicity campaign for the book and endorsements by such public religious luminaries as Norman Vincent Peale situated the book in this context, the astronomers and other scientists appropriated it within another contemporary debate: that over Joseph Stalin's imposition of Trofim Lysenko's antigenetic theories as obligatory orthodoxy in the Soviet Union and the simultaneous congressional persecution of left-leaning scientists in the United States.[6] The attack on Velikovsky, surprising in its intensity, was a by-product of a moment when paranoia and anxiety about ideology's possibly fatal effect on scientific inquiry gripped the American scientific community. The Velikovsky affair was absorbed into 1950's "silly season," and it seemed that his moment was past.[7]

And then, almost despite himself, Velikovsky rose once more. Amid the tumults that rocked American culture in the late 1960s, a marked enthusiasm for cosmic catastrophism was shaken loose. Velikovsky became, as one of his critics put it in 1977, "the grand curmudgeon of antiestablishment science."[8] For Harold Urey—winner of the 1934 Nobel Prize in Chemistry for his 1932 discovery of deuterium (a heavy isotope of hydrogen)—the potential reverberations of this sea change were no less catastrophic than a near collision with a comet. "Velikovsky is a most remarkable phenomenon of the last 20 years," he wrote in 1967 to University of Kansas chemist (and Velikovskian) Albert Burgstahler in answer to an earlier missive. "If someone of this kind should turn up in science once a year I think it would wreck science completely."[9]

The counterculture had found a scientific guru in a septuagenarian Russian-Israeli émigré. "The counterculture," of course, was not one thing but comprised a diverse array of peaceniks, New Age spiritual seekers, Black Power activists, the drug-addled, the musically hip, and those who just refused to keep on keeping on: "a culture so radically disaffiliated from the mainstream assumptions of our society that it scarcely looks to many as a culture at all, but takes on the alarming appearance of a barbaric intrusion."[10] This characterization came from Theodore Roszak, the writer whose 1969 book, *The Making of a Counter Culture*, gave the phenomenon its name. But Roszak also imbued the counterculture with his own agenda: a critique of scientific and technical objectivity. Many scientists saw the excitement surrounding Velikovsky's theories as more evidence of the same: "It is as if many Velikovskyites are saying 'Nya! Science isn't so great—look at all the things it can't explain!' . . . Many Velikovskyites, like many others who have no experience in research, betray a basic hostility to science per se."[11]

This interpretation is misleading. There were many ways to be hostile to science in the 1970s—getting stoned instead of going to class, joining a radical farming commune, or bombing a computer center, for instance—but it is not obvious that the detailed study of orbital paths, geological formations, ancient inscriptions, and the latest reports from Soviet Venus landers was among their number. Being interested in Velikovsky meant being interested *in* science, just science of a different sort. Rejected decisively by the Establishment, by the 1970s Immanuel Velikovsky had acquired, in a fit of absent-mindedness, a counter-Establishment: his books assigned in college courses, peer-reviewed journals dedicated to his theories, and countless invitations to address packed lecture halls.

This chapter uses the story of Velikovsky's final decade to trace two paths in the relationship between science and the counterculture. First, I will examine how the growing swath of Velikovsky fans perceived the scientific "Establishment," a fighting word. Scientists were feeling just as embattled as they had during the events that provoked the spasmodic reaction of 1950. The barbarians had come through the gates, they were sitting in classrooms, and they could vote. By 1972 Alvin Weinberg, director of the Oak Ridge National Laboratory, was deeply concerned: "Today, however, one wonders whether science can afford the loss in public confidence that the Velikovsky incident [of 1950] cost it. The republic of science can be destroyed more surely by withdrawal of public support for science than by intrusion of the public into its workings."[12] Mapping out precisely how both mainstream scientists and Velikovskians perceived the conflict between established doctrines and cosmic catastrophism thus delineates the fault lines between different conceptions of science.

If the first issue is *what* the several sides were fighting over, the second concern is *how* they did it. The countercultural Velikovskians were unable, by definition, to publish in mainstream venues like *Science* or *Astrophysical Journal*. To broadcast their arguments they had to use alternative modes of communication: special courses, lectures, student groups, and especially the emergence of specialized Velikovskian journals (principally *Pensée* and *Kronos*). The rise and fall of this alternative publishing forum shows us not only how alternative sciences were framed and promulgated but also what happened when Velikovsky's massive popularity among the youth began to strain his own self-conception of his ideas. Youth had appropriated him for its own reasons, not his, and Velikovsky neither liked nor trusted these camp followers of cosmic catastrophism. It is important to recall that in this age of gurus, the energy often emanated from the followers, not the charismatic leader. We will thus follow Velikovskianism more than Velikovsky, as his untimely season finally arrived.

Velikovsky 101

It was difficult to attend college in the 1970s without being somewhat aware of Immanuel Velikovsky and his revolutionary theories. The extent of his popularity is hard to measure, but there is no question that, when one tallies up the letters from fans across the country, the tremendous sales of his books (especially *Worlds in Collision*) in college bookstores, and the numerous invitations to lecture, Velikovsky was becoming something of a phenomenon—even, one might say, a celebrity. (Consider the fact that Peter Fonda, the easy-riding poster child of the counterculture, name-checked Velikovsky in an interview.)[13]

In retrospect, Velikovsky presented this transformation as the foreseen result of his own changed tactics: instead of diplomatic overtures to established scientists, he would focus on the young, who were less likely to be indoctrinated into uniformitarian dogma. Velikovsky declared in 1969 that a decade earlier "I evaluated my resources and concluded that I should not spread myself on all fronts but dedicate my efforts to the goal of reaching the young generation—college students and young professors."[14] Velikovsky was very concerned with youth, and he particularly enjoyed the contrast of his own aging frame with the boundless energy of his fans. Writing in the late 1970s, he gloried that "I, an octogenarian, stride with the young of mind. There is no cult of Velikovsky; there is only the cult of scientific and historical truth."[15]

And so Velikovsky looked to the students; or rather, the students looked to him. He experimented with a campus organization, Cosmos and Chronos, established in the mid-1960s at Princeton University, less than a mile from his home. By 1967 the fledgling clubs received a four-page mimeographed newsletter from the Princeton chapter of the "Campus Study Groups in Interdisciplinary Synthesis."[16] These intermittently produced newsletters mentioned recent pro-Velikovsky publications, confirmations of scientific predictions, and his impressive roster of upcoming talks. Between 1964 and 1969, by his own count, he had lectured at sixty college campuses of all types, seeding Cosmos and Chronos Groups along the way. On 27 April 1966 Velikovsky gave a talk at Yale University entitled "The Pyramids"; on 24 January 1968 he gave a lecture, "A Changing View of the Universe," at the Towne School of Civil and Mechanical Engineering at the University of Pennsylvania; and on 17 February 1972 he capped his tour of the Ivy League with "My Star Witnesses," presented by invitation before the Society of Engineers and Scientists of Harvard University. Elitist about many things, Velikovsky was assiduously democratic when it came to speaking about his theories. He accepted an invitation from the Forum for Free Speech at Swarthmore, and he did not shun San Fernando Valley State College or the University of North Texas. He even spoke at high schools.

"The new generation on campuses—in this country—is definitely following the heretic; the professors find themselves before unbelieving audiences," Velikovsky crowed. "My visits to campuses are triumphs. And more recently some large universities re-evaluate the entire situation; thus I was selected to address the Honors Day Convocation (June 3 [1967]) at the Washington University, St. Louis, over a two-times Nobel Prize winner (Lynus [sic] Pauling)."[17] On 14 April 1970, in celebratory anticipation of the first Earth Day the following week, Velikovsky achieved top billing at the Parsons School of Design with the talk "Is the Earth an Optimal Place to Live?" Stewart Brand, the editor of the *Whole Earth Catalog* and fixture of the counterculture, played backup.[18] The appeal spread northward. At McMaster University in Hamilton, Ontario, in 1974, Velikovsky drew a crowd of 1,100, and he received an honorary doctorate of arts and sciences in spring 1974 from the University of Lethbridge in Alberta, accompanied by a Velikovskian conference.[19]

The change was in the audience, not in Velikovsky: his claims remained almost identical to the position sketched out in 1950 in his best seller. Why were people lining up for him, and why *now*? His theories served as a middle ground for people of all political persuasions. He was an underdog in an age that had ceased to trust scientists (capturing the Left), but he also promoted deeper study of the Bible (seducing the Right) in a decade whose best-selling work was Hal Lindsey's *Late Great Planet Earth* (1970), an application of biblical eschatology to Cold War geopolitics (to which I will return in the conclusion). Velikovsky was anti-Establishment but not New Left, and thus shared affinities with strands of the counterculture that have dimmed in our memory today.[20] To a speaker at the 1974 Lethbridge conference, Velikovsky was the choice of a new generation:

> The veil of amnesia has been lifted, the result is the awakening of consciousness, whether the apocalyptic agent is perceived to be an extra-terrestrial jostling, or biospheric poisoning, atomic weaponry overkill, or overpopulation; or whether one has experienced the disintegration of his world view by chemical inducement—a magical mushroom or the fabled LSD. The generation of the *Whole Earth Catalogue* has experienced the catastrophe and, consistent with Dr. Velikovsky's amnesia theory, they no

longer itch to re-enact the primordial paroxysm that heralded
our present age—the bomb has gone off![21]

This view fits nicely with Roszak's antitechnocracy interpretation of the
youth movement and Gell-Mann's fears about a dawn of obscurantism.
Surely this was part of his growing appeal: bringing up Velikovsky in
class enraged science faculty. Yet explaining support of Velikovsky as
expressing "anti-Establishment" sympathies is no explanation at all.
In the 1970s *everyone* was opposed to the Establishment. As historian
Bruce Schulman has observed: "Richard Nixon hated the establishment. He loathed the prep school and private club set, the opera-goers
and intellectuals, the northeastern Ivy League elite."[22] When the president of the United States can claim anti-Establishment credentials, we
need a more nuanced framework. The point was not opposition to an
Establishment but what the Establishment signified to those who opposed it.

Support for Velikovsky concentrated among the lay public, humanists and social scientists, and, interestingly, scientists working for private
industry.[23] For Velikovsky and the inner circle, the youthful exuberance
for his doctrines was both flattering and a bit of an embarrassment.
As Chris Sherrerd, a peripheral member of his inner circle of devotees,
wrote to Velikovsky in 1968: "I suspect that much of the support you
are finding on college campuses is mot[i]vated not so much epistemo-
logically but rather socially: as part of a general revolt of today's youth
against 'the establishment.'"[24] If youth were following Velikovsky en
masse, the Velikovskians wanted it to be not because he was rejected by
the "Establishment" but because he was *right*.

The issue of correctness crops up repeatedly in pro-Velikovsky articles and underscores that this movement was not "antiscience" in
any straightforward way. One article from 1968 asserted that his resurgence "is due to one circumstance that the Scientific Establishment
did not foresee when it all but unanimously dismissed Velikovsky as
a crank and mocked his theories as ridiculous. With the accumula-
tion of new knowledge, especially that gathered in the last decade by
space probes"—such as the unexpectedly high temperature of Venus,
radio noises emitted from a cold Jupiter, and especially the (disputed)
detection of hydrocarbons in the Venusian atmosphere—"Velikovsky's
picture of the solar system has proved to be more accurate on many
important points than the theories embraced by the Establishment."[25]
(The first two of these findings are real phenomena, but astronomy both
in the 1970s and today attributes them to conventional mechanisms of

geophysics, such as the greenhouse effect for Venus; the hydrocarbon finding was later retracted as spurious.) A combination of excitement about new astrophysical discoveries, a chafing at the bonds of authority, and the widespread distribution of Velikovsky's works in paperback changed the climate.

Nowhere was this more visible than in the rise of college courses dedicated to exploring Velikovsky's work. Much as he had long predicted and fervently desired, *Worlds in Collision* became required reading in colleges across the United States and Canada, although not always in a manner that Velikovsky would have found flattering. For example, W. C. Straka, an assistant professor of astronomy at Boston University, taught a course called Science and Anti-science in Astronomy, where he assigned *Worlds in Collision* in order to debunk it.[26] More intriguing were courses that defended Velikovsky's theories to the young. Given the outsider status of Velikovskianism, it is not surprising to learn that many of these instructors were adjuncts at less prominent institutions, and even they had to fight to get their courses listed. A scheduled course at the University of Alabama was canceled at the last minute because of controversy over an advertisement for the class posted by the teacher.[27] A course at Penn State was also scrapped, with the argument that "students at the freshman and sophomore level can't judge what is correct or incorrect reasoning. We feel they should only be taught material that is correct beyond any doubt."[28] (The wrangling about the meaning of that statement occupied many pages of appeal and protest. Although the course was never reinstated, the professor was eventually granted tenure.) In 1971 Velikovsky's close acolyte C. J. Ransom successfully struggled to get a course on the theories accepted at Texas Christian University in Fort Worth.

This night course proved very popular, with an enrollment of twenty-nine students. "Overall the students agree with your theory," Ransom wrote to Velikovsky. "Most of the discussion concerns details, and no one seems opposed to the total concept."[29] Three days later he added: "To the young people, the theory seems quite logical and some do not understand why there is so much controversy."[30] Another close disciple, Lynn Rose, used some Velikovskian material in his philosophy of science class at SUNY-Buffalo (today the University at Buffalo) in 1971, and by 1973 he was teaching courses entirely devoted to *Worlds in Collision*.[31] Rose was tenured; he could do as he wished. Others had to exploit the makeshift experiments of the Age of Aquarius, such as the proliferation of "free universities" that paralleled established institutions of higher education. There were courses on Velikovsky at the Free University at the University of Pennsylvania, the University of Connecticut Free University, and even at the Medical College of Virginia.[32]

Counter-Establishment Science, in Print and in Public

Ben Bova, the editor of the science fiction magazine *Analog*, believed that the enthusiasm for Velikovsky was news and thus should be addressed, but he privately fumed against the Velikovskians. "Sometimes it's not your enemies that hurt you, it's your friends. The only thing more tedious, sententious and lacking in physical proof than Velikovsky's own writings are the writings of many of those who attempt to support his thesis," he wrote to Lynn Rose in 1974. "I'm not interested in counting alleged errors in articles either by or for Velikovsky. I am interested in physical evidence either for or against his ideas, the kind of evidence that one uses to decide the validity of any other physical theory."[33] And that evidence, to his mind, was sorely lacking: "Velikovsky's ideas hold about as much water as a well-worn piece of cheesecloth. They're the result of trying to find one sweeping explanation for every strange and wonderful event that confronts us; this is a syndrome that's very common in science fiction."[34] Nonetheless, in the spirit of fairness, he spent two years negotiating with Velikovsky's inner circle to get the master to write a piece for *Analog* to address negative articles that had appeared there. After countless stipulations about copyright, billing on the cover, space constraints, and more, Bova called the whole thing off in 1975. "This hardly seems like the attitude of a man who wants to use rational discourse to convince skeptics," he wrote to Frederic Jueneman. "He's acting like a petulant child."[35]

There was a moment when Velikovsky would have leapt at the chance to be published in a broad-circulation magazine like *Analog*, but not anymore, not during his season of grooviness. The 1970s saw the emergence of dedicated journals that promoted Velikovskianism, packed with articles bristling with footnotes, equations, and archeological evidence.[36] Though Velikovsky had not been accepted into the Establishment, he now found himself with a full-blown counter-Establishment. This was a sudden development. As recently as 1967, he felt so locked out of print venues for his ideas—aside from his books, of course, which continued to sell—that he even took special pains with an undergraduate periodical.

That April, *Yale Scientific Magazine*, "operated by undergraduates with complete editorial freedom" from the elite educational institution in New Haven, published a special issue focusing on a dispassionate scientific discussion of one aspect of Velikovsky's theories: the issues surrounding Venus, including recent discoveries from space probes. The editor, John W. Crowley, insisted that the magazine "does not pretend either to vindicate or to demolish Velikovsky's ideas in this issue; we seek only to present a paradigm for further discussion by avoiding the abusive tone" of prior discourse.[37] The centerpiece of the issue was Velikovsky's article "Venus—a Youthful Planet," which had been written in 1963 and submitted to the *Proceedings of the American Philosophical Society* by Princeton University geologist Harry Hess, a member of that august organization. The dispute over whether to publish it almost ruptured the journal's editorial board, so a separate panel was established to decide upon the fate of the piece. In January 1964 Velikovsky was informed that the article had been rejected, and it was subsequently also rejected by the *Bulletin of the Atomic Scientists*. *Yale Scientific Magazine* was to be its home.[38] It was followed by friendly and not-so-friendly critiques from University of Kansas chemist Albert Burgstahler and Columbia astronomer Lloyd Motz, respectively, both rebutted at length by Velikovsky.[39] In 1967, this was the best that he could manage; within five years, the situation had utterly changed.

The first Velikovskian journal initially had nothing to do with Velikovsky. It was called *Pensée* and was officially published in Portland, Oregon, through the Student Academic Freedom Forum of Lewis and Clark College. It is very difficult to reconstruct the early history of the journal. According to a press release for a Velikovsky symposium hosted by *Pensée* at that college in 1972, it was founded in 1966 by David Talbott, then an undergraduate at Portland State University.[40] No issues appear to survive from those early years. In the winter of 1970–71, the journal was reactivated under the editorship of David's brother, Stephen Talbott, a graduate of Wheaton College (where he had edited the school paper). Judging from the content of these early issues, *Pensée* was a rather-typical student journal in those countercultural days, with opinion battles pro and con on issues like Vietnam (June 1971), local environmental activism (November 1971), and abortion (January 1972). Of a medium-sized format akin to *Time* magazine, although much thinner and with less glossy, black-and-white pages, each month's installment began with an amusing series of sarcastic commentaries on national and local issues—often with a conservative bent—and signed pieces hailed from undergraduate and graduate students across the Portland region. The content was haphazard in both origin and quality, and the artwork also displayed some of the rushed character of many countercultural publications. Only rarely did Stephen (who characterized himself in his byline as "an on-and-off-again student") choose to pen a piece, as he did in June 1971, "The Population Crisis Is a Put-On," articulating a view that skewed slightly to the right in the wake of Paul Ehrlich and Anne Ehrlich's *The Population Bomb* (1968).[41] The journal had nothing to do with science and not even a hint of Velikovsky—no Venus or Egyptian king lists or boycott campaigns against Macmillan. In May 1972 the entire emphasis shifted. David Talbott suddenly appeared as the magazine's publisher (earlier it had been Robert G. Wallenstein) and Stephen remained the editor. The Talbotts launched a series entitled "Immanuel Velikovsky Reconsidered" to examine the debates over cosmic catastrophism, including some contributions from Velikovsky himself. It was a fateful decision: the circulation of the magazine spiked, its content became entirely dedicated to Velikovsky, and by 1974 the editorial board was populated by the inner circle, including Ralph Juergens, William Mullen, and C. J. Ransom. In appearance, too, *Pensée* began to look slightly glossier, the layout less jury-rigged and more professionally seamless (fig. 7.1). Some of that was the benefit of experience; the rest was due to the magazine's surprising success and improved financial situation.

The counterculture gave birth to *Pensée*, Velikovsky's theories gave the magazine a mission, and then *Pensée* returned the favor by bringing Velikovskianism to the counterculture. *Pensée* provided a forum for his supporters (and some critics) to discuss their thoughts, to puzzle through problems in the chronology of Egypt's Middle Kingdom or the orbital dampening of Venus—in short, to build a community. Circulation boomed: for the two and a half years of its Velikovskian adventure, *Pensée* had an annual circulation of 10,000–20,000, but the first issue in the Velikovsky series was reprinted twice, with a total run of 75,000 copies.[42] Not bad for a fly-by-night operation in Portland. Submissions flooded in, and the Talbott brothers (principally Stephen) had to develop a system to filter out the good from the bad. They borrowed a practice from the academic establishment, one that has often been held up as differentiating "real science" from "crackpot writings": peer review. Every submission to *Pensée*—with the important exception of those by Velikovsky—was reviewed, often by the inner circle. Not surprisingly, many of the critical anti-Velikovsky pieces were rejected or returned for revisions, usually because of logical flaws or problems with the empirical data.[43] (Several were published upon revision.)

1974: The High Tide

Velikovsky had for years been trying to get a hearing before a committee of scientists, an organized panel of diverse experts in Assyriology,

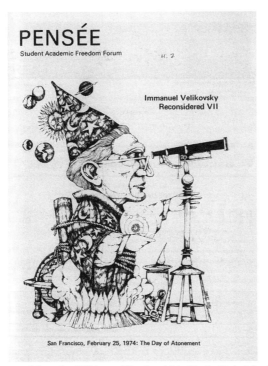

PENSÉE
Student Academic Freedom Forum

II. 2

Immanuel Velikovsky
Reconsidered VII

San Francisco, February 25, 1974: The Day of Atonement

FIGURE 7.1 The cover of the spring 1974 issue of *Pensée*, published by the Student Academic Freedom Forum of Lewis and Clark College in Portland, Oregon. This, the seventh issue in the wildly popular "Immanuel Velikovsky Reconsidered" series, chronicles in detail the confrontation between Velikovsky and his critics at the February 1974 meeting of the American Association for the Advancement of Science, held in San Francisco. The glossy quality and high production values of this series, which contrast with the earlier non-Velikovskian *Pensée*, are easily visible.

astrophysics, planetary science, and history. In 1966 he approached the American Association for the Advancement of Science (AAAS) and received no response.[44] Around the same time as the phenomenal success of the first Velikovskian journal, however, certain members of the academic community thought, for the first time since the uproar of 1950, that they should confront Velikovsky directly. There is no question that student interest in *Worlds in Collision* and his other books motivated these Establishment scientists; in the late 1960s Cornell astronomer Carl Sagan and a few others informally suggested that perhaps it was time to refute Velikovsky so that students would not be led astray by one-sided endorsements of cosmic catastrophism.[45] But although Sagan put together a symposium at the AAAS meeting in 1969 on UFOs, there was no movement on the Velikovsky issue.

Then in 1973, Ivan King, the chair of the astronomy section of the

AAAS, and Owen Gingerich, chair of the history of science section, decided to sponsor a symposium on Velikovsky for the 1974 annual meeting, scheduled for February. This event is, next to the 1950 boycott campaign, the single most discussed episode of Velikovsky's career. Everyone who has written about it has been eager to declare one side—the scientists or Velikovsky—the victor.[46] My goal here is different: to focus on the interpretations attached to this event, especially by the Velikovskians. The year 1974 featured not only the AAAS meeting; it was an annus mirabilis of four separate Velikovsky symposia, of which San Francisco was just the kickoff. Putting the controversial first symposium in context also highlights that the AAAS event was not so much an attempt to staunch the countercultural current on which Velikovsky was borne but more a gambit to join it in order to shape the flow.

It didn't work. The symposium took place at the St. Francis Hotel on 25 February 1974 before a tremendous audience (estimates vary between 500 and the room's capacity of 1,500, with the actual number likely closer to the latter) and lasted for seven hours—four hours in the morning and an additional three in the evening. The participants—sociologist Norman W. Storer, astronomer Carl Sagan, physicist J. Derral Mulholland, statistician and amateur Assyriologist Peter Huber, physicist Irving Michelson, and Velikovsky himself—were all supposed to speak within defined time periods and then address some questions from the audience. Sagan and Velikovsky both went significantly over their time allotments (necessitating the evening session), although Velikovsky more so, on the grounds that only he and Michelson were speaking on his behalf and that most of the audience had come to see him anyway (which happened to be true). The anti-Velikovsky papers were later published as a volume by Cornell University Press after negotiations broke down over the length of the rebuttals Velikovsky wished to include. He and Michelson published their presentations in *Pensée*.[47]

After the fact the discussion was reduced to a head-on confrontation between the seventy-nine-year-old catastrophist and America's most popular astronomer, Carl Sagan (fig. 7.2). Those looking for sharp verbal fireworks were mostly disappointed. Velikovsky spoke eloquently and at great length about his concepts, and after making some suggestions for the Viking probe to Mars based on his conclusions, he ended his speech by hurling down a gauntlet: "None of my critics can erase the magnetosphere, nobody can stop the noises of Jupiter, nobody can cool off Venus, and nobody can change a single sentence in my books."[48] Applause ensued, and so did questions, which Velikovsky addressed effusively, his responses to single inquiries at times occupying whole pages

FIGURE 7.2 Carl Sagan (left), in his trademark turtleneck and blazer, and Immanuel Velikovsky (right), in his no-less-characteristic dark suit and tie, on the dais at the AAAS annual meeting at the St. Francis Hotel in San Francisco, February 1974. This image, which nicely depicts the contrasting styles of the generational clash of the 1970s as well as defying expectations (the hipper Sagan is the less countercultural of the two), silently omits the other participants in the symposium, perpetuating the false impression that the meeting was a one-on-one confrontation between the two figures. Reprinted by permission of Dale Cruikshank.

of the transcripts and wandering across the range of human erudition (but not always answering the question).

Sagan's paper was also far too long for a twenty- to thirty-minute presentation, and he skimmed through it, reading segments here and there. Although he attempted a detailed refutation of several of Velikovsky's principal assertions, the rushed presentation and even the published version contained numerous errors, which harmed his arguments among the intended audience. Nonetheless, he did display moments of sparkling frustration, as in response to a question about the radio noises on Jupiter: "There is bound to be some residual magnetism everywhere. There is bound to be, just as in the Earth's oxidizing atmosphere there are today hydrocarbons. Methane is one part per million of the Earth's atmosphere. That has nothing to do with manna. It has nothing to do with any of this."[49] Likewise, Velikovsky, in the midst of a voluminous "short comment" on Norman Storer's presentation, had one of his several rhetorical triumphs: "But neutral is not objective.

You cannot be objective between evil and the victim of evil, neutral between the behavior of science—how it was and how it started from 1950 and continued till today, almost till today, till yesterday, better let us say."[50] But the net result was more confusion than enlightenment. Both the Velikovskians and their critics held unanimous views of who won—they just were different unanimous views. From these accounts, we should note the intense effort both sides expended in spinning the AAAS symposium, confirming that the event was more about public relations and propaganda than coming to a scientific evaluation of Velikovsky's theories.[51]

Norman Storer, whose opening paper at the symposium was decidedly lackluster, found the case more interesting in retrospect. "My private opinion is that the old guy is quite out of his tree, and I am much more negative about him after having seen him in action than I was before the San Francisco meeting," he wrote later. "But the interesting thing is his following—who are they, what structural circumstances might account for their 'faith,' and what sustains them in the face of overwhelming evidence to the contrary? Questions like this suggest a rich lode of sociological material for someone who wants to dig into 'em."[52] But there was to be no extensive study of Velikovsky's countercultural following nor any further attempts to confront Velikovsky directly on the scientists' own turf. Dennis Rawlins, a fellow of the Royal Astronomical Society, deftly noted the Catch-22: "If one simply ignores the crank, this is 'close-mindedness' or 'arrogance.' If one then instead agrees to meet him in debate, this is billed as showing that he is a serious scholar. (For why else would the lordly establishment agree even to discuss him?) Irksome *either* way."[53] So the 1974 experiment was never repeated. It had been neither success nor failure. It had raised the visibility of scientific opposition but had resolved nothing.

The Velikovskians, however, were on a roll. Lewis and Clark College, in Portland, had hosted the first Velikovskian conference on 16–18 August 1972 under the prodding of *Pensée* and its editors. But 1974 was different, studded with panels and discussions of Velikovsky. In May, Velikovsky received an honorary degree from Lethbridge University in Canada and then traveled to McMaster University in Hamilton, Ontario, for an oversubscribed conference on 17–19 June, entitled "Velikovsky and the Recent History of the Solar System." He had some time to rest before trekking to Duquesne University in Pittsburgh for a History Forum on 27 October to 2 November and immediately thereafter shuttling to a 2 November session of the Philosophy of Science Association meeting at Notre Dame, entitled "Velikovsky and the Politics of

Science," where the speakers were three Velikovskians (the man himself, Lynn Rose, and Rose's fellow Buffalo professor Antoinette M. Patterson) juxtaposed with one critical physicist (Michael W. Friedlander of Washington University in St. Louis).[54] Papers from all these venues appeared shortly afterward, indicating that the San Francisco meeting had done little to dampen enthusiasm for the author of *Worlds in Collision* and his theories.

And when those papers appeared in 1974, they usually appeared in *Pensée*, which had become—next to Velikovsky himself—the single most important clearing house for information about his concepts. Not imagining that they could write directly to the man in Princeton, people wrote to the journal, which they assumed had a direct connection with Velikovsky. That connection was in fact rather tenuous, and the tensions surrounding just how much control Velikovsky did or should have over the contents of *Pensée* would soon come to a head. The Talbott brothers were not in Velikovsky's inner circle, and the contributors for *Pensée* included critics of Velikovsky's theories. When Velikovsky attempted to assert control, he shattered the entire venture.

The Guru out of Season

Why was *Pensée* so popular? To judge from the correspondence of contributors and editors (the subscribers are, sadly, impossible to trace fully), the primary reason was that what it published seemed to be *real*: actual scholarship, actual science. The periodical had peer review, boasted a large readership, and featured detailed discussions on many sides of an issue. This couldn't be pseudoscience. True, many of the citations were to other articles in *Pensée*, but that was because Establishment journals had unfairly frozen out these discussions. According to Lewis Greenberg, writing to Velikovsky in 1974: "The credibility of *Pensée* depends upon its ability to remain positively inclined to your work without seeming to be too one-sided or partial."[55]

Not everyone thought the situation was clear-cut. While it was obviously beneficial for the journal not to appear as a propaganda organ, perhaps some skewing was permissible. "I do not fully agree that our effort is scholarship only and not propaganda, if the definition and not connotation of propaganda is used," C. J. Ransom wrote to Stephen Talbott in 1975. "We do not need to propagate discussion of the other side since they have their own propaganda machines. However, a certain amount of mixing would act as a catalyst to encouraging discussion of Velikovsky's works."[56] Talbott, in a slightly earlier letter to Lynn Rose discussing the same issues, disagreed, and his response is worth quoting at some length:

> Considerations centering on the overall balance among journals do not by themselves suggest any particular balance for *Pensee* . . . . [W]e have to deal realistically with the fact that, for example, the nearly universal opinion among "experts" is that available data do not allow for a recent Velikovskian episode on the lunar surface. If our more informed readers see us failing to interact with the weight of conventional opinion on the subject, they can only discount us. After all, there exists every sort of wierd [*sic*] publication and society, surviving merrily on in isolation, while boldly "challenging" accepted viewpoints in their widely unread pronouncements. Nobody bothers with them. The reason *Pensee* has achieved what it has is that it went straight into the scientific community (read: community of conventional thinkers) with its bold challenges, seeking in every way possible to avoid the isolation that normally would befall such an effort. That meant involving conventional antagonists.

The problem, he continued, was partially the perennial editorial conundrum of how to fill an issue: "It is my opinion that, up until now, and even now, we could not put together a presentable series of issues consisting solely of contributions by Velikovskians. The scholarship is simply not there; the result would look anemic. There are too few Velikovskian researchers, and not in enough fields."[57]

Velikovsky had been intimately involved with the editing of the journal for some time. For example, Thomas Ferté, of the Humanities Department at the Oregon College of Education, was incensed with Stephen Talbott as early as 1972 for allowing Velikovsky, he claimed, to edit his submission for *Pensée* for both content and style; he threatened legal action if the piece was published.[58] An article in *Science* on the AAAS meeting attributed the rise of Velikovskianism to *Pensée* and noted that while there was no financial connection between the two, "there is a kind of symbiotic relationship—he is good for circulation and circulation is good for him—and Velikovsky has, on occasion, exerted editorial influence." When interviewed for the piece, Velikovsky told the author that he had at one point given the editors an "ultimatum" when he wished to respond to a critical article in the same issue it appeared, as opposed to waiting for the next issue: "I said if they didn't do so, I would never write for them again."[59] *Pensée* backed down.

Ironically, the very success that prompted Velikovsky and his cohort to exert stronger control over the journal also induced the Talbotts to assert their autonomy. Relations soured among the principals, and Greenberg in January 1975 called Stephen Talbott "an inflexible, arrogant egomaniac who employs his editorial position as a dictator wields political power."[60] Still, in direct letters to the Talbotts, the tone remained civil. Ransom worriedly wrote to Stephen that if he did not continue to grant Velikovsky editorial say over the journal's contents, the catastrophist might walk away.[61] Talbott, in the same long letter to Rose quoted above, stood firm. "The 'proposed break' with Velikovsky, I trust you realize, is not anything I am proposing, but rather he has threatened," he wrote. Yet he insisted that *Pensée* was much more than a mouthpiece for Velikovsky, and that he did not believe "that the consideration of any single man's work is a sufficient base upon which to operate a journal." Furthermore, "it would surely be suicidal for us to commit ourselves *editorially* to the truth of his work. . . . *Pensee* cannot be '100% pro-Velikovsky'—or pro-Velikovsky at all, editorially. The commitment which 'has already been made' is—just as you remark—that 'Velikovsky was worth reconsidering'—neither more nor less. To commit ourselves further would be to remove ourselves from the ranks of truth-seeking journals to those of the axe-grinding journals."[62] It proved suicidal, however, to fail to take Velikovsky's threats seriously. After ten pro-Velikovsky issues, the catastrophist withdrew.[63] *Pensée* lay fallow in 1975, publishing no issues.

New Velikovskian publishing projects popped up. In 1974 there appeared a single issue of *Chiron: Journal of Interdisciplinary Studies*, run out of the Oregon College of Education by the same Thomas Ferté who had stormed out of *Pensée* in 1972. This journal was interested principally in the humanistic aspects of catastrophism and looked forward to compiling a special issue "on theoretical psychology (archetypes, collective amnesia, neo-Freudianism, three-brains hypothesis, etc.)" and also sought papers "on the work of Joseph Campbell, Cyrus Gordon, and Nikos Kazantzakis."[64] Lewis Greenberg attributed its speedy demise "to some rather inexplicable behavior on the part of Ferte [*sic*] who has failed to send out the first issue to many subscribers and chooses not to communicate with anyone since early October from his secluded retreat in Pocatello[,] Idaho."[65] Greenberg and Warner Sizemore seized the initiative and moved the journal east to Glassboro State College in New Jersey—where history professor Robert Hewsen announced the opening of a Center for Velikovskian Studies "as a focal point for the collection and dissemination of information relevant to the work of

Dr. Immanuel Velikovsky."[66] Contrary to his previous statement in favor of *Pensée*'s objectivity, in late 1974 Greenberg assured Velikovsky that the new journal would toe the party line: "I will promise you this right now. The new journal will have its doors solidly barred to any would-be critics. We are not interested in the open forum posture presently assumed by PENSEE. I should like to go on record with you personally on that account."[67] The new journal was *Kronos* and ran from spring 1975 until 1988.

Kronos was a smaller affair than *Pensée* by an order of magnitude: its circulation peaked at 2,400 in its second year, settled to roughly 1,500 by its tenth year. It was physically smaller than *Pensée* too, with matte pages and occasionally an almost mimeographed feel, sparse in images and heavy on footnotes. By 1980, it had subscribers in twenty-four foreign countries, all the while competing with the British "revisionist" journal *S.I.S. Review*. In the first issue, the editorial preface declared: "Thus we present KRONOS, a journal of interdisciplinary synthesis, whose initial contents are dedicated to *Immanuel Velikovsky*—progenitor and inspirational force for the ideas contained herein." But, careful to avoid angering Velikovsky by attributing to him positions he did not hold, they immediately added a footnote: "The views expressed by the authors in this journal are their own and do not necessarily reflect editorial opinion nor that of Dr. Velikovsky."[68] (They also added, later in the issue: "**KRONOS**, an independent journal, is in no way affiliated with **Chiron** or **Pensee** or the Student Academic Freedom Forum.")[69] Velikovsky's unhappiness with the Talbotts did not imply he would be kinder to Sizemore and Greenberg. He received the first issue after returning from a visit to the hospital for some medical treatments (he was eighty-one), and he wrote Greenberg a letter trashing *Kronos* and offering suggestions. Greenberg acceded to all of them.[70]

Back in Oregon, *Pensée* was about to fold, and the wrangling over the financial results of that event—selling mailing lists, possession of rights to the *Velikovsky Reconsidered* edited volume—occupies many pages of angry correspondence in Velikovsky's personal archive. It would be impossible to tease apart accusations and counteraccusations in a reasonable space, even if we had complete documentation from all sides (which we do not). Suffice it to say that in the end, *Pensée* shut up shop for good. Harvard astronomer Donald Menzel penned a gloating letter to Stephen expressing his delight in the news that *Pensée* was no more, since the "magazine is, in my opinion, detrimental to the best interests of science."[71] A bewildered Velikovsky supporter wrote Menzel to see whether this was a mistake or a prank. No such luck, Menzel retorted:

"Pensée, from its inception, has been primarily devoted to the glorification of Velikovsky, one of the greatest Cranks of modern times. . . . It is the magazine, Pensée, which is irresponsible."[72] And he would dance on its grave while he could.

Meanwhile, *Kronos* forged ahead, accepting and publishing articles on questions of planetary atmospheres, moon craters, and the exact dating of various Babylonian conflicts. Sizemore was pleased: "We now have an organ—a powerful organ—that will allow no distortion of your work to go unanswered."[73] But only if people read it. There was no getting around the issue of Velikovsky's control over the journal and the tension between allowing people to debate his views—letting them live in the conflict of scholarship—and having them remain faithful, even if the project died in the process. The very insurgent quality that had drawn the counterculture to Velikovsky made it want to push farther, question his conclusions, edge closer to finding the *real* truth, and he was no more going to allow that for *Kronos* than he had for *Pensée*.

In 1977 a man named Jerry Rosenthal inherited a bit of money and, having come to admire Velikovsky's theories and his tenacity in defending them, proposed donating some of those funds toward an expansion of Velikovsky's audience. The problem, as Rosenthal saw it, was not too little scientific research but rather a biased older generation. With better public relations, Velikovskians could win over the young—the scientists and decision makers of the next generation. "Your primary emphasis, as I talk with you," he wrote to Velikovsky, "is toward research and print media publication. This reaches only a small fraction of the public today as movies, TV, radio, even lectures influence many more people. Even young scientists today, because your books and theory are virtually blacklisted, cannot easily be introduced to you." But with the right medium, Velikovsky's reach could be extended dramatically. "Young people have a strong desire to know the facts. They are misled and feel empty with the pseudo-answers of the establishment," Rosenthal continued. "Some spring off into religious cults or into escapist philosophy; some remain in the establishment knowingly frustrated. There is also a large group of people interested in space, science fiction, and the sciences that would be potential customers of a Velikovsky media event. Young people must be made aware before they get a vested interest in the existing system."[74]

Rosenthal proposed bankrolling a documentary or a TV series. This was not a new idea: the Canadian Broadcasting Corporation had produced Henry Zemel's *Velikovsky: Bonds of the Past* in 1972, and the British Broadcasting Corporation released *Horizon: Worlds in Colli-*

sion that same year. Velikovsky suggested that Rosenthal give $4,000 to *Kronos* instead. The latter was unhappy. "I specifically stated that my continued support was dependent upon the expansion of the readership and the necessary broadening of the base for whom *Kronos* could be a useful journal," he complained. "There is a mass market for these ideas; Close Encounters, Star Wars, etc. all prove the mass appeal that popularly packaged Velikovsky would have. Talbott had 35,000 copies printed—over ten times what you might reach, and Sagan reaches millions!" If Velikovsky were not more cooperative, then Rosenthal would no longer give money to ventures like the special *Kronos* issue on the AAAS meeting, which he considered "an ego-boosting rejoinder to Sagan."[75] Velikovsky might be "a great scientist and researcher, but is a failure at leading a revolutionary movement," he wrote to Greenberg. He suggested that the "movement must be simplified, digested and regurgitated for the masses. They will then do the work you are trying to do—apply pressure on the scientists, universities, Congress. Then the power, money, success, research grants, books, television shows, expeditions and fame will be given to you."[76] Velikovsky, Greenberg, and the *Kronos* set would not cooperate, and so Rosenthal walked away, bewildered at the man who had captivated hundreds of thousands with his writings but now seemed willing to converse only with a faithful band of a few dozen. We have reached the end of the season.

Conclusion

Today, Velikovsky's name has fallen into obscurity, recalled only dimly by those who associate the name with late-night dorm room conversations in the glow of the lava lamp (and perhaps the bong). *Worlds in Collision*, that product of Freudianism and Zionism from the late 1930s, fanned into the public eye by the gusts of anti-Communism and anti-Lysenkoism in the early 1950s, bloomed into one of the totemic books of the early 1970s. Why? What made Velikovsky's cosmic catastrophism so beloved of segments of the counterculture—so "groovy"?

Although *Worlds in Collision* bears traces of its original contexts, to those ignorant of the history—and, in the period in question, just about all Velikovsky's young readers can be considered such—the book seemed very much of the moment. Consider two other major best sellers of the 1970s that have lived on in public consciousness a lot longer than Velikovsky's has, and the affinities are obvious.

The first, often juxtaposed at the time with *Worlds in Collision*, as much because of the author's alliterative foreign name as the obvious similarities in argument, was Erich von Däniken's megapopular *Chariots of the Gods?*, published in German in 1968 and soon translated into dozens of languages. Von Däniken claimed in this work that the wonders of ancient civilization were the products of "gods"—visitors from other worlds—who came to Earth in antiquity, interbred with the almost-simian humanoids, bequeathed civilization to their progeny, and introduced technological advances (pyramids, cities on mountaintops). Any reader can pick up on the similarities between von Däniken and Velikovsky. The former also read the Bible and mythology in his quest for ancient astronauts: "The almost uniform texts can stem only from facts, *i.e.*, from prehistoric events. They related what was actually there to see." Von Däniken even saw the two approaches as compatible: after the Venus catastrophe, perhaps the extent of planetary destruction witnessed from space prompted the Martians to visit Earth.[77] Von Däniken's theories skyrocketed in popularity, often among the same set of college enthusiasts who devoured the collected oeuvre of Immanuel Velikovsky. Velikovsky and the Velikovskians insisted on putting significant distance between their own claims and those of the godly charioteers. They maintained that Velikovsky was more careful with evidence, while von Däniken admitted to massaging facts to suit his narrative. Even critics, like science fiction writer Ben Bova, agreed that "[t]here is no question of fraud, or of winking at known facts, in Velikovsky's case."[78]

Today, von Däniken is substantially more recognizable than Velikovsky. This is in part because he is still publishing—his most recent book appeared in 2014[79]—although with markedly reduced sales, but probably more because his arguments and scenarios for ancient history have continued to be appropriated in science fiction blockbuster movies (*Indiana Jones and the Kingdom of the Crystal Skull, Alien vs. Predator, Transformers: Revenge of the Fallen,* and *Prometheus,* to name four recent ones). The continuing life of alien-astronaut theories should not obscure their very particular emergence in the countercultural soup of enthusiasm for the Space Age, the trippiness of astronomy, the quest for spirituality in ancient texts, and the desire for a universal explanation for everything.

The second interesting text for comparison is the top best-selling work of the 1970s: Hal Lindsey's *The Late Great Planet Earth.*[80] Lindsey represents another strand of the counterculture often considered antithetical to science: evangelical Christianity. While the manifestly countercultural Jesus Freaks of the 1970s are starting to receive scholarly attention, the much more studied right-wing evangelical revival is often considered a breed apart, largely on political grounds.[81] But Lindsey read his Bible (both Old and New Testaments) in a manner that resonated with Velikovsky—and surely did with his readers as well. For Velikovsky, the catastrophes depicted in the historical parts of the Bible were real, just couched in metaphorical language; for Lindsey, the prophetic ecstasies of the Bible were no less real, no less metaphorically presented. Gog and Magog, the Warsaw Pact, Red China, the Four Horsemen—Revelation was nothing more than plain speaking once the reader approached the text properly. And the book and its notions took wing, nourishing an apocalyptic dispensationalism in American spirituality that remains alive today.[82]

The unseasonable grooviness of Immanuel Velikovsky was not so unseasonable after all, despite the decided ungrooviness of the strikingly tall senior citizen with his dark suits, heavy Russian-Israeli accent, and dreams of scientific respectability. He remained more or less the same figure he had been as a struggling émigré in New York City during World War II. He did not turn himself into a countercultural guru; the youth did that for him, and Velikovsky both relished and chafed against the honor. They placed him on that pedestal because he represented a kind of science they felt spoke to them, and they propagated that science using a toolkit widespread among various strands of the counterculture— new journals, new courses, new conferences—repurposed to fit their age as much as von Däniken's astral optimism and Lindsey's eschatological pessimism were. It wasn't the adoration from certain segments of the counterculture that eventually spooked Velikovsky so that he retreated to his home at 78 Hartley Avenue in Princeton, New Jersey, until his death on 17 November 1979—it was their desire to take his new science and make it their own. That, he felt, was decidedly not in season.

Acknowledgments

Portions of this essay are adapted from chapter 6 of Michael D. Gordin, *The Pseudoscience Wars: Immanuel Velikovsky and the Birth of the Modern Fringe* (Chicago: University of Chicago Press, 2012). My thanks to David Kaiser and Patrick McCray for suggestions toward the revision in the present volume.

Notes

1. Murray Gell-Mann, "How Scientists Can Really Help," *Physics Today* 24 (May 1971): 23. On the obvious and rapid expansion of interest in the oc-

cult in this period, see Richard G. Kyle, *The New Age Movement in American Culture* (Lanham, MD: University Press of America, 1995), esp. 47; Steven Dutch, "Four Decades of Fringe Literature," *Skeptical Inquirer* 10 (1986): 342–51; Edward A. Tiryakian, "Toward the Sociology of Esoteric Culture," *American Journal of Sociology* 78 (1972): 491–512; and Andrew W. Greeley, "Superstition, Ecstasy and Tribal Consciousness," *Social Research* 37 (Summer 1970): 203–11.

2. Immanuel Velikovsky, *Worlds in Collision* (New York: Macmillan, 1950). A huge amount of literature was produced by and about Velikovsky during his lifetime. For surveys of the Velikovsky debates with differing degrees of partisanship, see Henry H. Bauer, *Beyond Velikovsky: The History of a Public Controversy* (Urbana: University of Illinois Press, 1984); Robert E. McAulay, "Substantive and Ideological Aspects of Science: An Analysis of the Velikovsky Controversy" (MA thesis, University of New Mexico, Albuquerque, 1975); Alfred de Grazia, ed., *The Velikovsky Affair: The Warfare of Science and Scientism* (New Hyde Park, NY: University Books, [1966]); and Duane Leroy Vorhees, "The 'Jewish Science' of Immanuel Velikovsky: Culture and Biography as Ideational Determinants" (PhD diss., Bowling Green State University, 1990).

3. A detailed account of these events can be found in Michael D. Gordin, *The Pseudoscience Wars: Immanuel Velikovsky and the Birth of the Modern Fringe* (Chicago: University of Chicago Press, 2012), chap. 1.

4. Sigmund Freud, *Moses and Monotheism*, trans. Katherine Jones (New York: Vintage Books, 1967); originally published in German in 1938 and original Jones translation published in 1939.

5. Gordin, *Pseudoscience Wars*, chap. 2.

6. Ibid., chap. 3.

7. Chester R. Longwell, "The 1950 Silly Season," *Science* 113, no. 2937 (13 April 1951): 418.

8. Donald Goldsmith, introduction to *Scientists Confront Velikovsky*, ed. Donald Goldsmith (New York: Norton, 1977), 20.

9. Harold Urey to Albert W. Burgstahler, 21 June 1967, Immanuel Velikovsky Papers, C0968, Firestone Library Special Collections, Princeton University (hereafter cited as IVP), 126:8.

10. Theodore Roszak, *The Making of a Counter Culture: Reflections on the Technocratic Society and Its Youthful Opposition* (Garden City, NY: Anchor / Doubleday, 1969), 42.

11. Robert J. Good, "It Can't Be Our Fault, Can It?," *Chemical and Engineering News* 51 (20 August 1973): 3. Historian of science Owen Gingerich agreed, maintaining that Velikovsky represented "a resurgence of interest associated with the disenchantment with science and the science establishment. It's part of a pattern, and has a great appeal to people looking for a literal explanation of miracles." Quoted in Robert Cooke, "Theory on Collision of Planets Sets Up a Few Earthly Ripples, by Jupiter . . . ," *Boston Globe*, 26 February 1974, 2.

12. Alvin M. Weinberg, "Science and Trans-science," *Minerva* 10 (1972): 222. On the new sense of vulnerability of the scientific establishment beginning

in the mid-1960s, see Kelly Moore, *Disrupting Science: Social Movements, American Scientists, and the Politics of the Military, 1945–1975* (Princeton, NJ: Princeton University Press, 2008), 37.

13. Peggy Constantine, "Peter Fonda Not Really a Hippie," *Los Angeles Times*, 19 September 1967, D13.

14. Velikovsky to Bruce Mainwaring and John Holbrook, 20 September 1969, IVP 88:6.

15. Immanuel Velikovsky and Lynn E. Rose, "The Sins of the Sons: A Critique of Velikovsky's A.A.A.S. Critics," [late 1970s], IVP 53:7, p. 3. Similar statements appear on pp. 181–82 and 185 of this manuscript, which is a detailed criticism of both the statements made against Velikovsky at the 1974 AAAS meeting and the ensuing publications by those critics in Goldsmith, *Scientists Confront Velikovsky*, and other venues.

16. *Cosmos and Chronos*, no. 4 (December 1967), IVP 126:3.

17. Velikovsky to Sune Hjorth, 7 September 1967, IVP 81:5.

18. Press release from Parsons School of Design, 14 April 1970, IVP 66:3.

19. See the documents about the honorary degree reproduced in E. R. Milton, ed., *Recollections of a Fallen Sky: Velikovsky and Cultural Amnesia; Papers Presented at the University of Lethbridge, May 9 and 10, 1974* (Lethbridge, AB: Unileth Press, 1978).

20. Daniel Cohen, *Myths of the Space Age* (1965; New York: Dodd, Mead, 1967), 191; and Bauer, *Beyond Velikovsky*, 207–8. On counterculture beyond the New Left, see Peter Braunstein and Michael William Doyle, eds., *Imagine Nation: The American Counterculture in the 1960s and 70s* (New York: Routledge, 2002); Michael William Doyle, "Debating the Counterculture: Ecstasy and Anxiety over the Hip Alternative," in *The Columbia Guide to America in the 1960s*, ed. David Farber and Beth Bailey (New York: Columbia University Press, 2001), 143–56; Van Gosse, *Rethinking the New Left: An Interpretative History* (New York: Palgrave Macmillan, 2005); and Christopher Gair, *The American Counterculture* (Edinburgh: Edinburgh University Press, 2007), 8.

21. Patrick Doran, "Living with Velikovsky: Catastrophism as World View," in Milton, *Recollections of a Fallen Sky*, 143.

22. Bruce J. Schulman, *The Seventies: The Great Shift in American Culture, Society, and Politics* (New York: Da Capo Press, 2001), 24.

23. This sociological splitting of Velikovsky's audience draws from Robert McAulay, "Velikovsky and the Infrastructure of Science: The Metaphysics of a Close Encounter," *Theory and Society* 6 (1978): 332. As a view from industry, consider Velikovsky's close supporter Frederic Jueneman, who wrote: "Personally, I don't have a quarrel or an ax to grind with the academic community *per se*; still, as a member of the industrial community I have a growing impatience with those members of the academic hierarchy who have taken it upon themselves to tell me what I should or shouldn't believe." Frederic Jueneman, "The Search for Truth," *Analog* 94, no. 2 (October 1974): 30. On humanists acting out resentment of the scientists, see George Grinnell (Department of History, McMaster University) to Velikovsky, 14 April 1972, IVP 80:6.

24. Chris Sherrerd to Velikovsky, 29 July 1968, IVP 97:3.

25. Charles H. McNamara, "The Persecution and Character Assassination

of Immanuel Velikovsky as Performed by the Inmates of the Scientific Establishment," *Philadelphia*, April 1968, 64.

26. W. C. Straka to editors of *Pensée*, 22 June 1972, IVP 128:13.

27. Guenter Koehler to Velikovsky and Elisheva Velikovsky, 20 July 1979, IVP 85:23.

28. R. H. Good Jr. to Dr. Hilton Hinderliter, 3 January 1974, IVP 81:4. Hinderliter responded angrily and at great length on 13 February. Ironically, Good had been dismissed from Berkeley in the early 1950s because his own work in abstract quantum field theory was not considered useful for graduate students and was thus pedagogically unacceptable. See David Kaiser, *American Physics and the Cold War Bubble* (Chicago: University of Chicago Press, forthcoming), chap. 4.

29. C. J. Ransom to Velikovsky, 4 October 1971, IVP 92:2.

30. C. J. Ransom to Velikovsky, 7 October 1971, IVP 92:2. Ransom later used the course to produce a popularized introduction to Velikovsky's entire thesis, published as C. J. Ransom, *The Age of Velikovsky* (Glassboro, NJ: Kronos Press, 1976).

31. Lynn E. Rose to Velikovsky, 14 November 1971 and 16 March 1973, IVP 93:1.

32. Jerry Rosenthal to Velikovsky, 18 February 1977, IVP 94:2.

33. Ben Bova to Lynn E. Rose, 20 November 1974, IVP 93:1.

34. Ben Bova, "'With Friends Like These . . . ,'" *Analog* 90, no. 5 (January 1973): 6.

35. Bova to Frederic Jueneman, 7 March 1975, IVP 82:17.

36. On the spectrum of Velikovskian journals, see Bauer, *Beyond Velikovsky*, 68–69.

37. John W. Crowley, "A Scientific Approach to Velikovsky," *Yale Scientific Magazine* 41, no. 7 (April 1967): 5.

38. Immanuel Velikovsky, "Venus—a Youthful Planet," *Yale Scientific Magazine* 41, no. 7 (April 1967): 8–11, 32.

39. Albert W. Burgstahler and Ernest E. Angino, "Venus—Young or Old?," *Yale Scientific Magazine* 41, no. 7 (April 1967): 18–19; Lloyd Motz, "Velikovsky—a Rebuttal," in ibid., 12–13; Immanuel Velikovsky, "A Rejoinder to Burgstahler and Angino," in ibid., 20–25, 32; Immanuel Velikovsky and Ralph E. Juergens, "A Rejoinder to Motz," in ibid., 14–16, 30; and Horace Kallen, "A Letter to the Editor," in ibid., 30. Burgstahler, almost certainly because of his prominent academic credentials, was granted a much greater degree of tolerance in criticizing Velikovsky's framework. Although Burgstahler disagreed with some of Velikovsky's positions, these were only at the level of detail; he endorsed the general picture. Furthermore, being an advocate of other fringe theories—such as his opposition to fluoridation of water or the conventional ascription of Shakespeare authorship—Burgstahler always levied his criticisms respectfully.

40. Press release for Velikovsky Symposium, Lewis and Clark College, [1972], IVP 65:2.

41. Stephen Talbott, "The Population Crisis Is a Put-On," *Pensée* 1, no. 4 (June 1971): 14–15, 30. For more on the debates over population and their

political valence, especially among science and technology enthusiasts on college campuses, see W. Patrick McCray, *The Visioneers: How an Elite Group of Scientists Pursued Space Colonies, Nanotechnologies, and a Limitless Future* (Princeton, NJ: Princeton University Press, 2012).

42. Philip M. Boffey, "'Worlds in Collision' Runs into Phalanx of Critics," *Chronicle of Higher Education* 8, no. 22 (4 March 1974): 7. On the enormous heterogeneity of publishing ventures spawned by countercultural enthusiasts in the 1970s, see Sam Binkley, *Getting Loose: Lifestyle Consumption in the 1970s* (Durham, NC: Duke University Press, 2007).

43. See Lynn Rose to David Morrison, 26 November 1973, IVP 93:1; and Ralph E. Juergens to Stephen L. Talbott, 30 December 1974, IVP 93:1. On the "scientization" of Velikovskianism in the *Pensée* period, see Judith Fox, "Immanuel Velikovsky and the Scientific Method," *Synthesis* 5 (1980): 49.

44. Albert Burgstahler was the intermediary. See Burgstahler to Velikovsky, 6 December 1966, IVP 71:10.

45. Carl Sagan, *Carl Sagan's Cosmic Connection: An Extraterrestrial Perspective* (1973; Cambridge: Cambridge University Press, 2000), 59; and Albert Shadowitz and Peter Walsh, *The Dark Side of Knowledge: Exploring the Occult* (Reading, MA: Addison-Wesley, 1976), 241.

46. For pro-Establishment reportage on the AAAS meeting, see Robert Gillette, "Velikovsky: AAAS Forum for a Mild Collision," *Science* 183 (15 March 1974): 1059–62; Boffey, "'Worlds in Collision' Runs into Phalanx of Critics"; Miranda Robertson, "Velikovsky in the Open," *Nature* 248 (15 March 1974): 190; George Alexander, "Controversial Author, Scientists in Collision," *Los Angeles Times*, 26 February 1974, A4; Walter Sullivan, "Writer Collides with Scientists," *New York Times*, 26 February 1974, 9; David F. Salisbury, "Velikovsky Cosmology: Theories in Collision," *Christian Science Monitor*, 12 March 1974, 14; and Graham Chedd, "Velikovsky in Chaos," *New Scientist*, 7 March 1974, 624–25. For pro-Velikovskian accounts, see n. 51 below.

47. The relevant sources are Goldsmith, *Scientists Confront Velikovsky*; Immanuel Velikovsky, "My Challenge to Conventional Views in Science," *Pensée* 4, no. 2 (Spring 1974): 10–14; Irving Michelson, "Mechanics Bears Witness," *Pensée* 4, no. 2 (Spring 1974): 15–21; and the verbatim transcripts of the discussions around each paper reproduced in Lynn E. Rose, ed., "Transcripts of the Morning and Evening Sessions of the A.A.A.S. Symposium on 'Velikovsky's Challenge to Science' Held on February 25, 1974," in *Stephen J. Gould and Immanuel Velikovsky: Essays in the Continuing Velikovsky Affair*, ed. Dale Ann Pearlman (Forest Hills, NY: Ivy Press, 1996), 727–95. Sagan's essay, long to begin with and expanded after the meeting, became the focal point of discussions later and was published as Carl Sagan, "An Analysis of *Worlds in Collision*," in Goldsmith, *Scientists Confront Velikovsky*, 41–104. A revised and corrected version is printed in Carl Sagan, *Broca's Brain: Reflections on the Romance of Science* (1974; New York: Presidio Press, 1979), chap. 7. One latter-day Velikovskian devoted an entire monograph to refuting Sagan's piece: Charles Ginenthal, *Carl Sagan and Immanuel Velikovsky*, 2nd ed. (Tempe, AZ: New Falcon Publications, 1995). The Goldsmith volume also includes an essay

by David Morrison ("Planetary Astronomy and Velikovsky's Catastrophism," 145–76) that was not part of the original symposium.

48. Velikovsky, "My Challenge to Conventional Views in Science," 14.

49. Sagan in Rose, "Transcripts of the Morning and Evening Sessions of the A.A.A.S. Symposium," 757.

50. Velikovsky in ibid., 772.

51. Velikovsky and Lynn Rose wrote a book-length manuscript dissecting the meeting and defending Velikovsky against his critics, especially Sagan; see their "The Sins of the Sons: A Critique of Velikovsky's A.A.A.S. Critics," [late 1970s], IVP 53:7. According to them (e.g., p. 2), the pro-Sagan press did not reflect actual events but was part of a conspiracy to taint Velikovsky. For additional negative reviews by the Velikovskians, see "Velikovsky's Challenge to Science," *Pensée* 4, no. 2 (Spring 1974): 23–44; George Grinnell, "Trying to Find the Truth about the Controversial Theories of Velikovsky," *Science Forum* 38 (April 1974): 3–5; Frederic B. Jueneman, "A Kick in the AAAS," *Industrial Research*, August 1976, 9; Charles Ginenthal, "The AAAS Symposium on Velikovsky," in Pearlman, *Stephen J. Gould and Immanuel Velikovsky*, 51–138; Lynn E. Rose, "The A.A.A.S. Affair: From Twenty Years After," in Pearlman, *Stephen J. Gould and Immanuel Velikovsky*, 139–85; George W. Early, "Velikovsky Confronts His Critics," *Fate* 32 (February 1979): 81–88; and Shane Mage, *Velikovsky and His Critics* (Grand Haven, MI: Cornelius Press, 1978).

52. Norman Storer to Sidney M. Wilhelm, 18 April 1974, IVP 93:3.

53. Dennis Rawlins, "Sagan and Velikovsky," *Science News* 105, no. 17 (27 April 1974): 267 (emphasis in original).

54. See the program for McMaster University in IVP 65:3 and the descriptions of the various conferences in Frederic B. Jueneman, *Velikovsky: A Personal View* (1975; Glassboro, NJ: Kronos Press, 1980), 42.

55. Lewis Greenberg to Velikovsky, 8 June 1974, IVP 79:14.

56. C. J. Ransom to Steve Talbott, 26 January 1975, IVP 92:4.

57. S. Talbott to Lynn E. Rose, 30 December 1974, IVP 93:1.

58. Thomas L. Ferté to S. Talbott, 10 March 1972, IVP 77:4; and S. Talbott to Ferté, 15 March 1972, IVP 77:4. For the legal proceedings, see letter from lawyer Scott McArthur to S. Talbott, 17 March 1972, IVP 77:4.

59. Quoted in Gillette, "Velikovsky," 1060.

60. Lewis Greenberg to Velikovsky, 29 January 1975, IVP 79:14.

61. C. J. Ransom to S. Talbott, 26 January 1975, IVP 92:4.

62. S. Talbott to Lynn Rose, 30 December 1974, IVP 93:1 (emphasis in original).

63. Velikovsky to Eddie Schorr, 19 May 1975, IVP 96:1.

64. Table of contents of *Chiron: Journal of Interdisciplinary Studies* 1 (Winter–Spring 1974). Winter 1974 also saw the appearance of volume 1, number 1, of *Chiron: The Velikovsky Newsletter*, which offered Velikovskians news of recent events. This folded equally quickly.

65. Lewis Greenberg to Velikovsky, 29 December 1974, IVP 79:14.

66. Robert Hewsen, preface to Alice Miller, *Index to the Works of Im-*

manuel Velikovsky, vol. 1 (Glassboro, NJ: Center for Velikovskian and Interdisciplinary Studies, 1977), i. See also "A Focal Point," *S.I.S. Review* 1, no. 3 (1976): 17–18; and Jueneman, *Velikovsky,* 52–53. Here is how Hewsen characterized the venture to Velikovsky: "I would say, then, that my role in the Center, of which I am the Director, will be largely organizational. It is our goal to make your work better known, to give it publicity, and to stimulate its discussion. . . . To accomplish this aim the Center, as proper with an academic institution, cannot and should not take a *public* stand for or against your views" (Robert Hewsen to Velikovsky, 17 April 1975, IVP 81:2). Kronos Press, based in Glassboro, also published several books, such as an un-Velikovskian initial publication, H. C. Dudley, *The Morality of Nuclear Planning?* (Glassboro, NJ: Kronos Press, 1976); most of their publications, however, referred to Velikovsky prominently, such as Miller's *Index.*

67. Lewis Greenberg to Velikovsky, 29 December 1974, IVP 79:14.

68. *Kronos* 1, no. 1 (Spring 1975): inside cover.

69. Ibid., 64n.

70. Velikovsky to Lewis Greenberg, 5 June [1975], and Greenberg to Velikovsky, 9 June [1975], IVP 79:14.

71. Donald H. Menzel to Stephen L. Talbott, 13 May 1975, IVP 124:2.

72. Donald H. Menzel to Frederic Jueneman, 23 July 1975, IVP 124:2. This was in response to Jueneman to Menzel, 16 June 1975, IVP 124:2.

73. Warner Sizemore to Velikovsky, 5 October 1975, IVP 97:10.

74. Jerry Rosenthal to Velikovsky, 28 June 1977, IVP 94:2.

75. This was Lewis Greenberg, ed., "Velikovsky and Establishment Science," special issue, *Kronos* 3, no. 2 (November 1977).

76. Jerry Rosenthal to Lewis Greenberg, 10 March 1978, IVP 94:2.

77. Erich von Däniken, *Chariots of the Gods? Unsolved Mysteries of the Past,* trans. Michael Heron (1968; New York: Berkley Books, 1999), 75 (quotation), 158. On sales figures, see Kenneth L. Feder, "Cult Archaeology and Creationism: A Coordinated Research Project," in *Cult Archaeology and Creationism: Understanding Pseudoscientific Beliefs about the Past,* exp. ed., ed. Francis B. Harrold and Raymond A. Eve (Iowa City: University of Iowa Press, 1995), 34–48. For a critique of von Däniken's theories, see William H. Stiebing Jr., "The Nature and Dangers of Cult Archaeology," in Harrold and Eve, *Cult Archaeology and Creationism,* 1–10.

78. Ben Bova, "The Whole Truth: Editorial," *Analog* 94, no. 2 (October 1974): 8. See also the vehement dissociation of the two in Ransom, *Age of Velikovsky,* 238. Interestingly, the most systematic investigation of Velikovsky's sources concluded that von Däniken was *more* careful with his evidence than Velikovsky: Bob Forrest, *Velikovsky's Sources,* vols. 1–6 (Manchester: printed by author, 1981–83), 1:5.

79. Erich von Däniken, *Remnants of the Gods: A Visual Tour of Alien Influence in Egypt, Spain, France, Turkey, and Italy* (Pompton Plains, NJ: Career Press, 2014).

80. Hal Lindsey with C. C. Carlson, *The Late Great Planet Earth* (1970; New York: Bantam Books, 1973).

81. On both strands of 1970s evangelical spirituality, see T. M. Luhrmann,

When God Talks Back: Understanding the American Evangelical Relationship with God (New York: Alfred A. Knopf, 2012).

82. Daniel Wojick, "Embracing Doomsday: Faith, Fatalism, and Apocalyptic Beliefs in the Nuclear Age," *Western Folklore* 55, no. 4 (Autumn 1996): 297–330; and Andrew J. Weigert, "Christian Eschatological Identities and the Nuclear Context," *Journal for the Scientific Study of Religion* 27, no. 2 (June 1988): 175–91.

8

Timothy Leary's Transhumanist SMI²LE

W. Patrick McCray

Timothy Leary—former Harvard professor, LSD advocate, proto–New Age guru, and prison escapee—took one last trip in April 1997. Engineers had bolted a canister with two dozen small aluminum containers inside the third stage of a fifty-five-foot rocket. One of the lipstick-sized capsules was packed with seven grams of Leary's ashes. "They look like little cocaine vials," one of Leary's friends watching the launch said, "which is kind of hysterical in Timothy's case."[1] A few minutes after the rocket motor ignited, Leary (or at least some of his ashes) orbited the planet.

Less than a year before his final flight, on 31 May 1996, Leary passed away (or, as he referred to it, "deanimated"). But, before Leary's death from cancer, there was considerable speculation about how he would spend his remaining time. Some journalists reported that he planned to broadcast the "world's first visible, interactive suicide" on his website with a running account of his last days.[2] "Dying," he quipped, "is a team sport." Meanwhile, news reports speculated that Leary would have his head frozen in the hope that future medical technologies might revive him.[3]

A once-promising researcher who abandoned the protocols of mainstream psychology for notoriety, Leary was the century's most visible advocate for psychedelic drug research. After his dismissal from Harvard in 1963, Leary and Richard Alpert (later known as Ram Dass) continued their experiments in Millbrook, New York.[4] Ensconced in a vast mansion on a sprawling estate, Leary imagined that the ever-changing cadre of truth seekers, hucksters, and hangers-on circulating around him were "anthropologists from the twenty-first century inhabiting a time module set somewhere in the dark ages of the 1960s. On this space colony we were attempting to create a new paganism and a new dedication to life as art."[5]

Leary had long professed a belief that chemicals could be tools for cognitive enhancement. In the summer of 1960, Leary began his research with psilocybin, combining observations of human test subjects with self-experimentation. Leary soon branched out to lysergic acid diethylamide (LSD). Following Leary's succinct injunction to "turn on, tune in, drop out," young adults embraced LSD, and President Nixon branded Leary "the most dangerous man in America."[6] Despite the inconvenient fact that possession of LSD became a federal offense in 1968, chemical technologies offered Leary and his followers an opportunity to refashion their own minds. Leary saw no coincidence in the fact that scientists had discovered the knowledge to build nuclear weapons and LSD's psychedelic properties within the same decade. One path led to annihilation; the other opened doors to revelation, transcendence, and self-improvement. "I look around us," he told an audience in 1966, "and I see metal—all living things and all my cells hate metal—and I see the pollution of the air and the poisoning of the rivers and the concrete over the earth, and I have to say 'Baby, it's time to mutate.'"[7]

Self-directed individual change—"mutation"—was also a goal endorsed by the transhumanism movement as it emerged in coastal California in the late 1980s. According to Humanity+ (once known as the World Transhumanist Association), transhumanism is about "understanding and evaluating the opportunities for enhancing the human condition and the human organism opened up by the advancement of technology."[8] Short for "transitional human," the word was suggested decades earlier by Julian Huxley, a British evolutionary biologist whose brother Aldous authored *Brave New World*. Huxley considered what would happen when humanity decided to "transcend itself . . . realizing new possibilities of and for his human nature." By embracing transhumanism and the "zestful but scientific exploration of possibilities," Huxley said humanity would finally "be consciously fulfilling its real destiny."[9] An essential idea among transhumanists was that existing and anticipated technologies might enable individuals, if not entire societies, to acquire new capabilities, augment their physical and mental power, and thereby transcend biological limitations. As one early advocate told a journalist in the early 1990s, "I enjoy being human but I am not content."[10]

Leary's program of using chemical technologies to alter and enhance people's mental capacities would itself mark him as a progenitor of to-day's transhumanist movement. But while he was incarcerated, radical new ideas about space settlements and technologically enabled life extension began to circulate throughout the American news media and popular culture. After California governor Jerry Brown paroled him in April 1976, Leary, now even more on the fringes of respectability, added these new ingredients to his evolving recipe for mutation. Ever adept at coining a catchy phrase, Leary cheerily christened his new plan SMI²LE: "Space Migration, Intelligence Increase, and Life Extension." In books, lectures, radio shows, and even comic books, Leary and a few close associates promoted SMI²LE to a small community of devotees (fig. 8.1). The well-publicized placement of Leary's ashes into orbit symbolized Leary's long-standing interest in space, immortality, and (to be fair) publicity.

While Leary served his prison time, Americans' love affair with technology reached a nadir of sorts. Loudly and sometimes violently,

FIGURE 8.1 Timothy Leary speaking to a group of students in New York about SMI²LE in October 1976. Photograph © Bettmann/CORBIS.

many (but not all) Americans challenged the trust society had placed in science and technology. Some Americans, for example, were concerned about the escalating arms race. Others worried about an increasingly polluted environment or questioned the values of a society that prized conformity, consumerism, and planned obsolescence.[11] Fears about the future—especially, the technological future—were not just the province of concerned citizens, campus intellectuals, and ecologists. In the early 1970s the Christian fundamentalist revival in the United States coincided with apocalyptic excitation about the future. The most visible author, at least on best-seller lists, was former tugboat captain Hal Lindsey, who prophesied doomsday in *The Late Great Planet Earth* (1970) and other books. Lindsey predicted that new technologies, including the "worldwide computer banking system" and "in-home computers," would allow the Antichrist to control the future.[12] By the time Richard Nixon resigned the presidency, many American citizens had some doubts about technology, anxieties which that era's intellectuals, such as Jacques Ellul, Lewis Mumford, Herbert Marcuse, and Theodore Roszak, spoke of in their books and articles.

Despite simplistic characterizations, that broadly defined intersection of social movements and demographics called the "counterculture" displayed a conflicted and complex relationship with science and technology.[13] This was the cohort, after all, that advocated throwing their bodies on the gears and levers of The Machine. But, mixing wariness with enthusiasm, some of these same people eschewed the wholesale condemnation of technology à la Mumford, Marcuse, and Ellul.[14] They joined the Homebrew Computer Club and integrated new technologies like strobes, synthesizers, and holograms into art and music.[15] Fringe scientific topics such as UFOs, the paranormal, and searches for Bigfoot reflected an ongoing interest in science, albeit of a flavor vastly different from what appeared in the pages of *Science*. Moreover, people like *Whole Earth Catalog* publisher Stewart Brand embraced ecology and cybernetics while promoting certain forms of technology as tools to build new communities and foster social experimentation.

Leary's SMI²LE reflected this optimistic view toward science and technology, especially that which was small in scale and existed outside the margins of mainstream research. The ideas that Leary presented with SMI²LE suggest that some Americans were willing to consider radical technologies in a positive light. As one former 1960s activist turned grassroots space booster predicted, "I'm going to become a billionaire. A lot of us are."[16] By tracking the development of Leary's SMI²LE, we see how he tapped into two different streams of 1970s "groovy science."

On the one hand, Leary borrowed eagerly from a good deal of serious research and speculation done by credentialed people on topics such as space habitation and biomedicine. At the same time, the ideas that Leary bundled together into SMI²LE suggested possibilities for personal improvement and expression, reflecting the larger zeitgeist with its ideals grounded in self-help movements, "*est*" workshops, and New Age seminars.

Initially, enthusiasts for the concepts that SMI²LE included, such as space settlements and life extension, were relatively isolated from one another. During the 1980s, however, advocates from these different areas of interest began to overlap, cross-pollinate, and cooperate while adding yet more new technologies to their futuristic mix. By the dot.com boom of the 1990s, some of these technological enthusiasts had coalesced around transhumanism. While the vision of individual improvement was not new, the paths leading to the goal—genetic modification, mind-enhancing drugs, cybernetic implants, nanotechnology, and artificial intelligence—present an amalgam of late twentieth-century technoscience.

Technological enthusiasts like Leary—especially Leary—were not immune to the lures of profit, celebrity, and sensationalism. These modern utopians imagined the technologies they advocated having a radical, deterministic, and transformative effect on society. Leary and the transhumanists he helped inspire imagined that the future would differ sharply from the past as humans mastered the ability to create new settlements away from Earth and overcome their own biological and cognitive limits.

A High Frontier

Leary's ashes were not the only ones blasted into orbit in April 1997. The same rocket also carried the cremated remains of Princeton physicist Gerard O'Neill. There is considerable irony to the fact that their ashes orbited Earth together until finally burning up (again) five years later over Australia. Although only seven years apart in age, the professional and personal differences between O'Neill and Leary were conspicuous, to say the least.

After his stint in the military, the straitlaced O'Neill divided his career time between Princeton and Palo Alto, where he built and designed equipment for experiments at the Stanford Linear Accelerator Center. His most notable accomplishment was building devices in which particle beams could be "stacked" to make them more dense and then kept circulating for long periods of time, allowing more collisions between them to take place.[17] Although trained as a physicist, O'Neill enjoyed and promoted big engineering projects. In interviews, O'Neill spoke in a near-monotone voice that invited reporters to compare him to *Star Trek*'s Mr. Spock. In dress and manner, O'Neill embodied rationality.

Leary was Leary—outrageous, extravagant, and effusive. Even before his incarceration, Leary performed his celebrity on the margins of American society. A photograph taken soon after his release from prison shows him looking ecstatic, his famous smile flashing back at the camera, while cutting it up at New York's Studio 54.

But in the 1970s, both O'Neill and Leary advocated human expansion into space. That they chose space is not surprising. The success of the Apollo program suggested that the human exploration of space would eventually extend far beyond the moon. For much of the twentieth century, space exploration was the archetypal technological frontier, the tabula rasa on which generations of engineers and schoolchildren projected their wildest dreams. When he became frustrated with his physics career, O'Neill himself tried to join NASA's Scientist-as-Astronaut program. Even for people who appreciated the simpler technologies advocated by E. F. Schumacher in his 1973 book *Small Is Beautiful*, space exploration still held a strong fascination.

For decades, of course, fiction writers and scientists harbored dreams of space-based settlements that might provide the basis for new, perhaps utopian, societies.[18] Believing that the "days of blind trust in science and in progress were past," O'Neill wanted to focus on problems relevant to the environment and "the amelioration of the human condition."[19] Predictions that there would eventually be restrictions on economic and population growth as well as personal liberties—ideas professed most famously in the 1972 report *The Limits to Growth*—motivated O'Neill. O'Neill imagined space settlements as a safety valve for human expansion, a place for social experimentation, and the next frontier for business and manufacturing.

However, O'Neill differed from earlier futurists and visionaries who offered only descriptions and vague speculations about their space-based utopias. O'Neill drew on his extensive engineering skills and experience with large-scale projects to produce detailed mathematical calculations and designs for self-contained worlds. Presented as microcosms of larger Earth-bound systems, O'Neill conceptualized them using informed extrapolations of existing technologies. What matured over time were designs for self-enclosed ecological systems that would

FIGURE 8.2 Interior view, as depicted by artist Rick Guidice in 1975, of an O'Neill-style space colony. Photograph courtesy of NASA Ames Research Center.

be maintained through established engineering and cybernetic feedback principles (fig. 8.2).[20]

When compared with the cost of the era's other megaprojects, O'Neill's estimate, based on NASA's own optimistic cost projections, that the first space colony could be built for $30 billion or less didn't seem wildly impossible. Experts predicted that the space shuttle system and the trans-Alaska pipeline would cost about the same, while Nixon-era plans for national energy self-sufficiency were far more lavish, running as high as $500 billion.[21] After spending the national purse to explore the moon, O'Neill claimed it was "now time to cash in" on what scientists and engineers had learned from Apollo.[22]

During the heyday of the Apollo moon landings, O'Neill presented his designs and ideas to campus groups and university colleagues. Seeking to expand his audience and knowledge base, O'Neill hosted a small conference on space colonization at Princeton in May 1974. Seed funding came from the Bay Area–based Point Foundation, a small nonprofit organization started by Stewart Brand, a former "Merry Prankster" and publisher of the *Whole Earth Catalog*. The Point Foundation's philanthropy blended politics and an enthusiasm for alternative, or "soft," technologies with "western libertarian sensibility" and an interest in experimental communities.[23] The *New York Times* put O'Neill's meeting on its front page, and media coverage of him blossomed in mainstream newspapers and magazines.[24] The attention only increased after *Physics Today* published an article by O'Neill in which he laid out technical details for building space settlements. Congress invited O'Neill to testify about his ideas for the future of the US space program. Meanwhile, regular profiles in mainstream publications and appearances on talk shows with Merv Griffin and Johnny Carson popularized the space colony "meme" further and made O'Neill a minor celebrity.[25] When his book *The High Frontier* appeared in 1977, it won Phi Beta Kappa's award for the best science book and fostered more discussion about how people might live and work in space in the future.[26]

What O'Neill called the "humanization of space" was an amalgam of his engineering-based conceptualizations coupled with decades-old ideas about technology, society, and their coproduction of modern America.[27] But what really sparked public and media reaction to O'Neill's ideas was his evocative depiction of space not as a government-run *program* but as a *place* to be settled by ordinary people, not astronaut-demigods. This critical shift in perspective—seeing space as the future home for humanity—also became a key element in Timothy Leary's SMI²LE.

H.O.M.E.s for Sale?

Whereas O'Neill's interest was in supplying detailed explanations for how a space colony could be built, Leary was intrigued about what the humanization of space would mean for humanity's continued advancement up the astrologically tinged path of development Leary called the "evolutionary sequence."[28] As Leary, now a self-proclaimed "hope dealer," told the *Los Angeles Times* after his parole, "only in space can we take the next steps in our evolution."[29]

"Intelligence increase"—which Leary defined as the "purposeful use of psychoactive drugs for reprogramming the brain"—was something he had studied and promoted for years.[30] This work, he said, had sparked the "inner consciousness movement" of the 1960s and marked the start of humanity's expansion into space and, eventually, toward self-directed evolution. As he put it, "You cannot create anything externally which you have not experienced internally."[31] Some progress toward intelligence increase, Leary said, had already been made. The "consciousness movement of the sixties is now the consciousness industry of the seventies," he noted in an interview. "People understand that they can grow and evolve personally."[32] But this evolution needed new ideas and new technologies if it were to continue.

In his autobiography (one should caution that the former professor's recollections, steeped as they are in performance and provocation, often must not be taken purely at face value) Leary recalled that the winter of 1975–76 meant "twelve hours a day reading and writing. Lots of science" while still confined to prison.[33] Through *CoEvolution Quarterly*, a new magazine published by counterculture icon Stewart Brand, Leary learned about O'Neill's "great space colony revelation," and he concluded that "the next step in human evolution was up."[34] Space migration, especially O'Neill's vision for the humanization of space, therefore became a major new ingredient Leary adopted as he conceived SMI²LE. Whether he truly believed, as he claimed, in the inevitability of space migration or was just riding the coattails of O'Neill's publicity is hard to tell. In any case, Leary called O'Neill the "most important human being alive today" and, channeling C. P. Snow, saw space colonies as a bridge across "the gap between the hardware/scientific and the humanistic/artistic."[35]

O'Neill's ideas encouraged speculation among anthropologists and sociologists as to what space-based communities might mean for future social, sexual, and economic relations. Leary's own musings continued in this philosophical and futurist-oriented vein. Like O'Neill, Leary preferred that space be settled with "no reliance on big government or big industry." But Leary expressed little interest in the technical details of how this might be done.[36] Instead his talks and writings presented abstract considerations of what would happen when people could escape the "nursery planet and find our adult evolutionary status" as spacefaring creatures.[37]

Exactly when and why Leary began to formulate SMI²LE is unclear. Jack Sarfatti, a physicist who started the Physics/Consciousness Research Group in Berkeley, claimed that Leary's inspiration came from an unexpected source: General Douglas MacArthur. Leary, Sarfatti said, was really MacArthur's "lovechild." The general-to-be often danced with Leary's mother when Leary was in utero, and Leary's father was an army dentist who supposedly had MacArthur as a patient.[38] And MacArthur, Sarfatti pointed out, made some astonishing prototranshumanist predictions near the end of his life. Tomorrow's cadets, the retired general told West Pointers in 1962, would experience humanity's "staggering evolution" as people harnessed "cosmic energy" and created "disease preventatives to expand life into the hundreds of years" and "space ships to the moon." All this was preparation for, MacArthur mused, some final apocalyptic conflict with "the sinister forces of some other planetary galaxy."[39] By Sarfatti's tortuous reasoning, Leary's parents knew MacArthur, Leary had (briefly) attended West Point, and MacArthur spoke about some prototranshumanist ideas, thus sparking Leary's imagination—QED.

Whether or not Dugout Doug inspired SMI²LE, Leary was certainly thinking about space migration while he languished in California's penal system. The 1973 arrival of comet Kohoutek, which some hippie cultists saw as a harbinger of doom, inspired Leary to write a short tract, *Starseed*, which he "transmitted" from the "black hole" of Folsom Prison. In *Starseed*, Leary claimed that he was preparing "a complete systematic philosophy: cosmology, politic, epistemology, ethic, aesthetic, ontology, and the most hopeful eschatology ever specified." Like astrologers of the Middle Ages, Leary claimed that Kohoutek (the "starseed" in Leary's title) meant a "higher Intelligence has already established itself on earth, writ its testament within our cells, decipherable by our nervous system. That it's about time to mutate. Create and transmit the new philosophy. . . . Starseed will turn-on the new network."[40]

Unfortunately, Kohoutek—dubbed by science writers as the "comet of the century"—failed to be as bright and prominent in the night sky as scientists had predicted. As the comet and hype over its appearance faded in early 1974, Leary exchanged letters with planetary astronomer Carl Sagan, who had seen a copy of *Starseed*. The two batted Leary's conjectures about space colonies, mutation, and self-directed evolution—what Leary called the "transgalactic gardening club"—back and forth.[41] Sagan even offered to visit Leary on his way to a conference to challenge Immanuel Velikovsky's curious theories of cosmic catastrophe (see chapter 7 in this volume). So, at the very same time, we have Sagan planning to debunk Velikovsky's interpretation of planetary origins while trading some rather far-out ideas about human evolution and space exploration with Leary. Such seeming contradictions just highlight how groovy science was such a fertile meeting ground of mainstream and alternative ideas.

Sagan was not the only person intrigued by *Starseed*'s evolutionary ideas. Soon after its publication, Bay Area writer Robert Anton Wilson helped organize a small group called The Network. Known for his *Illuminatus* books, which mixed occult themes, conspiracy theories, and wildly discontinuous narratives, Wilson became an aficionado of SMI²LE, and The Network's logo featured the words "neurologic-immortality-star flight" around a stylized yin-yang symbol. Wilson later coauthored several articles and essays with Leary, and he incorporated

"Leary's Exo-Psychology" into an "8 Circuit Model of Consciousness."[42] Wilson had a special reason to be drawn to Leary's techno-evolutionary optimism. After the murder of his teenage daughter, the bereaved writer had her brain cryogenically preserved—frozen—in the hopes that one day she could be cloned.[43]

With help from The Network, Leary started promoting SMI²LE. A postcard sent to the offices of *CoEvolution Quarterly* advertised a private workshop, led by Leary, on space colonization, life extension, and quantum physics.[44] Before he set out on a tour of college campuses around the country that year, Leary held a "Starseed Seminar" at Berkeley's Institute for the Study of Consciousness. Besides Leary and Wilson, the workshop featured longevity researchers and advocates from the Bay Area Cryonics Society. Jack Sarfatti and a few other underemployed physicists, intrigued by Leary's confluence of mysticism, space travel, and quantum theories, joined the two-day seminar and supplied their own riffs on Leary's radical technological enthusiasm. Leary continued his association with Sarfatti, and together they attended workshops at the Esalen Institute ("a Cape Canaveral of inner space"), nestled amid Big Sur's rugged beauty.[45]

Leary provided more details about SMI²LE in two books, both published in 1977: *Neuropolitics* and *Exo-psychology*.[46] Mixing philosophical musings with fiction and satire, *Neuropolitics* provided an unruly articulation of Leary's ever-evolving message. Although Leary's book is silent when it comes to eugenics, he depicted SMI²LE as a way for humans to improve themselves via "neurogenetic evolution pre-programmed by DNA."[47] Advances in genetic engineering (Bay Area molecular biologists had recently succeeded in splicing together DNA drawn from different species) were proof that humans would some-day soon consciously control their own evolution toward a "quantum model of consciousness." Scientific and technological advances were "making space migration a practical alternative to our polluted and overcrowded planet," said the introduction to *Neuropolitics*, "at a time when NASA is making Star Trek's *Enterprise* science faction" (fig. 8.3).

Central to all the evolutionary progress Leary imagined was the humanization of space à la O'Neill. He imagined that a citizen's space program could provide all sorts of opportunities so long as it wasn't an "insectoid bureaucracy, or another Alaskan pipeline controlled by the oil politicians."[48] Expansion into space would create space for more people, something that would be a necessity if future technologies gave people longer life-spans. Meanwhile, space itself would become a new Petri dish for social and biological experimentation. "O'Neill's space-

FIGURE 8.3 The very groovy cover of Timothy Leary's *Neurocomics*, a 1979 comic adapted from Leary's 1977 book *Neuropolitics*. The artwork is by Pete Von Sholly, and the image is used with permission of Last Gasp Books.

cities, housing not monastic astronauts but families, tribes, human communities," would begin the "greatest evolutionary mutation since the ascent from the ocean to land."[49] Leary's continued focus on mutation via technology presaged later transhumanist thinking which held that accelerating technological change would converge to create the next stage in human evolution.

Readers of *Neuropolitics* also learned, among other things, why "dopers seem to prefer sci-fi" and how space colonies would permit an almost endless variety of sexual experimentation.[50] At the same time, the book satirized Leary's legal problems and lampooned Watergate-era politics, with Richard Nixon, G. Gordon Liddy, and antifeminist writer (and future cheerleader for cyberspace) George Gilder all held up for ridicule. Replete with drug references and social observations wrapped in scientific-sounding jargon, *Neuropolitics* concluded with a fictional stock prospectus for Leary's version of space colonies, which he called "High Orbital Mini Earths" (i.e., H.O.M.E.).

Leary gave a fuller exposition of SMI²LE in *Exo-psychology*, a book he dedicated to "evolutionary agents, on this planet and elsewhere."[51] Leary mutated his own definition of exo-psychology throughout the book. It was a "Science which Studies the Evolution of the Nervous System in its Larval and post-terrestrial Phases," the "psychology of physics" (Psi Phy), as well as a "theory of Interstellar Neurogenetics." Juxtaposing his ideas with "pre-Einsteinian psychology," Leary claimed that astronautics, astrophysics, genetics, and nuclear science were all research areas with "significance for human destiny in the future."[52] The book goes on to describe the "eight circuits" of the human nervous system and the "twenty-four stages of Neural Evolution," which Leary likened to the periodic table. Resembling better-known books, such as Fritjof Capra's *The Tao of Physics*, *Exo-Psychology* cited the quantum musings of physicists like Jack Sarfatti and John A. Wheeler while criticizing the cynicism of Werner Erhard's *est* and "puritanical protestant-ethic manipulators" like B. F. Skinner.[53] Overall, *Exo-Psychology* blended a freewheeling pastiche of ideas from quantum physics and genetics with Vedic, Islamic, and Zen philosophies.

Neuropolitics and *Exo-Psychology* were clear signs that Leary had strayed far from O'Neill's comparatively straightforward ideas, which were grounded in optimistic yet measured extrapolations of 1970s technology. It's difficult to determine exactly how people responded to Leary's two books. Contemporary responses were relatively rare and memories today are hazy. But some indication can be seen via copies that circulated within the University of California's library system. These are heavily marked and annotated with phrases like "how to see the world," "TOWER OF BABEL," and "technological mysticism" scattered throughout.

From his own home, tucked away in one of Los Angeles County's steep, shaded canyons, "where the migrants and the mutants, and the future people come from, the end point of terrestrial migration," Leary spread his SMI²LE.[54] Leary's professed enthusiasm for "high orbital living" was proof that the "humanization of space" meme had migrated far from its origins in Ivy League classrooms and NASA workshops. And as the idea of space migration circulated between university campuses and sci-fi conventions and the coffee shops and hot tubs of coastal California, it continued to mutate in ways O'Neill would never have imagined.

Countercultural Consideration of Space Settlements

Opinions about space colonies were colorful, plentiful, and sharply divided. O'Neill's own colleagues in the staid Princeton Department of

Physics raised eyebrows, especially when he gave interviews for un-conventional venues such as *Penthouse*.[55] Meanwhile, Senator William Proxmire, famous for his "Golden Fleece" awards and an opponent of increased NASA funding in general, promised "not one cent for this nutty fantasy." Nonetheless, thousands of university students and other young adults were thrilled by the idea that someday they might live in space.[56]

Their interest helped spark a modest-sized grassroots movement as dozens of space-related interest groups formed across the United States, with a membership that peaked in the thousands.[57] The western United States was home to much of this organizational activity, with about a third of these groups based in California. Membership was especially strong in the Los Angeles and San Francisco areas. Of the many pro-space groups launched in the 1970s, the L5 Society claimed the largest membership and loudest voice. Keith and Carolyn Henson, two O'Neill devotees, started the organization in Tucson, Arizona. The Hensons op-posed the Vietnam War and supported the women's movement. They became friends with Leary, who put the couple in his "genetic hall of fame" in the 1979 book *The Intelligence Agents*. The name the Hensons chose for their pro-space group came from O'Neill's proposal to locate space settlements at one of the Lagrangian libration points. These five locations, named L1, L2, etc., are places in outer space where gravita-tional forces between the earth, moon, and sun are balanced so that objects placed there remain in a relatively stable position.[58]

The L5 Society adopted the slogan "L5 in '95" to express their ulti-mate goal of settling in space. Having accomplished this, they imagined one last meeting at their new L5 home, before disbanding the society. A song, "Home on LaGrange," conveyed the group's enthusiasm for a range of things, technological and otherwise:

> Oh, give me a locus where the gravitons focus
> Where the three-body problem is solved,
> Where the microwaves play down at three degrees K,
> And the cold virus never evolved.
> CHORUS: Home, home on LaGrange,
> Where the space debris always collects,
> We possess, so it seems, two of Man's greatest dreams:
> Solar power and zero-gee sex.[59]

Well into the 1980s, L5 members continued to debate space-related topics, including lunar mining, space colonies, missions to Mars, and the

Strategic Defense Initiative. L5's membership was relatively small, fewer than ten thousand or so.[60] But it was a vocal—at times argumentative—group bolstered by local chapters in the Bay Area, San Diego, and Los Angeles. One member described the community as a "post-graduate camp for space nuts." Members tended to be "young, well-educated, and receptive to new ideas."[61] More than two-thirds were under the age of thirty-five, most were single men, and about 70 percent had college degrees, typically in a technical field.[62]

L5's membership wasn't limited to college students. At John Muir High School in Pasadena, Taylor Dark III and a few friends established their own L5 chapter. The group grew to about thirty members (including Carl Feynman, son of Nobel laureate Richard Feynman) and took advantage of nearby Caltech and the Jet Propulsion Laboratory to find speakers on space-related topics.[63] One John Muir student filmed his own space-themed movie, and the teens convinced their principal to invite George Koopman, a contributor to *Neuropolitics*, to give a pro-space address to the entire school. The teens' mimeographed newsletter noted their plans to see a space shuttle prototype at a local aerospace company and their enthusiasm for solar power. Space hardware, environmental concerns, fund-raising, political activism: all were part of a broader techno-optimistic perspective that appealed to Dark and his friends.[64]

Like many other technological enthusiasts, L5 members found the dire predictions in books such as *The Population Bomb* and *The Limits to Growth* alarming while they disparaged NASA's diluted and directionless manned space program. Moreover, they believed that the humanization of space could produce opportunities for unfettered social experimentation. "There will probably be some creative anachronism people," Carolyn Henson predicted, imagining parallels with Amish communities or medieval reenactors, "who will fabricate their own little pocket of history to live in."[65]

Other motives drove them too. "We realized that if we waited around for the jobs to open up, it would never happen," Carolyn Henson told an interviewer in 1978. "The only way we're going to get jobs in space is to do it ourselves!" Moreover, life might simply get "very BORING if we stick around on this planet too long." Her husband put it more directly: "We want to GO! And become millionaires and come back to earth to visit!"[66] Such comments, one might surmise, reflect the indulgent, elitist, self-centered attitudes of baby boomers—the mainstay of the pro-space movement—that social critic Christopher Lasch critiqued in his best-selling 1979 book *The Culture of Narcissism*.

Initially, the grassroots pro-space community appreciated the sense of countercultural utopian and pan-spiritual ideals that Leary referred to in *Neuropolitics* and *Exo-Psychology*. Futurist Barbara Marx Hubbard, for example, endorsed a rejuvenated American space program throughout the 1970s. An heiress to a toy company fortune, Hubbard was a devotee of Pierre Teilhard de Chardin, the controversial French priest and philosopher who argued that humans and the universe were evolving together toward greater complexity and consciousness. Like Leary, Hubbard imagined space as a place for humanity's "conscious evolution." While witnessing the *Apollo 11* launch, Hubbard felt herself "rising in space . . . , the words 'freedom, freedom, freedom' pounding in my head."[67] The astronaut Russell "Rusty" Schweickart was noted for his "sensitivity to left-of-center issues," and after flying with the Apollo and Skylab programs, he spoke emotionally about the transcendental experience of being in space.[68]

The swell of public interest in space caught the attention of California governor Jerry Brown. Brown first met O'Neill at meetings facilitated by Rusty Schweickart. Aerospace jobs were vital to the Golden State's economy, and a California-NASA partnership to launch a remote-sensing satellite for the state's use intrigued Brown (and helped provide the nickname, Governor Moonbeam, that damaged his political career). In August 1977, as the movie *Star Wars* sold out theaters nationwide, California held its first Space Day at the Museum of Science and Industry in Los Angeles. The event blended O'Neill's techno-utopianism with presentations from major aerospace firms. Originally an advocate of "small is beautiful," Brown spoke as a convert to the humanization of space. "It is a world of limits but through respecting and reverencing the limits, endless possibilities emerge," he said. "As for space colonies, it's not a question of whether—only when and how."[69] Outside the packed lecture hall, Leary pushed his own psychedelic version of O'Neill's vision. "Now there is nowhere left for smart Americans to go but out into high orbit," he vamped to reporters. "I love that phrase—high orbit. We were talking about high orbit long before the space program."[70]

What did O'Neill think of all the citizen-based interest? On the one hand, the Princeton physicist recognized the value of grassroots activism and he encouraged citizens to get involved. On the other hand, O'Neill was cautious about the messianic fervor that some passionate L5'ers expressed. If O'Neill was discomfited by the isolated oddballs who represented the fringe of L5's membership, he was even more wary when Timothy Leary, ever the provocateur, publicly advocated space colonization. Even a few L5 members expressed mixed feelings about having Leary on board, fearing that space migration might appear as "a new way . . . to Turn On and Drop Out."[71] O'Neill eschewed direct associations with Leary but the disregard wasn't returned in kind. Leary dropped the Princeton physicist's name into almost every exposition of SMI²LE, claiming "sexy Gerard O'Neill" was proof that the days of the "retiring, square, fuddy-duddy scientist" were finally over and the next stage of human evolution could begin.[72]

While O'Neill's vision for space settlements was what initially attracted many L5 members, some also expressed interest in Leary's SMI²LE. Space, perhaps, was a way to continue experimenting with counterculture ideas. In San Francisco, the "April Coalition," for example, mixed pro-space advocacy with antinuclear and environmental issues. "There is no time to waste uniting the progressive political groups of the United States to wrest control of the national destiny from the militarists," organizers claimed. "Space is the place for the NEW human race!"[73] Groups such as these were relatively small, with influence that was modest at best.

The same could not be said for Stewart Brand, who, like Leary, became a vocal advocate for the physicist's ideas. Brand, who had once lobbied NASA for a picture of the "whole earth" from space, shifted from "mild interest . . . to obsession."[74] Brand was enthused by the idea that astronauts and cosmonauts blended the "unfashionable aesthetic" of astounding technological prowess with a personal voyage of discovery. "We were wrong," he wrote in his diary, "in perceiving the astronauts as crew cut robots."[75] Eager to initiate discussion of O'Neill's ideas, Brand began publishing stories about space colonization in his new magazine, *CoEvolution Quarterly*.

The idea that a reconciliation of consumerism, environmentalism, and technology was possible proved a dominant theme in *CoEvolution*'s eclectic pages.[76] Concerned that scientists with ideas that were out of step with the mainstream had no forum to express them, Brand envisioned *CoEvolution* as the West Coast's alternative to traditional "establishment" journals like *Science* and *Nature*.[77] He even went so far as to ask MIT's Philip Morrison, *Scientific American*'s book reviewer, for "good stuff you can't use."[78] This broad-minded approach sometimes created odd confluences. One *CoEvolution* reader, for example, proposed "studying and obtaining legal protection for the North American sasquatch."[79]

In 1968 Brand made a bold statement about people and their relationship to technology: "We are as gods and might as well get good at it."[80] Fostering a debate over space colonies was a way to test this premise. If nothing else, Brand thought a conversation about O'Neill's ideas could provide a "fresh angle on old problems" such as population growth, resource depletion, the energy crisis, and perhaps even what sorts of technology were indeed "appropriate."[81] Like O'Neill, Brand believed that spending a fraction of what the United States was preparing to invest in energy independence on other options might be worthwhile, especially if it offered an escape from the dire predictions of the era's doomsday scenarios and counterculture ecopuritans. "The manmade idyll is too man-made, too idyllic or too ecologically unlikely—say the ired," Brand wrote. "It's a general representation of the natural scale of life attainable in a large rotating environment—say the inspired. Either way, it makes people jump."[82]

Brand's admitted obsession certainly made his mail bin fill up. For more than two years, *CoEvolution* published articles and letters attacking or praising O'Neill's ideas. Colorful space-inspired images appeared on two of its covers. Brand himself tried to remain neutral during debate but his diary suggests where his loyalty lay. "Technology, kiddo," he wrote after seeing the first space shuttle. "This is to today what the great sailing ships were to their day." Making a not-so-subtle dig at overzealous adherents to E. F. Schumacher's "small is beautiful" philosophy, Brand said, "Get with the program or stick to your spinning wheel."[83]

Of the scores of letters Brand's office received, about four of every five writers viewed the humanization of space favorably. To some *CoEvolution* readers, living in space seemed a natural extension of a "back-to-earth" lifestyle, which eschewed crowded urban environments for communes and wilderness preservation. Some readers expressed escapist yearnings. "Whatever I can do," said one, "may help my beautiful daughter to slip away from this failing civilization here on Earth."[84]

Space colonies provoked outrage and ignited fears of technocracy among other readers as well as intellectuals like Lewis Mumford ("infantile fantasies"), Wendell Berry ("shallow and gullible"), and John Holt (only for the "starving and desperate").[85] O'Neill's ideas violated the ideals of small-scale appropriate technology many admired. Others detected the scent of massive federal programs and the military-industrial complex ("the same old technological whiz-bang and dreary imperialism").[86] At times, critics aimed their barbs directly at Brand for his supposed betrayal of the counterculture ideals embodied in the pages of the *Whole Earth Catalog*. One *CoEvolution* reader penned this caustic remark:

It seems to be no secret
Where all the flowers have went
Yesterday's counterculture
Is today's establishment.[87]

But Brand wasn't the only counterculture icon to be bashed by some *CoEvolution* readers. In a critique of Leary's SMI²LE, a writer for one 1970s Bay Area 'zine opined: "The I AM NOT A CROOK of the counterculture has escapologised himself into another day-glo corner." Leary's proconsumption, protechnology message meant "NO MORE JOURNEYS TO THE CENTER OF THE HEAD, just EASY LIVING with the elite in the ULTIMATE IN SUBURBAN LIVING, the SWIMMING POOL IN SPACE." All this made Leary a "FRONT MAN FOR NASA and AMERIKAN INDUSTRIAL HARDWARE, SALESMAN for SPACE CARS."[88]

Whether one supported or opposed the humanization of space, 1977–78 marked the idea's apogee. Debates over space-based weapons and other Reagan-era space initiatives eventually divided the L5 Society, while Gerard O'Neill gave up his Princeton professorship to become a high-tech entrepreneur. Space migration, however, remained intrinsic to Leary's prototranshumanist thinking into the early 1980s, a broad agenda into which he integrated other radical new ideas.

Many Are Cold, Few Are Frozen

Leary incorporated another fringy ingredient besides space settlements and drug-enhanced mental capacity into his formulation for SMI²LE. Just as Leary tapped into the enthusiasm that O'Neill's ideas generated, he also noticed a growing interest, especially in California, in ideas for life extension. For some, this meant enhancing longevity via nontechnological techniques such as a lower caloric intake.[89] Others preferred the more direct approach of cryonics—the preservation of one's body or brain at liquid-nitrogen temperatures in the hope that future medical advances might be able to bring about revival. Whatever the path, Leary and other enthusiasts on the technological borderlands saw death, like space, as yet another frontier to be overcome.

Space exploration and life extension, of course, have a long association. A quarter century before Nazi rockets fell on London and Antwerp, American engineer Robert Goddard speculated about spacefaring "generation ships" that might carry people, their life functions suspended, beyond our solar system. "It has long been known," he wrote, "that protoplasm can remain inanimate for great periods of time, and can also

withstand great cold, if in the granular state."⁹⁰ Popular culture reinforced the connection between space and life extension/suspension. For
example, "Space Seed," a 1967 episode of the television show *Star Trek*
freely mixed space travel, genetic enhancement, eugenics, and cryonic
suspension. This fiction reflected ideas afloat among some researchers.

Even before Yuri Gagarin and Alan Shepard left the earth's atmosphere, medical doctors discussed the possibility of lowering a person's
metabolism to facilitate long-term space travel, coining the word "cyborg" in the process. A subsequent NASA study called "Engineering
Man for Space" suggested a range of technologies—drugs, artificial
organs, suspended animation, and psychological modification via sensory deprivation—which all showed up later in transhumanist think-
ing.⁹¹ *Physics Today* even published an article entitled "Physics and Life
Prolongation" by Gerald Feinberg. A respected physicist at Columbia
University whose theoretical research included hypothetical faster-than-
light particles called tachyons, Feinberg noted that "freezing and storing
at low temperatures might lead to many new potentialities for the human race." Fellow physicist Leo Szilard played with a similar idea in his
story "The Mark Gable Foundation."⁹² While these publications generated publicity and bemused interest, mainstream scientists still gave
cryonics a chilly reception.⁹³

Proponents readily acknowledged that cryonics was speculative.
Not surprisingly, however, when reporters learned that a Californian
(not Walt Disney) had been frozen for a "future revival experiment,"
cryonics became front-page news. The man in question was James H.
Bedford, a retired psychology professor from Glendale.⁹⁴ After a doctor
pronounced Bedford dead in January 1967 (his last words reportedly
were "I'm feeling better"), members of the Cryonics Society of California injected him with an anticoagulant and placed him in a metal "cryo-
capsule" filled with liquid nitrogen.⁹⁵

Although medical doctors dismissed the experiment, new organizations sprang up to support the nascent life extension community.⁹⁶
For example, Fred Chamberlain, an employee at NASA's Jet Propulsion
Laboratory, started a company called Manrise and self-published one of
the first manuals for cryopreservation.⁹⁷ In 1972 Chamberlain formed
the Alcor Society for Solid State Hypothermia. By the mid-1970s California hosted a small but active community devoted to life extension.

This was another subculture with which Leary connected as he promoted SMI²LE. And, of course, if one believed that space settlement
was a real possibility, then life extension made even more sense. Leary
explained the logic in a 1976 interview. "Life extension before space mi-

gration was an impossible nightmare," he said. "We'd just clutter up the planet. But, once space migration has begun and we start moving out into space and to the stars, life extension is the technical tool to make the long voyages feasible."[98]

In *Neuropolitics*, Leary included life extension as part of humanity's directed evolution. With so many advances happening in genetics, biology, and other areas of science, longevity and, eventually, immortality "can't be much further away."[99] Leary spoke at Alcor events and wove life extension into his prototranshumanist public talks. Taylor Dark recalled even getting some money from his high school to attend a conference on life extension, where he saw Timothy Leary bounding onstage singing the Bee Gees' hit song "Stayin' Alive."[100]

The technologies that constituted Leary's SMI²LE program started to reach a larger audience in the late 1970s. One pathway for this was *Omni* magazine. Slickly produced, lavishly illustrated, and bankrolled by soft-core porn king Bob Guccione, *Omni* presented the cutting edge of radical scientific research and new technologies to mainstream America with optimistic articles and colorful photo-essays. The topics Guccione and his partner/copublisher Kathy Keeton chose reflected their interest in all kinds of futuristic technologies, including space exploration, genetic engineering, nanotechnology, and human cognition. In addition, *Omni* presented articles about research that straddled the borders of fiction and fact. Extrasensory perception, alien life, human-dolphin communication, cattle mutilations, consciousness studies, and UFOs were all popular subjects.

The possibility of life extension especially piqued Guccione and Keeton's interest. *Omni*'s inaugural issue featured a nonfiction piece called "Some of Us May Never Die," cryonics appeared in numerous short stories, and Keeton published a book called *Longevity: The Science of Staying Young*. Even one of *Omni*'s advertisers got in on the act. Champion International, a traditional wood and paper company, ran a two-page spread that discussed a range of possible life extension options from organ transplants to cryonic preservation. Featuring a pair of "healthy, hearty, 'Methuselahs,'" it was a small counterweight to *Omni*'s pervasive pitches for cigarettes, top-shelf liquors, and sports cars.[101]

Exactly how frozen "cryonauts" might experience their high-tech revival was largely left unexplained by Leary and other life extension advocates. However, a potential answer to this question appeared in the mid-1980s: nanotechnology. The cryonics community was among the first to enthusiastically accept the speculative ideas about molecular engineering put forth by former MIT student, O'Neill acolyte, and early nano advocate K. Eric Drexler. In the future, Drexler suggested, molecular machines might be able to repair the frozen or damaged tissue of people who had "de-animated," an idea Leary repeated in his 1983 autobiography *Flashbacks*.[102]

Spurred by enthusiasm for nanotechnology, Alcor's membership had tripled in size by 1990.[103] The possibility of life extension via nanotechnology and cryonics continued to find an especially favorable reception among the Silicon Valley's high-tech culture. As the *San Jose Mercury News* reported, tech-savvy people will "accept and dive into things before it is widely accepted."[104] Leary proudly proclaimed himself as one of these future "nerds on ice." Cryonics, he said, is "the second dumbest idea in the world. Number one is letting them put you in a coffin and letting the worms eat you."[105] In 1988 he reportedly paid Alcor $35,000 to ensure that his head would be removed, frozen, and preserved after his death. Leary eventually changed his mind about this plan; one story holds that, by the time he died in 1996, he thought even cryonics was becoming too Establishment.[106] Despite the furor of speculation in the weeks before his death, Leary's passing, at home and in the company of friends, was anything but radical. Leary's last words—"Why not?"[107]

Keep on SMI²LEing

In the fall of 1988, a new fanzine began to circulate among California's technohipsters. Put out by two philosophy graduate students at the University of Southern California—Max T. O'Connor and Tom W. Bell, or, as they called themselves, "Max More" and "Tom Morrow"— *Extropy* was pitched as the "Vaccine for Future Shock." O'Connor and Bell defined "extropy" as the optimistic opposite of entropy; it meant increasing "order, usable energy, and information." Their magazine covered the technological topics that "promise to radically transform virtually every aspect of our existence." A more elaborate and politically informed version of Leary's SMI²LE, extropy included "artificial intelligence, . . . intelligence-increase technologies, life extension, cryonics and biostasis, nanotechnology . . . , space colonization, economics and politics (especially libertarian) . . . [and the] intelligent use of psychochemicals."[108]

An even slicker exposition of these ideas, titled *Mondo 2000*, appeared the following year. The irregularly published magazine was the brainchild of Ken Goffman (aka R. U. Sirius), a Bay Area denizen who professed a long fascination with the "neuro-futurisms" of Leary and Robert Anton Wilson.[109] Lavishly illustrated, *Mondo 2000* caught the cyberpunk wave started a few years earlier by authors like William Gibson and Bruce Sterling, a wave Leary happily rode as well. Continuing where *Omni* stopped, *Mondo 2000*'s considerable audience wanted to know about virtual reality, hacker culture, smart drugs, life extension, and nanotechnologies, all seen as part of *Mondo*'s take on "fringe science."[110]

From modest origins in coastal California, Extropians and their brand of transhumanism and techno-optimism mutated and moved to the glossy pages of mainstream magazines like *Wired*. Riffing on ideas that Leary had sketched with SMI²LE, this new cohort of technoenthusiasts "promiscuously combine[d] the free-wheeling spirit of the hippies" with "the entrepreneurial zeal of the yuppies."[111] This came with a disdain for the material world and even the human body itself. As one of the seminal cybermanifestos declared, "The central event of the 20th century is the overthrow of matter. . . . The powers of mind are ascendant over the brute force of things."[112] What some critics called the "Californian ideology" was grounded in a "profound faith in the emancipatory potential" of new cybertechnologies. Having forsaken ecotopia for the electronic agora, the hippies and their younger cyberpunk brethren would create a new techno-utopia where "everybody will be both hip and rich."[113]

Hip and rich certainly appealed to Leary. At the end of his life he became a godfather of sorts for the transhumanists and digerati who read *Mondo 2000* and *Wired*. His 1994 book *Chaos and Cyberculture* collected his various writings on computers and cyberspace and attempted to cash in on popular interest in trendy topics like artificial life, chaos theory, and immersive computing.[114] By this point, Leary had all but dropped his transcendental references to space migration as the next step in human evolution. Cyberspace, not outer or inner space, was the new frontier. "The PC is the LSD of the 1990s," Leary said, while his advice for achieving a longer life was now "To immortalize: digitize."[115]

After his death, the tenets Leary proposed in the 1970s with SMI²LE appeared in some discussions about the so-called Singularity.[116] Associated most notably with Raymond Kurzweil, an MIT-trained engineer, National Medal of Technology recipient, and highly visible futurist, the Singularity refers to a time when "machines become more like humans— programmed with replicated brain synapses that re-create the ability to respond appropriately to human emotion, and humans become more like machines."[117] With Moore's law providing both precedent and a metronomic beat for the future, Kurzweil and his followers posited their own "law of accelerating returns," with the exponential growth of technologies radically reshaping the capabilities of humans.

In the decades between Leary's sometimes playful and abstract ideas and Kurzweil's detailed articulations of accelerating technological change, visions of the technological future changed markedly.[118] Leary's own commitment to SMI²LE is itself somewhat debatable given his not-inconsiderable desire for the spotlight. In his autobiography, he admitted that after his 1976 parole he was "with no home, no job, no credit, and little credibility. . . . It was a great time to start a new career."[119] It would be giving Leary too much credit to say that SMI²LE led directly to transhumanism, let alone today's ideas about the "Singularity." But Leary's focus on personal metamorphosis and enhancement of the individual through technology places SMI²LE in conversation with ideas that transhumanists have long trafficked in. What SMI²LE didn't express, however, is the overt millennialism—the "rapture of the geeks"—that typically accompanies contemporary debates and predictions about technological "singularities."[120]

Was Leary's SMI²LE program an example of 1970s "groovy science"? Can we even call it "scientific"? Leary presented few technical details, provided no blueprints for its realization, and shrouded his ideas in cryptic references to quantum fields and neurological circuits of consciousness. His public persona was invested in a provocateur's persistent desire to jab at the Establishment. In these ways, he differs sharply from "visioneers" like O'Neill who grounded their ideas about the technological future on detailed engineering studies and who published and occasionally presented research in professional scientific venues.[121]

In keeping with the spirit of the "Me Decade," Leary's SMI²LE was much more about improving, changing, and enhancing the individual. Just as dropping LSD wasn't for everyone, Leary's writings divided willing and enlightened "techno-creatures" from members of the "anti-change factions."[122] And given the era's bleak economic times, predictions of environmental collapse, and ever-present threat of nuclear war, perhaps Leary's advocacy of life extension was a logical conclusion to people's anxiety about survival, while space migration offered some hope for self-preservation.[123] "The meek shall inherit the earth," Leary predicted in *Neuropolitics*, "and domesticate it totally in Maoist-insectoid fashion. The bold shall migrate to High Orbital Mini Earths aloft."[124]

However, the elitism such statements implied became one of the criticisms leveled against proponents of space settlements, which were portrayed as a high-tech form of "white flight."[125] Indeed, Leary's antics, despite his (mis)adventures with the Black Panthers, appealed to a rather-narrow college-age demographic. When underground comic artist R. Crumb accepted an invitation from Stewart Brand to attend the first Space Day ("or whatever the hell it was called," Crumb sniped), he came away infuriated and described space advocates as just a "smug bunch of hypocrites."[126] Fifteen years later, critics derided the libertarian-infused ideology of the dot.com cohort as selfish, elitist, and childish.[127] Critics have likewise attacked transhumanism. "The Singularity is not the great vision for society," one observer said. "It is rich people building a lifeboat and getting off the ship."[128]

Leary's ideas tapped into a potpourri of fringe sciences, including *est*, quantum consciousness, space habitation, and other topics that spanned physics, psychology, and the paranormal. Meanwhile, Leary catered to the "emotionally expressive, hedonistic, and firmly this-worldly" Americans who made up what Tom Wolfe called the "Third Great Awakening."[129] Like Woody Allen's Zelig character, Leary appears everywhere promiscuously embracing all sorts of radical technologies, from psychopharmacology and space settlements to life extension, nanotechnology, and cyberspace. Leary and those charmed by his SMI²LE are one bridge between the groovy science of the 1970s and the transhumanist communities and concepts that emerged a decade later. In all these scenarios, the future would differ sharply from the past as humans built and migrated to new worlds, mastered atomic-scale engineering, and overcame their own biological limits. Today, similar techno-optimistic schemes circulate (minus the trappings of Leary's 1970s neuropolitics and exo-psychology) for geoengineering and synthetic biology. For all these techno-optimists, the present serves merely as a prototype of what could become a groovy future.

Notes

1. Marlise Simons, "A Final Turn-On Lifts Timothy Leary Off," *New York Times*, 22 April 1997, A1.

2. Edward Rothstein, "On the Web, Tuning In to Timothy Leary's Last Trip . . . ," *New York Times*, 29 April 1996, D23.

3. "Dr. Leary Joins Up," *Cryonics*, September 1988, 6; and Laura Mansnerus, "At Death's Door, the Message Is Tune In, Turn On, Drop In," *New York Times*, 26 November 1995, E7. A sequence from the 1996 fictional film *Timothy Leary's Dead*, directed by Paul Davids, shows Leary's friends, including Ram Dass, watching as his head is removed and frozen.

4. A recent view of this history is Don Lattin, *The Harvard Psychedelic*

Club: How Timothy Leary, Ram Dass, Huston Smith, and Andrew Weil Killed the Fifties and Ushered in a New Age for America (New York: HarperOne, 2011).

5. Jay Stevens, *Storming Heaven: LSD and the American Dream* (New York: Grove Press, 1998), 208. The history of psychedelic drug use from the 1950s onward, in both its medical and its countercultural context, has been well documented. Besides Stevens's book, see Steven J. Novak, "LSD before Leary: Sidney Cohen's Critique of 1950s Psychedelic Drug Research," *Isis* 88, no. 1 (1997): 87–110; Martin A. Lee and Bruce Shlain, *Acid Dreams: The CIA, LSD, and the Sixties Rebellion* (New York: Grove Press, 1985); and Erika Dyck, *Psychedelic Psychiatry: LSD from Clinic to Campus* (Baltimore: Johns Hopkins University Press, 2008).

6. As quoted in Ron Chepesiuk, *The War on Drugs: An International Encyclopedia* (Santa Barbara, CA: ABC-CLIO, 1999), 118.

7. "New York: Time to Mutate," *Time*, 29 April 1966, 30–31.

8. From the website of Humanity+, accessed June 2010, http://humanityplus.org/.

9. Julian Huxley, *Religion without Revelation* (1927; London: Harper and Brothers, 1967), 195.

10. Ed Regis, "Meet the Extropians," *Wired*, October 1994, accessed June 2010, http://www.wired.com/wired/archive/2.10/extropians.html.

11. Timothy Moy, "The End of Enthusiasm: Science and Technology," in *The Columbia Guide to America in the 1960s*, ed. David Farber and Beth Bailey (New York: Columbia University Press, 2001), 305–11.

12. Hal Lindsey, *There's a New World Coming* (Irving, CA: Harvest House, 1973), 93. See also Paul Boyer, *When Time Shall Be No More: Prophecy Belief in Modern American Culture* (Cambridge, MA: Harvard University Press, 1992).

13. John Markoff, *What the Dormouse Said: How the Sixties Counterculture Shaped the Personal Computer Industry* (New York: Penguin Books, 2005); Fred Turner, *From Counterculture to Cyberculture: Stewart Brand, the Whole Earth Network, and the Rise of Digital Utopianism* (Chicago: University of Chicago Press, 2006); and Andrew Kirk, *Counterculture Green: The "Whole Earth Catalog" and American Environmentalism* (Lawrence: University Press of Kansas, 2007).

14. See Moy, "End of Enthusiasm"; and Timothy Moy, "Culture, Technology, and the Cult of Tech in the 1970s," in *America in the Seventies*, ed. Beth Bailey and David Farber (Lawrence: University Press of Kansas, 2004), 208–27. See also Everett Mendelsohn, "The Politics of Pessimism: Science and Technology circa 1968," in *Technology, Pessimism, and Postmodernism*, ed. Yaron Ezrahi, Everett Mendelsohn, and Howard Segal (Dordrecht: Kluwer Academic, 1994), 151–74.

15. See, e.g., Sean F. Johnson, *Holographic Visions: A History of a New Science* (New York: Oxford University Press, 2006); and Trevor Pinch and Frank Trocco, *Analog Days: The Invention and Impact of the Moog Synthesizer* (Cambridge, MA: Harvard University Press, 2002).

16. Quotation from Keith Henson, a controversial protechnology activist,

in Ned Scharff, "Too Crowded Here? Why Not Fly into Space?," *Washington Star*, 3 November 1977.

17. Elizabeth Paris, "Ringing in the New Physics: The Politics and Technology of Electron Colliders in the United States, 1956–1972" (PhD diss., University of Pittsburgh, 1999); and Gerard K. O'Neill, "Storage Rings," *Science* 141, no. 3583 (1963): 679–86.

18. For an example, see Asif Siddiqi, "Imagining the Cosmos: Utopians, Mystics, and the Popular Culture of Spaceflight in Revolutionary Russia," *Osiris* 23 (2008): 260–88.

19. Gerard K. O'Neill, *The High Frontier: Human Colonies in Space* (New York: William Morrow, 1977), 235.

20. Peder Anker, "The Ecological Colonization of Space," *Environmental History* 10, no. 2 (2005): 239–68.

21. "Transcript of State of the Union Address," *New York Times*, 31 January 1974, 20.

22. From 23 July 1975 testimony O'Neill gave to the Congressional Subcommittee on Space Science and Applications; reprinted in Stewart Brand, ed., *Space Colonies: A Coevolution Book* (San Francisco: POINT, 1977).

23. Kirk, *Counterculture Green*, 164.

24. Walter Sullivan, "Proposal for Human Colonies in Space Is Hailed by Scientists as Feasible Now," *New York Times*, 13 May 1974, A1.

25. Richard Dawkins's book *The Selfish Gene* appeared in 1976. Dawkins's suggestion that "memes" were the basic unit that transmitted cultural ideas became popular among some members of the pro-space movement that O'Neill helped spark.

26. Further discussion of O'Neill's ideas can be found in De Witt Douglas Kilgore, *Astrofuturism: Science, Race, and Visions of Utopia in Space* (Philadelphia: University of Pennsylvania Press, 2003), chap. 5. Michael A. G. Michaud's *Reaching for the High Frontier: The American Pro-space Movement, 1972–1984* (New York: Praeger, 1986) presents excellent information on O'Neill and pro-space in general but without much analysis.

27. The phrase "humanization of space" appears in a number of articles by and about O'Neill throughout the 1970s; e.g., Richard K. Rein, "Maybe We Are Alone," *People*, 12 December 1977, 122–35.

28. Timothy Leary, *Neurocomics* (San Francisco: Last Gasp Eco-funnies, 1979).

29. William Overend, "Timothy Leary: Messenger of Evolution," *Los Angeles Times*, 30 January 1977; and Robin Snelson, "Interview: Timothy Leary," *Future Life*, May 1979, 33–34, 66.

30. Timothy Leary, *Flashbacks: An Autobiography* (Los Angeles: J. P. Tarcher, 1983), 372.

31. Elizabeth Robinson, "Movement into Space: A View from Two Worlds, Pt. 1," *L5 News*, December 1976, 7.

32. Elizabeth Robinson, "Movement into Space: A View from Two Worlds, Pt. 2," *L5 News*, January 1977, 7.

33. Leary, *Flashbacks*, 365.

34. Ibid.

35. Robinson, "Movement into Space: A View from Two Worlds, Pt. 2," 3.

36. Ibid., 10.

37. Ibid., 6.

38. Jack Sarfatti, *Destiny Matrix* (Bloomington, IN: AuthorHouse, 2002), 16.

39. General Douglas MacArthur, Farewell Speech, 12 May 1962, West Point, NY, accessed June 2010, http://www.nationalcenter.org/MacArthurFarewell .html. The similarities to scenarios spun by science fiction writer Robert Heinlein, who published promilitary novels such as *Space Cadet* and *Starship Troopers*, are interesting, as is the fact that Heinlein's older brother Lawrence was a MacArthur aide during the occupation of Japan.

40. Dr. Timothy Leary, *Starseed* (San Francisco: Level Press, 1973), accessed November 2010, http://www.lycaeum.org/books/books/starseed/starseed.shtml. This essay was reprinted in Timothy Leary, with contributions from Robert Anton Wilson and George Koopman, *Neuropolitics: The Sociobiology of Human Metamorphosis* (Culver City, CA: Starseed / Peace Press, 1977).

41. This correspondence can be found at http://www.timothylearyarchives.org/carl-sagans-letters-to-timothy-leary-1974/ (accessed August 2014); copies in the author's possession.

42. Flyer for 16–17 April 1977 talk by Wilson, in author's personal collection. Also described in Wilson's book *Prometheus Rising* (Reno, NV: New Falcon, 1983).

43. Described at http://www.cryonics.org/luna.html (accessed May 2011).

44. Postcard dated August 1975, folder 5, box 3, Whole Earth Access/ Co-Evolution Quarterly Records (M1045), Stanford University Archives (hereafter cited as CQ/SA).

45. Douglas Martin, "George Leonard, Voice of '60s Counterculture Dies at 86," *New York Times*, 18 January 2010, A22. See also David Kaiser, *How the Hippies Saved Physics: Science, Counterculture, and the Quantum Revival* (New York: W. W. Norton, 2011).

46. For *Neuropolitics*, see n. 40 above. Timothy Leary, *Exo-psychology: A Manual on the Use of the Human Nervous System according to the Instructions of the Manufacturers* (Los Angeles: Starseed / Peace Press, 1977).

47. Leary, *Neuropolitics*, 135.

48. Ibid., 49; this essay first appeared in the August 1976 issue of *L5 News*.

49. Leary, *Neuropolitics*, 100.

50. Ibid., 83.

51. Leary, *Exo-psychology*, frontispiece.

52. Ibid., 2.

53. Ibid., 30, 48–49.

54. Overend, "Timothy Leary: Messenger of Evolution."

55. David Gross, who shared the 2004 Nobel Prize in Physics, was in Princeton's Department of Physics with O'Neill in the 1970s. He recalled, "Not being an experimenter myself I was more tolerant of his ideas than many of my colleagues at Princeton, who thought he was too far out and had deserted 'real science'" (e-mail correspondence with the author, 22 December 2009).

56. Further details are given in W. Patrick McCray, "From L5 to X Prize:

California's Alternative Space Movement," in *Blue Sky Metropolis: The Aerospace Century in Southern California*, ed. Peter J. Westwick (Berkeley: University of California Press, 2012), 171–93; as well as in my book *The Visioneers: How a Group of Elite Scientists Pursued Space Colonies, Nanotechnologies, and a Limitless Future* (Princeton, NJ: Princeton University Press, 2013), chap. 3.

57. Trudy Bell, "Space Activism," *Omni*, February 1981, 50–54.

58. Gerard K. O'Neill, "A Lagrangian Community," *Nature* 250 (1974): 636.

59. © 1978 by William S. Higgins and Barry D. Gehm. The song originally appeared in *CoEvolution Quarterly* in 1978.

60. Michaud gives a figure of some 9,500 members in 1984 (*Reaching for the High Frontier*, 96).

61. Hugh Millward, "Where Is the Interest in Space Settlement?," *L5 News*, January 1980, 8–10.

62. Michaud, *Reaching for the High Frontier*, 112–14.

63. *L5 News*, November 1977, 9.

64. Taylor Dark III, interview by the author, 11 March 2010; as well as materials Dark shared with me.

65. Robin Snelson, "Space Now! An Interview with Keith and Carolyn Henson," *Future* 4 (1978): 53–57.

66. Ibid., 57.

67. Barbara Marx Hubbard, *The Hunger of Eve: One Woman's Odyssey toward the Future*, 2nd ed. (1976; Eastbound, WA: Sweet Forever Publishing, 1989), 106, 93 (quotation).

68. Francis French and Colin Burgess, *In the Shadow of the Moon: A Challenging Journey to Tranquility, 1965–1969* (Lincoln: University of Nebraska Press, 2007), 334.

69. Jerry Brown's 11 August 1977 speech, folder 10, box 10, CQ/SA.

70. Roger Rapoport, *California Dreaming: The Political Odyssey of Pat and Jerry Brown* (Berkeley: Nolo Press, 1982), 186–92.

71. *L5 News*, September 1976, 19.

72. Timothy Leary, "Scientist Superstars," *Future Life*, February 1981, 70–73.

73. From 21 and 28 November 1977 letters to Governor Jerry Brown and Stewart Brand, folder 3, box 12, CQ/SA.

74. Brand, *Space Colonies*, 8.

75. 21 January 1976 diary note, from 1975–76 notebooks, Stewart Brand Papers (M1237), Stanford University Archives (hereafter cited as SB/SA).

76. Kirk, *Counterculture Green*.

77. Chapter 3 of Sam Binkley's *Getting Loose: Lifestyle Consumption in the 1970s* (Durham, NC: Duke University Press, 2007) discusses East Coast versus West Coast publishing styles.

78. Stewart Brand to Philip Morrison, 25 January 1975, folder 5, box 2, CQ/SA; and draft letter from Brand, 24 June 1975, folder 9, box 5, CQ/SA.

79. 1978 letter, folder 9, box 12, CQ/SA.

80. Stewart Brand, *Whole Earth Catalog* (New York: Portola Institute, distributed by Random House, 1968).

81. Michael Allen, "'I Just Want to Be a Cosmic Cowboy': Hippies, Cowboy Code, and the Culture of a Counterculture," *Western Historical Quarterly* 36 (Autumn 2005): 275–99; and Stewart Brand, "Free Space," *CoEvolution Quarterly*, Fall 1975, 4 (quotations). Brand had extensive correspondence with biologists, such as Gaia proponent Lynn Margulis, as to whether outer space might be a place to test their ideas.

82. Brand, *Space Colonies*, 3.

83. Stewart Brand, mid-1977 journal entry, SB/SA.

84. Brand, *Space Colonies*, 42.

85. Ibid., 34, 36.

86. *CoEvolution Quarterly*, Spring 1976, 5, 47, 48, 52.

87. Diane Engle to Stewart Brand, 1 December 1977, folder 11, box 1, CQ/SA.

88. Brand, *Space Colonies*, 96.

89. One outspoken proponent of this idea in the 1970s and 1980s was UCLA researcher Roy Walford; Walford later had a chance to put his ideas to the test when he was a member of the Biosphere II crew from 1991 to 1993. A look at earlier research is Hyung Wook Park, "Longevity, Aging, and Caloric Restriction: Clive Maine McCay and the Construction of a Multidisciplinary Research Program," *Historical Studies in the Natural Sciences* 40, no. 1 (2010): 79–124.

90. "The Great Migration," in *The Papers of Robert H. Goddard*, ed. Robert Goddard and G. E. Pendray, vol. 3 (New York: McGraw Hill, 1970), 1611–12. Soviet philosophers of Goddard's era had similar ideas; see Asif A. Siddiqi, "Imagining the Cosmos: Utopians, Mystics, and the Popular Culture of Spaceflight in Revolutionary Russia," *Osiris* 23 (2008): 260–88.

91. Manfred E. Clynes and Nathan S. Kline, "Cyborgs and Space," *Astronautics*, 1960, 26–27, 74–75; "Spaceman Is Seen as Man-Machine," *New York Times*, 22 May 1960, 31; and "Engineering Man for Space: The Cyborg Study," 15 May 1963 Final Report, NASw-512 (Washington, DC: NASA, 1963). For discussion, see Ronald Kline, "Where Are the Cyborgs in Cybernetics?," *Social Studies of Science* 39, no. 3 (2009): 331–62.

92. Gerald Feinberg, "Physics and Life Prolongation," *Physics Today* 19 (1966): 45–48. Szilard's story appeared in his *The Voice of the Dolphin* (New York: Simon and Schuster, 1961).

93. D. E. Goldman, "American Way of Life?," *Science* 145 (1964): 475–76.

94. David Larsen, "Cancer Victim's Body Frozen for Future Revival Experiment," *Los Angeles Times*, 19 January 1967. The story was also covered in the *New York Times*, and the incident was described by the Cryonics Society of California in a self-published book called *We Froze the First Man* (1968).

95. Bedford is still in "suspension" and his body is located at the Alcor Life Extension Foundation in Scottsdale, Arizona.

96. Homer Bigart, "Group Advocates Freezing of the Dead," *New York Times*, 29 January 1967, 68; and Arlene Sheskin, *Cryonics: A Sociology of Death and Bereavement* (New York: Irvington, 1979).

97. Titled "Instructions for the Induction of Solid-State Hypothermia in Humans," it is available (as of August 2014) at http://www.lifepact.com/mm/mrm001.htm.

98. Robinson, "Movement into Space: A View from Two Worlds, Pt. 1," 7.

99. Leary, *Neuropolitics*, 100.

100. Dark, interview by the author, 11 March 2010.

101. *Omni*, October 1979, 40–41.

102. Drexler discussed "the eventual development of the ability to repair freezing damage [which] has consequences for the preservation of biological material today" in "Molecular Engineering: An Approach to the Development of General Capabilities for Molecular Manipulation," *Proceedings of the National Academy of Sciences* 78, no. 9 (1981): 5275–78. For Leary's take, see *Flashbacks*, 372. See also Timothy Leary and R. U. Sirius, *Design for Dying* (New York: HarperCollins, 1998).

103. Membership in Alcor, for instance, tripled between 1987 and 1990. According to a 1990 article, of the twenty-six people frozen in the United States, twenty-four of them were in California; Michael Cieply, "They Freeze Death if Not Taxes," *Los Angeles Times*, 9 September 1990.

104. Tim Larimer, "The Next Ice Age," *West* (magazine supplement to the *San Jose Mercury News*) 9 (December 1990): 17–26.

105. Robb Fulcher, "1960's Guru Turns On to Frozen Head Idea," *Riverside (California) Press-Enterprise*, 16 September 1988.

106. Brian Shock, review of *Design for Dying*, by Timothy Leary and R. U. Sirius, *Cryonics* 18, no. 4 (1997): 42.

107. Laura Mansnerus, "Timothy Leary, Pied Piper of the Psychedelic 60's, Dies at 75," *New York Times*, 1 June 1996, A1.

108. Quotations from *Extropy* 1 (Fall 1988); copy in author's possession.

109. Background on *Mondo 2000* comes from two main sources: a web-based history under construction in early 2011 (http://www.mondo2000history.com/history/) and a 1995 article in *SF Weekly* by Jack Boulware entitled "Mondo 1995: Up and Down with the Next Millennium's First Magazine," http://www.suck.com/daily/95/11/07/mondo1995.html (both accessed December 2010).

110. Rudy Rucker, R. U. Sirius, and Queen Mu, eds., *Mondo 2000: A User's Guide to the New Edge* (New York: HarperPerennial, 1992), 116.

111. Richard Barbrook and Andy Cameron, "The California Ideology," *Mute*, Autumn 1995, 3–8, accessed June 2014, http://www.hrc.wmin.ac.uk/theory-californianideology.html. See also Turner, *From Counterculture to Cyberculture*.

112. Esther Dyson, George Gilder, George Keyworth, and Alvin Toffler, "Cyberspace and the American Dream: A Magna Carta for the Knowledge Age," Release 1.2, 1994, accessed June 2014, http://www.pff.org/issues-pubs/futureinsights/fi1.2magnacarta.html.

113. Barbrook and Cameron, "California Ideology."

114. Timothy Leary, *Chaos and Cyberculture* (Berkeley: Ronin Publishing, 1994).

115. Ibid., frontispiece and p. 202. Also, for links between the cyber-

denizens of the early 1990s and their interest in "smart drugs" (and Leary), see Douglas Rushkoff, *Cyberia: Life in the Trenches of Hyperspace* (New York: HarperCollins, 1994).

116. Initially proposed in Vernor Vinge, "First Word," *Omni*, January 1983, 10. Vinge later acknowledged that this use of the term "Singularity" originated, so far as he knew, in a tribute that Stanislaw Ulam gave for his colleague John von Neumann. See Ray Kurzweil, *The Singularity Is Near: When Humans Transcend Biology* (New York: Viking, 2005).

117. Raymond Kurzweil, "Live Forever," *Psychology Today*, January/February 2000, 66–71, https://www.psychologytoday.com/articles/200001/live-forever.

118. Richard Dooling, *Rapture of the Geeks: When AI Outsmarts IQ* (New York: Harmony, 2008).

119. Leary, *Flashbacks*, 369.

120. The phrase "rapture of the geeks" was used in the introduction of a special issue of *IEEE Spectrum*, from June 2008, devoted to exploring (and debunking) Singularity-oriented ideas. See also John M. Bozeman, "Technological Millenarianism in the United States," in *Millennium, Messiahs, and Mayhem: Contemporary Apocalyptic Movements*, ed. Thomas Robbins and Susan J. Palmer (New York: Routledge, 1997), 139–58.

121. McCray, *Visioneers*.

122. Leary, *Neurocomics*, 9.

123. In chapter 8 of *Something Happened: A Political and Cultural Overview of the Seventies* (New York: Columbia University Press, 2006), Edward D. Berkowitz argues that self-preservation and enriching oneself in a zero-sum society were two hallmarks of the seventies.

124. Leary, *Neuropolitics*, 111.

125. Kilgore, *Astrofuturism*, 150–85.

126. R. Crumb, "Space Day Symposium," *CoEvolution Quarterly*, Fall 1977, 48–51.

127. See, e.g., Paulina Borsook, *Cyberselfish: A Critical Romp through the Terribly Libertarian Culture of High Tech* (New York: Public Affairs, 2000).

128. Andrew Orlowski quoted in Ashlee Vance, "Merely Human? That's So Yesterday," *New York Times*, 11 June 2010, B1.

129. Steven Sutcliffe, *Children of the New Age: A History of Spiritual Practices* (New York: Routledge, 2003), 3 (quotation); and Tom Wolfe, "The 'Me' Decade and the Third Great Awakening," *New York*, 23 August 1976, 26–40. See also Bruce J. Schulman, *The Seventies: The Great Shift in American Culture, Society, and Politics* (New York: Da Capo Press, 2001).

9

Science of the Sexy Beast: Biological Masculinities and the *Playboy* Lifestyle

Erika Lorraine Milam

In January 1968, before a single copy of Desmond Morris's *The Naked Ape: A Zoologist's View of the Human Animal* had been sold in the United States, it made national news.[1] A review in *Book World*, the shared book supplement of the *Chicago Tribune* and the *Washington Post*, used the word "penis" when discussing the anatomy of chimpanzees. The editors of these papers, who had not seen the text of the review until it was already in print, judged this language "in bad taste," and they recalled and reprinted the supplement—a costly decision that ultimately drew more attention to the book than the review itself would likely have merited.[2] Later that month, on Morris's American book tour, timed to coincide with the week of *The Naked Ape*'s release by McGraw-Hill, he mentioned the incident to Johnny Carson while being interviewed on *The Tonight Show*. Carson matter-of-factly replied, "You discuss the fact that man is one of the primates. You talked about his penis. What other word could you use for that?"[3] Despite Carson's calm demeanor, he had just spoken the word "penis" for the first time on (live) American television and (again) raised quite a few eyebrows.

When Morris stopped in Chicago a few days later, a

representative of *Playboy* enterprises, the appropriately named Mr. Rav-age, called and offered—at the behest of Hugh Hefner himself—to en-sure that Morris's stay was "a pleasurable one" and offered him "any-thing you want (pause) *anything at all*."[4] By April, film producer Zev Bufman and director Donald Driver signed a contract for three films with Universal Pictures, including the movie rights to Morris's now best-selling book.[5] Morris began to receive annual gifts imprinted with *Playboy*'s iconic bunny logo. Two years later, Hefner, whose interest in Morris had not waned in the meantime, agreed to back the project to the tune of $1,100,000.[6]

The filmic version of *The Naked Ape* didn't appear until 1973 (fig. 9.1). "Part live action, part animation and all banality," one re-

FIGURE 9.1 Advertising image from the Pressbook for *The Naked Ape* (Universal Pictures and Playboy Productions, 1973), directed by Donald Driver and based on the book by Desmond Morris. The Pressbook described the film as "a story about man's evolution, [that] traces his development from 10 million years ago to tomorrow with the integration of live action and animation." Photograph courtesy of Universal Studios Licensing LLC.

viewer claimed. "The film has its greatest potential among the high-school set."[7] A young Victoria Principal later lamented her involvement, claiming, "that movie almost ruined my career. . . . It was worse than terrible. The popcorn was better than the picture!" She added, "part of the deal was that I pose nude for *Playboy* Magazine. I still regret that."[8] Morris, for his part, mistakenly recalled, "The film was so awful that it never received a general release."[9] After Hefner's financial loss with Roman Polanski's *Macbeth, The Naked Ape* drained his coffers even more.[10] Morris stopped receiving his bunny-imprinted gifts.

The history of *Playboy* and discussions about the science of human behavior were far more entwined than this brief anecdote reveals. As part of the magazine's anti–Main Street renunciation of familial mas-culinity, editors included numerous articles on the sciences of sexual-ity and animal behavior, asserting through repetition that men were naturally composed of equal parts sexual prowess and aggression.[11] Although *Playboy* was neither an underground countercultural produc-tion nor a popular-science magazine, it provides a valuable space for tracing how these threads actively interwove in media that circulated through middle-class society in the era of groovy science.[12]

In the aftermath of World War II, American cultural anthropologists and biologists had sought to promote the idea of a universal human na-ture, by uniting all peoples under a single rubric—the family of man.[13] Within this framework, scientists looked to three primary sources of information to identify social behaviors common to all humans: the evolutionary past of humanity itself (paleoanthropology and archeol-ogy), the characteristics of human cultures ostensibly uncorrupted by access to space-age technologies (cultural anthropology), and the so-cial behavior of humanity's closest living relatives (animal behavior, or ethology).[14] The result was a morass of conflicting morals. Some archeologists, like Raymond Dart, suggested that humans became hu-man the moment our ancestors picked up a broken bone or piece of sharp rock and used it to kill.[15] The first true human was not Adam but Cain. Other archeologists remained skeptical of both Dart's evidence and his champion—playwright Robert Ardrey, whose attention turned to nonfiction in 1961 with his popularization of Dart's theories in the book *African Genesis*.[16] Paleoanthropologist Louis Leakey, on the other hand, preferred to think of humanity's origins in the *manufacture* of tools.[17] Both Leakey and Dart agreed, however, on the importance of cooperative hunting in driving the evolution of social cohesion and the human capacity for language. Hunting behaviors fascinated cultural anthropologists as well, and full-color pictures of "primitive" men with spears or bows and arrows graced many pages of *National Geographic* magazine.[18] Although each culture offered a different lesson to the reading public, several of the so-called primitive tribes seemed far gentler in their nature than modern Americans. The Bushmen of the Kalahari, the Australian Aborigines, and the peaceful Hopi all spoke to a more placid past. At the same time, New Guinea harbored fierce warriors and barebreasted women who were reputed to nurse their prize piglets themselves.[19] Even the early field studies of primates were divided; the peaceful chimpanzees popularized by primatologist Jane Goodall provided a stark contrast to the warring baboons studied by anthropologists such as Irven DeVore and Richard Lee.[20]

Only a small subset of this vast literature on human nature inspired the content editors at *Playboy* magazine. Nestled among sexy cartoons, advertisements for alcohol and inflatable chairs, and the obligatory pictures of naked women, a slow stream of references to the biological underpinnings of masculinity naturalized the *Playboy* lifestyle. The relationship between science and pornography extends at least as far back as the Enlightenment, when natural philosophers used the tropes and titillations of sexually explicit literary conventions to excite and entertain their readers. As historian Mary Terrall has suggested, the advantages for natural philosophers were clear (as they were for Morris); the eroticized language of their publications increased readership and sensationalized the ideas they hoped to convey.[21] Articles by and about scientists reciprocally served a legitimizing function for *Playboy* and helped the magazine establish its aspirational reputation among readers as an upmarket, intellectual publication.

However groovy the magazine may have seemed in the first decade of its publication, however, male virility came packaged with female acquiescence. By the late 1960s *Playboy*'s parallel construction of female sexuality landed it in trouble with some members of the women's movement.[22] Theories of human nature that emphasized male intellectual acumen and sexual prowess tended to ignore female contributions to the evolution of humanity. When the apostles and converts of pop ethology stridently invoked a naturalistic basis for differentiated sex roles in society, their critics believed such theories disavowed the importance of culture in understanding humanity and associated this move with a reactionary political agenda. Sexual mores that had appeared radical in the early 1960s counted as retrograde a decade later, as did the scientific theories that had supported them.

Sexual Revolutionaries?

Throughout the 1960s *Playboy* marketed itself as a key component of the burgeoning sexual revolution. Hefner's magazine not so subtly implied that the "girl next door"—the wholesome object of young American boys' affections—was not only beautiful but also sexually available. By the middle of the decade, millions of women were on the pill, as oral contraceptives became the most popular form of birth control in the country.[23] To many men, it felt as if the rules of courtship and relationships had changed. During this decade, *Playboy* helped to define a new masculine identity that ran counter to stereotypical gender norms of the 1950s. Monthly centerfolds depicted young, at least semi-naked women in soft lighting, posed amid casual evidence of their recent encounter with a man (a tie draped across the back of a chair, his shirt next to her on the couch). Advertisements and articles described new furniture, stereo equipment, even kitchen appliances, that cleverly evoked a masculine domestic space in which beautiful women could be entertained and then dismissed without fear of a lingering feminine presence.

The consumerism inherent to the celebrated lifestyle of the playboy, and the virile masculine identity that accompanied it, were rooted in postwar American affluence and reflected the hope of many Americans that money and perseverance would allow them to transform their role in society by remolding their character and appearance, inside and out.[24] In the decades following World War II, masculine stereotypes were increasingly up for grabs. Movies like *Easy Rider* portrayed male members of the counterculture as young hippies who grew their hair, donned colorful clothes, and became (in the disapproving eyes of more conservative members of society) indistinguishable from their female friends and lovers.[25] At the same time, Clint Eastwood and Richard Roundtree (to name only two popular actors of the period) each embodied a taciturn, muscular masculinity. They took law into their own hands, made women swoon and other men jealous. These stereotypes were equally anti-Establishment in their rejection of earlier social norms, in which masculinity had typically been defined through marriage and family,[26] and both relied on a form of lifestyle consumption.[27] Whereas social anxieties over the sexual identity of male hippies centered on their appropriation of traditionally feminine forms of dress and appearance, and tensions over aggressive men centered instead on their predilection for homosocial company and ephemeral liaisons with women, both enjoyed substantial overlap.[28] Within this matrix, as a magazine for men containing notes on interior decorating and fashion advice (tradition-

ally feminine activities), *Playboy* unquestionably asserted the hetero-
sexual identity of its male readers by hawking intimate encounters with
women.

Although the first issue of *Playboy* was produced on Hefner's kitchen
table in 1953 (and printed without a date because he wasn't sure when
he'd have enough money to produce a second issue), by 1965 he was
thirty-nine years old and already a multimillionaire (fig. 9.2). The previ-

FIGURE 9.2 Hugh Hefner, dressed in his iconic pajamas, draped in naked "bunnies," and contemplating
the works of Veblen, Kafka, Kierkegaard, Freud, and Plato (among others). Of the figures around him, I
recognize only Shel Silverstein, the bearded pool player looking to sink his shot in Hefner's lap and a
mainstay at the Playboy mansion in Chicago. The image accompanied Richard Gehman's "The Private Life
of Hugh Hefner" in the short-lived New York magazine *Fact* (*Fact* 2, no. 4 [1965]: 50–57, on 50).

ous year, the magazine had grossed $21 million. Additionally, Hefner
had opened fifteen Playboy Clubs (grossing a combined $12.6 million)
and had instigated other various bunny-imprinted merchandising projects (grossing $5 million).[29] Hefner successfully branded *Playboy* as a
magazine that redefined urban consumerism as an enterprise fit for men
by coupling leisure and domesticity with sex.[30] The magazine promised
to "crackle . . . with masculinity from bold and bracing fiction and nonfiction to the fairest of femmes," all for $8 a year.[31] *Playboy*'s circulation
had passed that of its inspiration, *Esquire*, by 1958; neared one million
copies by 1960; and reached seven million copies in 1972.[32] According
to their own marketing polls, 43 percent of households subscribing to
the magazine had annual incomes of over $10,000, and 35 percent spent
a "hefty $500 a year on apparel."[33] The idealized reader was a white-
collar man, "sensitive to pleasure," who "can live life to the hilt."[34]

With part of his new fortune, in 1960 Hefner purchased (for
$400,000) and renovated (an additional $350,000 at least) what became known as the Playboy House in Chicago.[35] According to architectural historian Beatriz Preciado, the expensive renovations—including
closed-circuit television cameras in most rooms—transformed the inner
spaces of the mansion into functionally public ones, creating a kind of
"non-domestic interiority."[36] If the seemingly private spaces of the mansion were anything but, the virtual consumption of self-identity that
Hefner created in the pages of his magazine offered readers the possibility of vicariously embodying a public masculine persona without
leaving their homes. A man need not star in critically acclaimed films or
physically go on a safari to Africa to kill a lion, he need only *read* about
such activities to enjoy their masculinizing benefits.[37]

Hefner's logic—that masculinity could be purchased—sat awkwardly
with his embrace of a countercultural sexuality. Because of *Playboy*'s
blatant consumerism, Theodore Roszak, in his quintessential book that
named the counterculture, identified the magazine as the antithesis of the
movement. Whereas *Playboy* depicted unobtainable artificiality, Roszak
argued, the counterculture sought to understand an authentic nature
through visceral and totalizing experiences.[38] He chided actress Vanessa
Redgrave, a cultural and political activist, for contributing to the "glossy
Playboy pornography of films like *Blow-Up*" (released in 1966) and be-
rated Eldridge Cleaver (interviewed by *Playboy* in December 1969), a
leading member of the Black Panthers and author of *Soul on Ice* (1968),
for conceiving of "the struggle for liberation as the province of manly
men who must prove themselves by 'laying their balls on the line.'"[39] Too
often, Roszak continued, these stereotypes suggested "that the female of

the species must content herself with keeping the home fires burning for her battle-scarred champion or joining the struggle as a camp follower." Either way, he contended, "the community is being saved *for* her, not *by* her as well."[40] Reciprocally, Roszak described the poetry of countercultural icon Allen Ginsberg as "an oracular outpouring . . . that reaches back to the rhapsodic prophets of Israel (and beyond them perhaps to the shamanism of the Stone Age)."[41] He also included a review of the musical stylings of the Doors, the dark underbelly of the counterculture, which, he believed, reflected an unhealthy and crude fascination with explicit sexuality: "The Doors are carnivores in a land of musical vegetarians," "their talons, fangs, and folded wings are seldom out of view, but if they leave us crotch raw and exhausted, at least they leave us aware of our own aliveness."[42] In short, Roszak argued that by embracing a nonrational and emotionally authentic reality, Americans would expose *Playboy*'s slick consumerism for what it was: an artificial dream. He further acknowledged that both gritty naturalism and slick consumerism contained at their roots a totalizing masculinity underwritten by popular theories of human nature.

As a magazine and franchise, *Playboy* had self-promoting reasons for embracing and commodifying a countercultural masculinity that emphasized male-male social bonding and a promiscuous outlook on life (in theory, for both sexes).[43] Thanks to its intellectual aspirations in the 1960s and early 1970s, *Playboy* also provides historians with a useful resource with which to track the mobilization of popular scientific theories in defense of these new masculine tropes.[44] By tracing the fate of these ideas through the pages of *Playboy*, we can see how the magazine appropriated and defended a vision of men as sexual hunters and women as mere signs of their success. Although in earlier decades constructions of men as evolutionary animals had resonated with a liberal commitment to seeing all human cultures as equal, by the early 1970s assertions of a biologically determined human nature instead resonated with a conservative defense of the status quo.[45]

The Science of Sex

Playboy drew on a wide array of sciences that described and endorsed a sexually active lifestyle. At first, this meant psychologists and sexologists. For example, in 1962 the magazine convened a panel to discuss the "womanization of America."[46] Eight men—Edward Bernays (public relations guru and nephew of Sigmund Freud), Ernest Dichter (psychologist), Alexander King (media personality), Norman Mailer (author, cofounder of the *Village Voice*, and later winner of two Pulitzer Prizes for Fiction), Herbert Mayes (successful magazine editor of *Good Housekeeping* and then *McCall's*), Ashley Montagu (anthropologist and architect of the first UNESCO Statement on Race), Theodor Reik (prominent psychoanalyst who trained with Freud in Vienna), and Mort Sahl (politically oriented comedian and personal friend of Hefner's[47])— discussed the reasons behind the newfound "dominance" of women in society and the threat women might (or might not) pose to traditional manhood. Of these luminaries, Montagu stood out as the sole anthropologist; everyone else on the panel with medical or scientific training was psychologically inclined. The article concluded by noting the increasing similarity of what it meant to be male and female and postulated a psychological shift in modern, urban youths:

> There is a new spirit on the land. . . . [T]he men are increasingly aware that one can be masculine without being hairy-chested and muscular: the women, that one can be intelligent and sensitive—and witty and wise—and at the same time completely feminine. Perhaps this is a new wave: perhaps it is merely a growing expressiveness—an acting out at last of latent, pent up feeling—in both sexes.[48]

In the following decade, the scientific ideas depicted in *Playboy* emphasized the perils confronting modern American culture but reinforced precisely the "hairy-chested and muscular" masculinity scorned in 1962.

Playboy's appeal to evolutionary arguments began only in the late 1960s. Here, *Playboy* drew support for its masculine ideal from several threads of public science.[49] The first was the animalistic basis of human nature—the innate drives that lurk inside all of us—taken from, for example, the publications of the aforementioned Robert Ardrey and the well-known animal behavior researcher Konrad Lorenz.[50] This line of thinking emphasized the importance of modern man as a hunter and, ultimately, a killer.[51] The second was Desmond Morris's contention that "the naked ape is the sexiest primate alive."[52] More than in any other species, Morris suggested, human social bonding resulted from sexual attraction and interactions. In both cases, *Playboy*, along with much of the popular media, focused on the evolutionary construction of men as sexual hunters and alliances with women as indicators of male social success.

Additionally, *Playboy* included a variety of other articles providing scientific support for humans as inherently sexy creatures. Wardell

Pomeroy, a New York–based sex therapist and coauthor of the Kinsey reports on human sexual behavior, reassured *Playboy*'s readers that all kinds of sexual activity were normal.[53] Polygamy, masturbation, homosexuality, mouth-genital stimulation, and face-to-face intercourse topped his list of behaviors exhibited by other animals. Even sexual relations with inanimate objects were more common than most people assumed. He continued, "the difference between humans and other mammals, therefore, is one of degree and not of kind."[54] Pomeroy found a simple comparison of humans to other animals insufficient to define typical sexual behavior, however, because rape, incest, and sadism also occur in mammalian species. So he adopted a series of criteria by which readers might define normal sexual behavior: statistical (which behaviors are common among humans?), phylogenetic (do animals do it?), moral (should we engage in these behaviors?), legal (what does the current law prohibit?), and social (do such behaviors hurt other people?). Given the wide variety of things people meant by "normal," Pomeroy argued that "normality" itself was a useless metric by which to judge sexual behavior, and he concluded that readers should feel free to engage in all kinds of sexual activities, as long as their actions did not interfere with anyone else's liberty.

Playboy also published an entire book written by editor Nat Lehrman summarizing the major findings of controversial sexologists William Masters and Virginia Johnson, and the magazine even funded its own survey of the sexual and social attitudes of American men.[55] In an article on the infamous "Sex Institute" (properly, the Institute for Sex Research, directed by Alfred Kinsey) located in Indiana's "quiet groves of academe," journalist Ernest Havemann began by recalling the horrifying years of his pre-Kinsey youth, when one of his classmates actually believed he might go crazy because of masturbating and another was forever labeled "queer" because of one sexual encounter in high school. Havemann heaped praise on the institute and its research but claimed it had not gone far enough. Although Kinsey had done a great job of asking "how much" and "how often," he failed to ask "how" or for more descriptive details. Havemann attributed Kinsey's reticence to prudishness about sex and described him as "a stern, grim and totally humorless man," who was so "mathematically-minded" that he "went about the business of tallying human sexual experiences in the same cold and mechanical way he might have counted the number of gall wasps landing on an oak leaf."[56] The vision of Kinsey as a disinterested scientist helped sustain the credibility of his research (surely the only other option was prurience) but sat at odds with the typical masculinity *Playboy* often projected. According to Havemann, Kinsey had begun to drink only when his doctors suggested it would be good for his ailing heart, and even then he preferred "glasses of syrupy liqueurs" sure to "gag his more sophisticated guests." He further reported that Kinsey's favorite form of entertainment consisted of sitting in "stiff-backed chairs" and listening to musicals, preceded by Kinsey's own "formally delivered program notes."[57] One gets the sense that the author had been invited to such an evening event and, expecting a risqué experience, left sorely disappointed. Although Wardell Pomeroy wrote in defense of Kinsey's reputation as far less "square" and "humorless" than Havemann had described, he couldn't erase the impression that any red-blooded reader of *Playboy* would have found Kinsey a terrible bore.[58]

The compassionate discussion of a range of sexual behaviors, including homosexuality, in the magazine's coverage of Kinsey and sexology research more generally reflected Hefner's long-standing commitment to sexual freedom.[59] When it came to coverage of animal behavior, however, naturalized gender roles leapt far more prevalently into the picture.

Man and Beast

Morton Hunt's 1970 article entitled "Man and Beast" discussed the popularity of books by Ardrey, Lorenz, and Morris—all based at least in part on the lessons one could learn about humanity from the study of animal behavior (fig. 9.3).[60] A prolific nonfiction writer, Hunt was especially concerned with the nature of American men and women. As early as 1962, he argued that all humans were a "hopeless tangle of heredity and environment," a position he continued to advocate through the 1970s.[61] Despite the widespread acclaim with which these recent books on human nature were received, Hunt refused to regard Ardrey, Lorenz, and Morris as "scientific prophets" and remained deeply skeptical of their conclusions. Lumping their research together—common by 1970 despite rather-dramatic differences—Hunt characterized these authors as epitomizing man as genetically destined to be "the most brutal and uninhibitedly aggressive of all animals."[62]

By way of contrast, so-called "primitive" or "Stone Age" cultures served as the focus of only one article in *Playboy* during these decades—Lewis Cotlow's "Twilight of the Primitive."[63] In 1961 Cotlow had released a surprising box office success, a documentary about the peoples of New Guinea called *Primitive Paradise*. Movie posters proclaimed that viewers would witness "[a] dangerous expedition into a Stone Age World," which would be "overwhelming, unbelievable . . . but REAL!"

a reasoned criticism of the fashionable contention that ethologists can unerringly understand and predict human behavior by observing that of lower animals

article By MORTON HUNT

FIGURE 9.3 The double-page spread illustrating Morton Hunt's "Man and Beast" essay in *Playboy*, July 1970, 80–81. The lede below the image reads, "A reasoned criticism of the fashionable contention that ethologists can unerringly understand and predict human behavior by observing that of lower animals."

"SEE Courtship rites of maidens in all their native glory." "SEE Dehydrated dead husband and widow!" "SEE Woman feeds suckling piglet!" "SEE Cannibalistic Dani Tribe feasting!"[64] In both *Primitive Paradise* and his contribution to *Playboy*, Cotlow's primary preoccupation was the destruction of other cultures, landscapes, and environments. In the last few paragraphs of his article, Cotlow's lesson for readers became clear: in the process of destroying others, we are destroying ourselves.[65] He never addressed questions of human nature. Given the lack of cultural anthropology and archeology in the content of the magazine, animal behavior research formed the primary basis by which readers would have encountered scientific theories of human nature. But before we turn to how such studies of animal behavior were appropriated by *Playboy*, we need to look at the theories of Ardrey, Lorenz, and Morris in a little more detail.

Robert Ardrey had been a playwriting student with Thornton Wilder at the University of Chicago and won a Guggenheim Fellowship in 1937 on the strength of his dramatic work. When he published *African Genesis* in 1961, he was a successful playwright and Hollywood screenwriter who turned to nonfiction partly because he reasoned that as a writer he was deeply attuned to questions of human nature—perhaps more so than scientists themselves.[66] If *African Genesis* had provided readers with an introduction to Raymond Dart's killer apes, then in *The Territorial Imperative* Ardrey explicitly articulated his belief that territoriality, or the defense of one's home against intruders, lay at the root of human aggression and fell on the shoulders of men.[67] "I can discover no qualitative break between the moral nature of an animal and the moral nature of man. This moral imperative lies more heavily on the human male than does sex itself."[68] Ardrey argued that whereas the capacity to kill defined the origins of modern man, territoriality had driven the evolution of his mental acuity. He drew a parallel between the mating habits of humans and competitions for territorial advantage in the mating arenas of kob (a species of African antelope). "In a city of human beings, real estate values increase block by block to the city's core; so on the kob stomping ground do values increase from the suburban periphery to the flashing excitements of Times Square. . . . The female wants her affection, but she wants it at a good address."[69] In this way, Ardrey insisted, males competed for real estate, not for females themselves. "It may come to us as the strangest of thoughts that the bond between a man and the soil he walks on should be more powerful than his bond with his wife. But how many men have you known of, in your lifetime, who died for their country? And how many for a woman?"[70] This question would have struck home for younger readers who questioned the recent escalation of US military activity in Vietnam and for older men who had served in World War II.

Konrad Lorenz, on the other hand, was a credentialed scientist through and through. He had been introduced to American audiences as the authoritative, witty author of *King Solomon's Ring: A New Light on Animal Ways*.[71] When Lorenz published an English translation (*On Aggression*) of his *Das sogenannte Böse: Zur Naturgeschichte der Agression* in 1966, it initially resonated with antiwar activists because he argued that humans are by nature dovelike—we lack natural weapons like fangs or claws—and it was the mechanized weapons of our own manufacture that drove our tendency to engage in world wars and outstripped our instinctual capacity to defuse modern warfare.[72] He opened a 1970 article in the *Bulletin of the Atomic Scientists* with the words, "Killing members of one's own kind, whether animal or human, is behavior ill-suited to the preservation of the species."[73] By placing humanity in a natural-historical context, he suggested that as early humans developed the capacity to escape the worst ravages of their environment and outsmart their natural predators, their primary concern became population growth and competition with their (most likely hostile) neighbors. Thus, natural selection favored the evolution of modern militarism from putative "ancient hordes" that he compared to some tribes in New Guinea who lived in a "constant state of war, with headhunting and cannibalism."[74] The bulk of his examples, however, came from animals and Lorenz quickly pointed to the variety of mechanisms that stopped animals from seriously injuring other members of the same species. What, then, went wrong with humans that allowed us to engage in such self-destructive behavior as global war? The problem, Lorenz suggested, was that we no longer knew our enemies, and worse, we had developed weapons that allowed us to kill at a distance. It wasn't that humans possessed a killer "instinct" but rather that we had never evolved inhibitions against killing with weapons.[75] Recognizing his solution as an oft-repeated adage, he nevertheless invoked "love of mankind and rational morals" as the necessary basis of political interactions designed to maintain world peace. The implied human nature in Lorenz's argument thus differed dramatically from Ardrey's, and he would later bemoan his use of the word "aggression" because he felt it obscured these distinctions.[76]

Lorenz's intention was even clearer in an interview he conducted with the French magazine *L'Express*, later translated into English and published in the *New York Times*. When asked whether *On Aggression* "justified" human aggression, he became quite upset: "Excuse aggression? Defend violence? I was trying to do just the opposite! I filled 499 pages in an attempt to explain that violence and war are a derailment of the normal instinct. I tried to show the existence of internal forces that man must know in order to master. I said that reason could conquer aggression."[77] Early reviewers of the books seemed quite aware of his message and its associations with an antiwar counterculture. For example, one reviewer noted that the "rebellious young . . . loved it when he [Lorenz] told them they were storming in the wrong direction."[78] Similarly, the 1966 review of *On Aggression* in the *New York Times* broadly claimed, "Man Has No 'Killer' Instinct," and went on to argue, "Even idiotic slogans such as 'Make love, not war' (as if the two activities had ever been incompatible!) and the use of drugs make the same point. Mankind is safer when men seek pleasure than when they seek the power and the glory."[79]

Desmond Morris emphasized the primacy of this pleasure-seeking aspect of human nature. Morris had earned his DPhil under Nikolaas Tinbergen at Oxford, but rather than follow a traditional academic career, for years he hosted the popular British television show *Zootime* and published an astounding number of popular-science books. In *The Naked Ape*, still his most famous book, Morris provocatively suggested that humans lost the fur that once covered their bodies and that still covers the bodies of most other mammals because its loss facilitated intimate caresses and made possible the development of other, now more accessible sexual signals: the rounded breasts and buttocks of women, the larger size of the penis in comparison to other ape species, female orgasm, and increased sensitivity of human nipples and genitals. All told, these features made sex a lot more fun for humans and rewarded pair-bonding with increased pleasure.[80] Morris's *Naked Ape* concentrated on those aspects of the human physique that caused individual attraction and desire, noting that human sexual partnerships formed the basis of our more general sociality.[81] He reasoned that a sexually active species, like humans, would be more altruistic than a less sexualized species.[82]

Not surprisingly, *Playboy* was quite interested in Morris's work and published an essay outlining the argument of his next book, *The Human Zoo*.[83] In the book and the essay, he no longer focused on the sexual habits of individuals but concentrated on the difficulties besetting urban populations around the globe. Were cities like jungles? No, Morris insisted; cities were like cages. When concentrated in large, overpopulated urban centers, the normal dominance relations that governed human social interactions broke down. "[T]he leaders of the packs, prides, colonies, or tribes come under severe strain. . . . So much time has to be spent sorting out the unnaturally complex status relationships that other aspects of social life, such as parental care, become seriously and damagingly neglected."[84] Morris spent the bulk of his *Playboy* essay providing ten lessons for managing human patterns of power, a vital skill in the unnaturally crowded environments of city life:

1. You must clearly display the trappings, postures and gestures of dominance. . . .
2. In moments of active rivalry, you must threaten your subordinates aggressively. . . .
3. In moments of physical challenge, you (or your delegates) must be able to forcibly overpower your subordinates. . . .
4. If a challenge involves brain rather than brawn, you must be able to outwit your subordinates. . . .
5. You must suppress squabbles that break out among your subordinates. . . .
6. You must reward your immediate subordinates by permitting them to enjoy the benefits of their high ranks. . . .
7. You must protect the weaker members of the group from undue persecution. . . .

8. You must make decisions concerning the social activities of
   your group. . . .
9. You must reassure your extreme subordinates from time to
   time. . . .
10. You must take the initiative in repelling threats or attacks
    arising from outside your group. . . .[85]

Morris's interests were moving with the times, and he provided his own
answer to the same problem of dominance that Ardrey had been eager
to sort out—a zoological manual for winning the human rat race based
on a comparison of social interactions in humans and other primates.

In his "Man and Beast" article, Hunt fixated on the aggressive drive
he believed was common to all such theories of human nature derived
from observations of animal behavior and expressed dismay at readers'
fascination and sympathy with "this depressing news." He implied that
the appeal of books by authors like Ardrey, Lorenz, and Morris lay in
their promise of a new basis for understanding human nature where
social scientists appeared to have failed. On the opposite end of the
intellectual spectrum, Hunt identified such widely read social scientists
as Ashley Montagu, well-known anthropologist Margaret Mead, and
philosopher Susanne Langer, who as a whole argued that humans had
no instinctive behaviors at all (much less an aggressive drive).[86] Between
these poles, Hunt suggested, lay most researchers interested in animal
behavior, who insisted that "the entire nature-nurture, innate-learned,
instinct-experience issue is outmoded, if not meaningless. . . . [B]y far
the largest part of it [i.e., behavior] results from interactions between
genotypic tendencies and environmental influences."[87]

Hunt's lesson for the readers? "For better or for worse, that's the
way it's going to be. The study of animal behavior will never again
be a quiet backwater of zoology. Men now fervently hope, and almost
demand, that animal-behavior researchers help them understand themselves and one another; and, given the present human condition, who
can blame them?" What was needed, in Hunt's view, was more careful
study of the functions of animal behavior without resorting to overly
simplistic zoomorphism.[88] "Man does have an aggressive instinct, but
it is not naturally or inevitably directed to killing his own kind. He is a
beast and perhaps at times the cruelest beast of all—but sometimes he is
also the kindest beast of all. He is not all good and not perfectible, but
he is not all bad and not wholly unchangeable or unimprovable. That is
the only basis on which one can have hope for him; but it is enough."[89]

Hunt's article attracted surprising attention from professional scientists. For example, Evelyn Shaw, psychologist and curator of animal behavior at the American Museum of Natural History, wrote a letter to the editor hailing Hunt's piece as a "brilliant rebuttal" to the simplistic determinism of Lorenz, Ardrey, and Morris. Julian Huxley, former director of UNESCO and student of animal behavior in his own right, described "Man and Beast" as "interesting and rather provocative," while biological anthropologist Irven DeVore sent compliments on Hunt's "provocative and judicious treatment" of animal behavior. Even Sally Carrighar, a prolific wildlife writer, added her praise.[90] After reading the piece, Montagu wrote to Hunt asking to reprint the article in a forthcoming edited collection.[91] Support for the piece, then, came from a variety of scientific communities. Even so, the nuance of Hunt's argument would quickly disappear from the pages of the magazine.

Playboys and Cowboys

Discussions of the aggressive and sexual nature of humans, particularly men, were not restricted to articles specifically devoted to these topics. In order to appreciate the wide appropriation of animal metaphors in justifying the *Playboy* lifestyle, I now turn to other sections of the magazine, especially the interviews conducted with public figures, many of which, like the description of Kinsey, commented on the physical appearance and demeanor of the (usually but not always) male interviewee.[92] *Playboy* interviewed one celebrity in each issue of the magazine. Despite the wide variety of personal styles and appearances, *Playboy*'s descriptions of these men often resonated deeply with a biological vision of masculinity as a combination of social control, physical strength, and sex appeal.[93] As an interview with Jesse Jackson noted, "Biologist Desmond Morris has written that a leader never scrabbles, twitches, fidgets or falters, and Jackson qualifies."[94] So did everyone else.

Well, almost everyone. Clint Eastwood stood out by defying expected masculine tropes, and for this reason his interview provides us with an opportunity to investigate the overlapping rhetorics of masculinity, sexuality, and aggression. The interviewer noted that "it's difficult to reconcile the real Clint Eastwood—gentle, soft-spoken, self-effacing—with the violent men he's played onscreen, men who were ready to shoot first and talk later, if at all."[95] Eastwood described himself as working on more of an emotional, "animal level" than an intellectual one.[96] When asked what he would do if he saw a woman, like Kitty Genovese, being hurt and no one coming to her rescue, Eastwood responded coolly.[97] "I don't know. I would hope that I would, at a minimum, raise the telephone and notify the police. At a maximum, wipe the guy out. I mean, people are capable of heroic action in life, but nobody knows what he'd do before the occasion arises." When pressed to further blur the distinction between his on-screen and off-screen lives, however, Eastwood resisted. He cracked, "I'm sure that if somebody were pointing a gun at me and I was standing there with a six-pack, I'd say, 'Care for one?'" Not to be outplayed, the interviewer persisted, comparing Eastwood's answer instead to the "realistic" antihero he had portrayed in Sergio Leone's spaghetti westerns (*Fistful of Dollars, For a Few Dollars More*, and *The Good, the Bad, and the Ugly*).[98] In the end, Eastwood came across as not that dissimilar from his on-screen personalities. He would shoot anyone who violated his home territory, believed that a fascination with guns was an innate male characteristic, and would stick up for the rights of the unprotected, but not at the expense of his own life.[99] He was indeed Ardrey's and Morris's man in total control of the situation (if not entirely in charge of the interview).

Perhaps the most revealing encapsulation of the biological underpinnings of the *Playboy* lifestyle, however, comes from the filmic adaptation of Morris's *Naked Ape*, financed by Playboy Productions and released in 1973. The movie combined live action and cartoon montages and took care to repeat much of Morris in his own words. For example, at one point in the film, a professor asks each student in his course to bring in an example of erotic classic literature. After embarking on a series of extended daydreams about dating fellow student Cathy (played by Victoria Principal), Lee (Johnny Crawford) apologizes for not having completed the assignment and instead reads a passage from Morris's *Naked Ape* about biologically natural attitudes toward sexual experimentation. His performance is such a success that the whole class breaks into spontaneous applause, and the professor can do little but nod in approval and remark, "Right."

To convey even more of Morris's message, the comic relief of the movie, Arnie (Dennis Olivieri), reads out loud two letters he wants to write to the US president from the front lines of Vietnam, where he had been sent along with Lee. In the second of these letters, he laments the lack of a true pair-bond and wonders about the fate of the girls GIs had left behind. As Arnie begins to read, he and the steamy jungle disappear, replaced by an idyllic Henri Rousseauesque cartoon:

> Dear Mr. President, for millions of years, guys like me have gone off to hunt or to wars, worrying about his woman back home. Then, true love . . . gave us males assurance our wives wouldn't

have thoughts of other guys, but attend his happy hearth at home, his unassailable haven from the horny. Ha ha! Why is it berries always seem riper on another man's bush? This fear, plus the need to stick around until the kids were grown, is why love was developed.

Meanwhile, in the cartoon jungle, an early human male is hiding in the bushes, watching a naked woman pick berries and frolic with butterflies, alone because her mate had just embarked on a hunting trip. As Arnie finishes his point about the evolution of true love, the man sneaks out of the bushes to chase the beautiful, abandoned woman. After a giggling chase through the lush foliage, the couple disappears from view, but the sounds of their lovemaking can be heard, as creatures of the forest turn out to watch and listen, including an idiotically grinning lion. Arnie has been talking all the while and speculates that "there's a creeping doubt" the pair-bonding process "was ever really perfected." According to Morris, this imperfect pair-bond allowed both men and women to seek sexual comfort in the arms of multiple others.[100] Of course, that wasn't to say that jealousy in the form of Ardrey's territorial (and sexual) defense wouldn't arise. In the movie, when the hunter returns to find his partner with the interloper, he roars with rage, and a close-up of his left pupil reveals a fanged beast leaping toward the camera as he chases the other man away. A female voice-over (Principal once again) picks up the thread of the visual argument: "So there he is, our vertical, hunting, territorial, brainy naked ape. The naked ape is a new experimental departure and new models frequently have imperfections. For instance, sexually the naked ape finds himself today in a somewhat confusing situation."

The film combines two images of masculinity—sexual prowess and aggression—into a single descriptor of innate human nature: the "sexual hunter." In the process of becoming fully human, the film implied, our ancestors had transformed from fruit-pickers to hunters and killers. The term "fruit-picker," used several times in the movie to refer to our prehuman arboreal ancestors, not so subtly associated vegetarianism with primitivity and meat-eating with the truly human. (Roszak's description of the Doors echoed a similar sentiment.)

Playboy's incorporation of these lessons from animal behavior for understanding human nature during the late 1960s and early 1970s complemented a social-scientific perspective on masculinity (provided, e.g., by Pomeroy) and highlights an almost-total disregard for other common sources of information on human nature, including archeology, cultural anthropology, and linguistics.[101] By the mid-1970s, however, the same research that had been used to justify a countercultural sexual revolution carried overtones of conservative politics. One cause of this shift lies in the vociferous reaction against conceptions of desirable femininity that were integral to images of the masculine sexual hunter.

In his interview with *Playboy*, Sam Peckinpah, the director of *The Wild Bunch* and *Straw Dogs* and devotee of Robert Ardrey's books, dismissed concern about the violence in his films as the result of overly sensitive feminists not in tune with the truths of biology. *Playboy* asked him about film critic Pauline Kael's review in particular—she had claimed Peckinpah "enshrined the territorial imperative" and was "out to spread the Neanderthal word."[102] Not swayed in the least, Peckinpah responded, "More, more, I love it!"[103] When asked whether or not his films caused violent behavior in his audiences, he explained that watching a violent film was a kind of physical release. It allowed people to experience violence safely within the movie theater, much like sports fans.[104] "Do you think people watch the Super Bowl because they think football is a beautiful sport? Bullshit! They're committing violence vicariously." When pushed even harder about the role of violence in films and its relation to human nature, Peckinpah explicitly invoked Ardrey's arguments. "I think it's wrong—and dangerous—to refuse to acknowledge the animal nature of man. That's what Robert Ardrey is talking about in those three great books of his, *African Genesis, The Territorial Imperative* and *The Social Contract*. Ardrey's the only prophet alive today." Peckinpah added that he had first come into contact with *African Genesis* while working on *The Wild Bunch*, released to critical acclaim in 1969. A friend had passed it along, suggesting that Peckinpah and Ardrey were "both on the same track." After finishing *The Wild Bunch*, Peckinpah finally had time to read it. In recalling the experience, he described Ardrey as a kind of kindred spirit. "I thought, wow, here's somebody who knows a couple of nasty secrets about us." Similarly, he suggested that *Straw Dogs* examined the life of a "guy who found out a couple of nasty secrets about himself—about his marriage, about where he is, about the world around him."[105] Ardrey, for his part, was so taken with at least this portion of Peckinpah's interview that he wrote to "Dear Playboy," saying, "Like him, I believe that until we have the courage to grasp the whole of human reality—namely, our propensity for violence—we possess small hope for improvement of our lot."[106]

In *Playboy*'s interview with Eastwood, the question of violence in his films had also arisen. Eastwood's reaction was similar to Peckinpah's. He, too, argued that fictional on-screen violence could reduce pent-up aggression in viewers. Eastwood admitted that he "knew they were tough films" but hoped the violence was sufficiently satirical that the films could serve as a catharsis for viewers. He continued, "I'm not a person who advocates violence in real life, and if I thought I'd made a film in which the violence inspired people to go out and commit more violence, I wouldn't make those films." When asked to elaborate, Eastwood added that the movies were a form of escapism, and he related a story he remembered reading in the *Los Angeles Times* in which a journalist reported that inmates at San Quentin said they loved Clint Eastwood westerns. Why? According to Eastwood, "any pent-up emotions they had were released when they saw those films. After they'd see one, everything would be very calm in the prison for the next few weeks." (In the original article, the inmate informant followed his remark that Eastwood's westerns functioned as a "tension release" with "one year they showed seven comedies and eight guys got stabbed.")[107] Eastwood, however, dismissed *The Wild Bunch* as part of a recent snowballing of excessive violence in films. Peckinpah, he suggested, simply "wanted to make a superviolent flick" to one-up other directors, to show "how *beautiful* it [violence] is, with slow-motion cameras and everything."[108] Eastwood hoped that members of the audience wouldn't be permanently brutalized by the violence of such recent films by becoming inured to it.

When it came to understanding the differences between men and women, Peckinpah (like Ardrey) insistently framed his answers in gendered universals. Men, he suggested, were turned on by a woman's physical appearance, maybe her beauty or the way she moved, and women were more interested in the material security that a man could provide. "It's the most basic and fascinating evolutionary process there is." The interviewer proposed that although experts on animal behavior would probably be amenable to his ideas, feminists would of course object. Peckinpah dismissively answered, "I don't care what goes on in people's heads; we are physically constructed in a certain way and we've been handed a set of instincts to go with the machinery. Tell that to any of these women's lib freaks and they'll swear you're a male chauvinist pig."[109] Indeed. Toward the end of the interview, Peckinpah attested to his own masculinity, describing a lifestyle consistent with the *Playboy* image: "I live plenty. I like good drink, good food, comfortable clothes and fancy women."[110]

Peckinpah's movies were part of a much larger reworking of the studio system in Hollywood that allowed a new generation of ultraviolent films, many of which embodied similarly gendered stereotypes—

taciturn men and women as weathervanes of male power. According to Pauline Kael, "Whatever their individual qualities, such films as 'Bonnie and Clyde,' 'The Graduate,' 'Easy Rider,' 'Five Easy Pieces,' 'Joe,' 'M*A*S*H,' 'Little Big Man,' 'Midnight Cowboy,' and 'They Shoot Horses, Don't They?' all helped to form the counterculture." Writing in the summer of 1975, she attributed to past "young, anti-draft, anti-Vietnam audiences" a willingness to watch these new kinds of cinematic experiments that were largely shunned by older audiences.[111] Although the fallout of events like Watergate and Vietnam was impossible to determine, Kael noted that earlier, studio films "were predicated on an implied system of values," which had been transmuted in New Hollywood films, if it survived at all, into the "corrupt, vigilante" justice embodied by Eastwood's Dirty Harry.[112] "The counterculture films," she argued, "made corruption seem inevitable and hence something you learn to live with"—either as unadulterated horror or through the lens of slapstick comedy. Kael lamented the "case-hardened audience" who viewed the violence so common in these films with ironic distance and "a new complacency."[113]

On cellulose acetate and paper, *Playboy* broadcast images of the male "sexual hunter." These stories encoded masculinity as a trait embodied in an individual but made visible through interactions with others, combining elements of private and public life, solitary and group activities, sexual and aggressive behavior. In turn, elements of a naturalized masculinity designed to demarcate a certain kind of man reciprocally contributed to social constructions of normative sexuality and behavior for women.[114]

Legacy of the Sexual Hunter

In the summer of 1975, the same summer Kael penned her thoughts on the ironically violent predilections of countercultural moviegoers, myrmecologist Edward O. Wilson of Harvard University published *Sociobiology: A New Synthesis*.[115] The debate over Wilson's political leanings followed the pattern established by earlier critiques of pop ethologists like Ardrey and his converts, such as Peckinpah.[116] Wilson argued that biological theory (especially evolution) constituted the best tool for understanding human nature. Other scientists who adopted this perspective soon came to be known as "sociobiologists," and as a community they sought to discredit Ardrey, Lorenz, and Morris as having fundamentally misconstrued natural selection as acting at the level of the group rather than individuals (or even genes).[117] Even so, more culturally minded natural and social scientists attacked Wilson and other sociobiologists as advancing an antifeminist, conservative perspective that let their political beliefs get in the way of objectively analyzing the facts.[118] For sociobiology's critics, Wilson's easy traffic between non-human and human societies seemed all too reminiscent of these earlier theories in emphasizing the natural aggression of men. These critics argued that the underlying evolutionary logic of sociobiology differed from previous theories only in nuance—its animalistic allegories and gendered conclusions remained the same.

After the publication of Wilson's *Sociobiology*, *Playboy*'s approach to human nature similarly changed in terms of the science it invoked but not in tone or meaning. The editors now called on Charles Darwin's theory of sexual selection, as articulated by sociobiologists. In a 1978 essay entitled "Darwin and the Double Standard," Scot Morris (no relation to Desmond) drew direct analogies between the sexual and social behavior of animal species and that of humans.[119] "It has been said that a man will try to make it with anything that moves—and a woman won't. Now the startling new science of sociobiology tells us why."[120] Although males (and, by implication, men) fight over females (and women), the article insisted, they do so "not because they like to or are innately aggressive but because that is what impressed their great-great-grandmothers."[121] Although the article provided readers with an absolute assurance that females controlled the timing and frequency of sex—in other words, rape was unnatural—it also proclaimed, even less subtly than before, that feminists were defying their biological heritage. "Recent scientific theory suggests that there are innate differences between the sexes and that what's right for the gander is wrong for the goose."[122] "Darwin and the Double Standard" also conjured a nonnegotiable genetic determinism: "If you get caught fooling around, don't say the Devil made you do it. It's the devil in your DNA."[123] The rhetoric had changed but the stereotype of a masculine sexual hunter remained the same. Each of these points built on a long tradition of *Playboy*'s use of science to bolster the lifestyle it sought to promote.

Throughout the 1970s, however, the early intellectual aspirations of the magazine were replaced by a need to compete with *Penthouse*.[124] The tone and content of the magazine became more sexually explicit and the style decidedly less upmarket. Accompanying "Darwin and the Double Standard," editors inserted a double-page spread illustrating "x-rated" examples of sexual behavior in a wide variety of species: lesbianism in seagulls, prostitution in hummingbirds, muscle beach parties in damselfish, cheating in elephant seals, and so on.[125] No scientists wrote in to express their dismay at this characterization. Their silence spoke volumes to the changing demographics of the magazine's audience.[126]

This story of a public conversation concerning our private constitutions raises questions about the genre of popular science in the age of the counterculture. Consider the imbricated histories of science and pornography and the various ways authors have (intentionally or not) generated frisson for their work by either discussing salacious topics or being associated with publication venues of questionable reputation.[127] Perhaps most obviously, authors benefited from the eroticized context of their arguments but inevitably opened themselves to critique by courting the edge of professional respectability.[128] Desmond Morris additionally published a book called *Intimate Behaviour* in 1971 and agreed to an interview with the even more sexually explicit *Gallery* magazine in 1978.[129] Even coverage of *The Naked Ape* in middlebrow *Life* magazine contained both suggestive subheadings and double entendres.[130] Much like his Enlightenment predecessors, Morris believed that sex and reproduction were natural processes and therefore should be discussed without shame.

Just as *Playboy* supported the publication of books disseminating recent research in "forbidden sciences" like sexology, including that of Kinsey, Masters, and Johnson, articles on human nature, and the worrisome health of the planet,[131] Kathy Keeton and Bob Guccione (the man behind *Penthouse*, whom she later married) cofounded the popular-science magazine *Omni*, which ran from 1978 to 1998.[132] Montcalm Publishing, the company behind *Gallery* (founded in 1972), also produced *The Twilight Zone Magazine* (founded in 1981). Even *Playboy*'s answer to *Penthouse—Oui* (also founded in 1972)—contained an occasional scientific article on quantum theory or interview with Carl Sagan.[133] Scientists were willing to publish in these magazines because of their wide readership. The magazines, for their part, sought intellectual legitimacy and believed (apparently, with good reason) that their target male audience enjoyed reading about both sex and science.

Acknowledgments

Sincere thanks to Margot Canaday, Bruce Lewenstein, and the participants of the "Groovy Science" workshop for helpful feedback on early drafts of this chapter, and to David Kaiser, Patrick McCray, Miranda Waggoner, and two anonymous reviewers for fruitful comments on later versions.

Notes

1. Desmond Morris, *The Naked Ape: A Zoologist's View of the Human Animal* (New York: McGraw-Hill, 1967).

2. Henry Raymont, "Review of 'Naked Ape' Causes 2 Papers to Call Back Sections," *New York Times*, 21 January 1968, 76.

3. "Programming: Reasonable v. Raunchy," *Time*, 9 February 1968, 67–68. See also Desmond Morris, *Watching: Encounters with Humans and Other Animals* (London: Max Press, 2006), 313.

4. D. Morris, *Watching*, 314–15 (emphasis in original).

5. Despite the 1967 copyright date on the book, *The Naked Ape* became available for sale in the United States in the third week of January 1968. For details on the movie contract, see A. H. Weiler, "Everybody's Going Ape," *New York Times*, 14 April 1968, D15; and Bill Edwards, "Bufman, Driver Strike It Rich in Univ. Deal," *Daily Variety*, 26 February 1969, 1, 13.

6. *Daily Variety*, 27 February 1974, 20.

7. Murf. [Arthur Murphy], "The Naked Ape," *Variety* (weekly), 15 August 1973, 12.

8. Vernon Scott, "The Girl Who Has Everything: Victoria Principal," *Good Housekeeping*, June 1989, 114.

9. D. Morris, *Watching*, 314.

10. *Daily Variety*, 27 February 1974, 20; and Frank Brady, *Hefner: An Unauthorized Biography* (New York: Macmillan, 1974), 212.

11. Such formulations of masculinity were certainly not new (Ernest Hemingway and Theodore Roosevelt are obvious earlier examples); nor have they disappeared (just think of the retrosexual charisma of the "Most Interesting Man in the World" in commercials for Dos Equis beer). At the time, critics abounded. For example, see Gloria Steinem's 1963 exposé of life as a waitress in the New York Playboy Club: "I Was a Playboy Bunny," reprinted in *Outrageous Acts and Everyday Rebellions*, 2nd ed. (New York: H. Holt, 1995), 32–75.

12. As Beth Bailey has argued, the sexual revolution took many forms and its origins lie not just in the radical fringes of society but also in mainstream middle-class cultures throughout the United States. See Beth Bailey, *Sex in the Heartland* (Cambridge, MA: Harvard University Press, 1999).

13. E.g., Michelle Brattain, "Race, Racism, and Antiracism: UNESCO and the Politics of Presenting Science to the Postwar Public," *American Historical Review* 112, no. 5 (2007): 1386–1413; Michael A. Little, "Human Population Biology in the Second Half of the Twentieth Century," *Current Anthropology* 53, no. S5 (2012): S126–S138; and Jonathan Marks, "The Origins of Anthropological Genetics," *Current Anthropology* 53, no. S5 (2012): S161–S172.

14. For an exploration of American research into the human animal throughout the twentieth century, see Donna Haraway, *Primate Visions: Gender, Race, and Nature in the World of Modern Science* (New York: Routledge, 1989).

15. Raymond Dart, *Adventures with the Missing Link* (New York: Harper, 1959).

16. Robert Ardrey, *African Genesis: A Personal Investigation into the Animal Origins and Nature of Man* (New York: Atheneum, 1961). For criticism, see Ashley Montagu, "The New Litany of Innate Depravity, or Original Sin Revisited," in *Man and Aggression*, ed. Ashley Montagu (New York: Oxford University Press, 1968), 3–17; and the other essays collected in the same volume.

17. National Geographic Specials, *Dr. Leakey and the Dawn of Man* (National Geographic, 1966); and Louis S. B. Leakey, "Exploring 1,750,000 Years into Man's Past," *National Geographic Magazine*, October 1961, 564–89.

18. Catherine H. Berndt and Ronald M. Berndt, "Australian Aborigines: Blending Past and Present," in *Vanishing Peoples of the Earth*, ed. R. L. Breeden (Washington, DC: National Geographic Society, 1968), 114–31; Orlando Boas and Claudio Villas Boas, "Saving Brazil's Stone Age Tribes from Extinction," *National Geographic Magazine* 134, no. 2 (1968): 424–44; and Frederick A. Milan, "The Indomitable Eskimo: Master of a Frozen World," in Breeden, *Vanishing Peoples of the Earth*, 132–51.

19. Father Alphonse Sowada, "New Guinea's Fierce Asmat: A Heritage of Headhunting," in Breeden, *Vanishing Peoples of the Earth*, 186–203; and Robert Gardner and Karl G. Heider, *Gardens of War: Life and Death in the New Guinea Stone Age* (New York: Random House, 1969).

20. Goodall's revelations about the warring, cannibalistic behaviors of chimpanzees would not come until 1974. Notice the dramatic differences between Jane Goodall, "My Life among Wild Chimpanzees," *National Geographic Magazine*, August 1963, 272–308, and "Life and Death at Gombe," *National Geographic Magazine*, May 1979, 592–621. See also Irven DeVore, The Primates (New York: TimeLife Books, 1965); and Richard B. Lee and Irven DeVore, eds., *Man the Hunter* (New York: Aldine de Gruyter, 1968).

21. Mary Terrall, "Salon, Academy, and Boudoir: Generation and Desire in Maupertuis's Science of Life," *Isis* 87 (1996): 217–29. See also Robert Darnton, *The Literary Underground of the Old Regime* (Cambridge, MA: Harvard University Press, 1982); and Robert Darnton, *The Forbidden Best-Sellers of Pre-revolutionary France* (New York: Norton, 1995).

22. Elizabeth Fraterrigo, *"Playboy" and the Making of the Good Life in Modern America* (New York: Oxford University Press, 2009), esp. "'Casualties of the Lifestyle Revolution': *Playboy*, the Permissive Society, and Women's Liberation," 167–204.

23. Elaine Tyler May, *America and the Pill: A History of Promise, Peril, and Liberation* (New York: Perseus, 2010); Lara V. Marks, *Sexual Chemistry: A History of the Contraceptive Pill* (New Haven, CT: Yale University Press, 2001); and Elizabeth Watkins, *On the Pill: A Social History of Oral Contraceptives, 1950–70* (Baltimore: Johns Hopkins University Press, 1998).

24. On the masculine consumerism embodied by *Esquire* (Hugh Hefner's model for *Playboy*), see Kenon Breazeale, "In Spite of Women: 'Esquire' Magazine and the Construction of the Male Consumer," *Signs* 20, no. 1 (1994): 1–22; and Beatriz Preciado, "Pornotopia," in *Cold War Hothouses: Inventing Postwar Culture from Cockpit to Playboy*, ed. Beatriz Colomina, Annmarie Brennan, and Jeannie Kim (Princeton, NJ: Princeton University Press, 2004),

216–53. On consumerism and sexual identity, see David Serlin, *Replaceable You: Engineering the Body in Postwar America* (Chicago: University of Chicago Press, 2004).

25. Geoffrey Gorer, "Man Has No 'Killer' Instinct," *New York Times*, 27 November 1966, SM24ff.

26. Wendy Kline, *Building a Better Race: Gender, Sexuality, and Eugenics from the Turn of the Century to the Baby Boom* (Berkeley: University of California Press, 2001), esp. "'Marriage Is Not Complete without Children': Positive Eugenics, 1930–1960," 124–56.

27. Thomas Frank, *The Conquest of Cool: Business Culture, Counterculture, and the Rise of Hip Consumerism* (Chicago: University of Chicago Press, 1997); and Sam Binkley, *Getting Loose: Lifestyle Consumption in the 1970s* (Durham, NC: Duke University Press, 2007).

28. Barbara Ehrenreich, *The Hearts of Men: American Dreams and the Flight from Commitment* (New York: Doubleday, 1983).

29. Charles McGrath, "How Hef Got His Groove Back," *New York Times*, 3 February 2011; and Richard Gehman, "The Private Life of Hugh Hefner," *Fact* 2, no. 4 (1965): 50–57.

30. Claire Hines, "'Entertainment for Men': Uncovering the *Playboy* Bond," in *The James Bond Phenomenon*, 2nd ed. (Manchester: Manchester University Press, 2009), 89–105; and Thomas Weyr, *Reaching for Paradise: The Playboy Vision of America* (New York: Time Books, 1978).

31. An insert in the November 1965 issue of *Playboy* advertised yearlong subscription rates of $8 (a savings of $2 off the cover price of $0.75 for a regular issue, and $1.25 for the January and December double issues), the equivalent of $58 in 2012.

32. Fraterrigo, *"Playboy" and the Making of the Good Life*, 25, 134, 167. The early 1970s represented the heyday of *Playboy*'s iconic status, after which circulation numbers began to drop due to competition from more sexually explicit magazines; Carrie Pitzulo, *Bachelors and Bunnies: The Sexual Politics of Playboy* (Chicago: University of Chicago Press, 2011), 12.

33. "What Sort of Man Reads *Playboy*?" *Playboy*, November 1965, 85, drawing on research from "1965 Starch Consumer Report Magazine." In 2015 dollars, that would mean $75,758 and $3,788, respectively (calculated according to the US Consumer Price Index).

34. "What Is a Playboy?," *Playboy*, April 1956, 73.

35. Preciado, "Pornotopia," 220; Elizabeth Fraterrigo, "Pads and Penthouses: *Playboy*'s Urban Answer to Suburbanization," in Fraterrigo, *"Playboy" and the Making of the Good Life*, 80–104; and Gehman, "Private Life," 52.

36. Preciado, "Pornotopia," 219.

37. Robert Ruark, "Far-Out Safari," *Playboy*, March 1965, 84ff.; and Richard Warren Lewis, "Playboy Interview: Jack Nicholson," *Playboy*, April 1972, 75ff.

38. Theodore Roszak, *The Making of a Counter Culture: Reflections on the Technocratic Society and Its Youthful Opposition* (New York: Doubleday, 1969); and Fraterrigo, *"Playboy" and the Making of the Good Life*, 163–65.

39. Roszak, *Making of a Counter Culture*, 65, 71; and Eldridge Cleaver, *Soul on Ice* (New York: McGraw Hill, 1968).

40. Roszak, *Making of a Counter Culture*, 65. In making this last point, he cited an article by his wife: Betty Roszak, "Sex and Caste," *Liberation*, December 1966, 28–31. See also their coauthored foreword to *Masculine/Feminine: Readings in Sexual Mythology and the Liberation of Women*, ed. Betty Roszak and Theodore Roszak (New York: Harper Colophon, 1969), vii–xii, in which "the sexual liberation of women is carefully distinguished from the phony (and male exploitative) freedom of *Playboy* permissiveness" (xii).

41. Roszak, *Making of a Counter Culture*, 128.

42. Ibid., 75.

43. On women's roles, see Fraterrigo, *"Playboy" and the Making of the Good Life*, 105–33; and Helen Gurley Brown, *Sex and the Single Girl* (New York: Random House, 1962).

44. *Playboy* sold especially well among men enrolled in college; Fraterrigo, *"Playboy" and the Making of the Good Life*, 135–38.

45. Robert N. Proctor, "Three Roots of Human Recency: Molecular Anthropology, the Refigured Acheulean, and the UNESCO Response to Auschwitz," *Current Anthropology* 44, no. 3 (2003): 213–39; Fred Turner, "*The Family of Man* and the Politics of Attention in Cold War America," *Public Culture* 24, no. 1 (2012): 55–84; and Erika Lorraine Milam, "Public Science of the Savage Mind: Contesting Cultural Anthropology in the Cold War Classroom," *Journal of the History of the Behavioral Sciences* 49, no. 3 (2013): 306–30.

46. Playboy Panel, "The Womanization of America," *Playboy*, June 1962, 43–50, 133–44. The panel took as their inspiration an article by Philip Wylie published in *Playboy* (September 1958) with the same title; see also Fraterrigo, *"Playboy" and the Making of the Good Life*, 34–36, 124–26.

47. In 1967 Sahl married China Lee, who appeared as the *Playboy* centerfold of August 1964.

48. Playboy Panel, "Womanization of America," 144.

49. On the value of integrating our vision of "professional" and "popular" science, see Katherine Pandora, "Popular Science in National and Transnational Perspective: Suggestions from the American Context," *Isis* 100, no. 2 (2009): 346–58; and James A. Secord, "Knowledge in Transit," *Isis* 95, no. 4 (2004): 654–72.

50. Konrad Lorenz, *On Aggression*, trans. Marjorie Kerr (New York: Harcourt, Brace, and World, 1966).

51. Morton Hunt, "Man and Beast," *Playboy*, July 1970, 82.

52. D. Morris, *Naked Ape*, 63.

53. Wardell B. Pomeroy, "What Is Normal?," *Playboy*, March 1965, 97, 174–76. See also Ernest Havemann, "The Sex Institute," *Playboy*, September 1965, 139ff.

54. Pomeroy, "What Is Normal?," 175.

55. William H. Masters and Virginia E. Johnson, *Human Sexual Response* (New York: Bantam Books, 1966); Nat Lehrman, *Masters and Johnson Explained* (Chicago: Playboy Press, 1970); and *The Playboy Report on American*

Men (Chicago: Playboy Press, 1979). For more on Lehrman, see Josh Lambert, "My Son, the Pornographer," *Tablet Magazine*, 24 February 2010. On the mutually beneficial relationship between Masters and Johnson and *Playboy*, see Pitzulo, *Bachelors and Bunnies*, 123–24. Pitzulo notes that the Playboy Foundation contributed over $100,000 to their research efforts.

56. Havemann, "Sex Institute," 194, 200. For more on Kinsey's reputation as an analytical scientist, see Miriam G. Reumann, *American Sexual Character: Sex, Gender, and National Identity in the Kinsey Reports* (Berkeley: University of California Press, 2005). By all accounts, Kinsey's private life was far less straitlaced. See James Jones, *Alfred C. Kinsey: A Public/Private Life* (New York: W. W. Norton, 1997); and Jonathan Gathorne-Hardy, *Alfred C. Kinsey: Sex the Measure of All Things; A Biography* (London: Chatto and Windus, 1998).

57. Havemann, "Sex Institute," 194.

58. Dear Playboy, *Playboy*, December 1965, 20.

59. Pitzulo, *Bachelors and Bunnies*, 109. In her chapter on sexology, Pitzulo engagingly explores how the magazine provided an important venue for discussing gay and lesbian issues throughout the 1960s; see esp. 109–18.

60. Hunt, "Man and Beast," 81.

61. Morton Hunt, *Her Infinite Variety: The American Woman as Lover, Mate, and Rival* (New York: Harper and Row, 1962), 36. Hunt's other publications include *The Natural History of Love* (1959; Washington, DC: Minerva Press, 1970), *The Affair: A Portrait of Extra-marital Love in Contemporary America* (New York: World Publishing, 1969), and *Sexual Behavior in the 1970s* (Chicago: Playboy Press, 1974). Hunt intended the last of these to be a popularization and contextualization of Kinsey's results, updated for a new era, and he cited none of the contemporary popular work on human nature.

62. Hunt, "Man and Beast," 82.

63. Lewis Cotlow, "Twilight of the Primitive," *Playboy*, October 1971, 134–36, 158, 254–58. This article was tied to the release of his new book, *The Twilight of the Primitive* (New York: Macmillan, 1971).

64. Lewis Cotlow, *Primitive Paradise* (The Lewis Cotlow New Guinea Expedition and Excelsior Pictures Corp., 1961), 56 min.; and Lewis Cotlow, *In Search of the Primitive* (Boston: Little, Brown, 1966).

65. Similar concerns over environmental destruction were raised by Paul Ehrlich in his *Playboy* interview (August 1970, 55–66, 150–54). Ehrlich advocated more extensive use of birth control, including vasectomies for men, and reforming US law to guarantee access to safe and cheap abortions for women of all classes.

66. Nadine Weidman, "Popularizing the Ancestry of Man: Ardrey, Dart, and the Killer Instinct," *Isis* 102, no. 2 (June 2011): 269–99.

67. Robert Ardrey, *Territorial Imperative* (New York: Atheneum, 1966).

68. Robert Ardrey, "The Drive for Territory," *Life* 61, no. 9 (August 1966): 47.

69. Ibid., 49.

70. Ibid., 47.

71. Konrad Lorenz, *King Solomon's Ring: A New Light on Animal Ways*,

illustrated by the author and with a foreword by Julian Huxley, trans. Marjorie Kerr Wilson (New York: Crowell, 1952). See also Richard W. Burkhardt Jr., *Patterns of Behavior: Konrad Lorenz, Niko Tinbergen, and the Founding of Ethology* (Chicago: University of Chicago Press, 2005).

72. Konrad Lorenz, "On Killing Members of One's Own Species," *Bulletin of the Atomic Scientists* 26, no. 8 (October 1970): 2–5, 51–56; Nikolaas Tinbergen, "On War and Peace in Animals and Men," *Science* 160 (1968): 1411–18; and Geoffrey Gorer, "Man Has No 'Killer' Instinct," *New York Times*, 27 November 1966, SM24ff.

73. Lorenz, "On Killing Members of One's Own Species," 3.

74. Ibid., 4.

75. Ibid., 55.

76. B. Lawren, "Interview with Konrad Lorenz," *Omni* 9 (April 1987): 86.

77. Frédéric de Towarnicki and Konrad Lorenz, "A Talk with Konrad Lorenz," trans. Stanley Hochman, *New York Times*, 5 July 1970, 121.

78. Konrad Lorenz, "People, the Plant Eaters," *Vogue*, April 1974, 153 (editor's note).

79. Gorer, "Man Has No 'Killer' Instinct," SM109; and Marston Bates, "A Naturalist at Large," *Natural History*, 1967, 14–18.

80. John Hurrell Crook, "Sexual Selection, Dimorphism, and Social Organization in the Primates," in *Sexual Selection and the Descent of Man*, ed. Bernard Campbell (Chicago: Aldine, 1972), 231–81, esp. 249–50.

81. D. Morris, "Sex," in *Naked Ape*, 50–102. Morris hoped to provide an evolutionary account of Masters and Johnson's *Human Sexual Response*.

82. Less risqué portions of the book were excerpted in *Life* (just as the magazine had done with Ardrey's *Territorial Imperative*); Desmond Morris, "The Naked Ape," *Life* 63, no. 25 (22 December 1967): 94D–108.

83. Desmond Morris, *The Human Zoo* (New York: McGraw-Hill, 1969); and Desmond Morris, "Status and Superstatus in *The Human Zoo*," *Playboy*, September 1969, 122ff.

84. D. Morris, "Status and Superstatus," 123.

85. Ibid., 124, 202–4. Each point was followed by multiple paragraphs of zoological justification.

86. Hunt, "Man and Beast," 82; Margaret Mead, *Coming of Age in Samoa: A Psychological Study of Primitive Youth for Western Civilisation* (New York: W. Morrow, 1928); and Susanne Langer, *Philosophy in a New Key: A Study in the Symbolism of Reason, Rite, and Art* (Cambridge, MA: Harvard University Press, 1942).

87. Hunt, "Man and Beast," 181–82.

88. Ibid., 114.

89. Ibid., 183.

90. Dear Playboy, *Playboy*, October 1970, 14. E.g., see Sally Carrighar, *One Day at Teton Marsh*, illus. George and Patricia Mattson (New York: A. A. Knopf, 1947).

91. Morton Hunt, "Man and Beast," in *Man and Aggression*, 2nd ed., ed. Ashley Montagu (Oxford: Oxford University Press, 1973), 19–38.

92. *Playboy* rarely interviewed two or more people together. When

the magazine did feature cointerviews, it was almost always a man and a woman—Jane Fonda and Tom Hayden (April 1974), for example, or William Masters and Virginia Johnson (May 1968 and November 1979). The only male duo interviewed in the 1960s and 1970s was the comic team of Don Rowen and Dick Martin (October 1969).

93. Of the interviews I have read, such physical descriptions did not accompany the *Playboy* "candid conversations" with men known primarily as intellectuals, such as Dr. Martin Luther King Jr. (January 1965), Paul Ehrlich (August 1970), or Barry Commoner (July 1974).

94. Arthur Kretchmer, "Playboy Interview: Jesse Jackson," *Playboy*, November 1969, 86. On the history of race and *Playboy*, see Elizabeth Fraterrigo, "The Racial Limits of 'The Playboy Life,'" in Fraterrigo, *"Playboy" and the Making of the Good Life*, 138–49.

95. Arthur Knight, "Playboy Interview: Clint Eastwood," *Playboy*, February 1974, 58.

96. Ibid., 70.

97. On the public horror surrounding Catherine Genovese's murder, as witnessed or overheard by thirty-eight neighbors, none of whom helped her or called the police, see "Study of the Sickness Called Apathy," *New York Times*, 3 May 1964, SM24ff.

98. Knight, "Clint Eastwood," 70.

99. Ibid., 170.

100. Morris's conception of an "imperfect" pair-bond differed from Hefner's earlier views of sex relations in early man. In a 1962 interview, for example, Hefner suggested that sex roles used to be quite easy and only became complicated with the influence of modern civilization: "You know it goes back to the beginning of time. The man goes out and kills a saber-toothed tiger while the woman stays at home and washes out the pots." See Simon Nathan, "About the Nudes in *Playboy*," *U.S. Camera*, April 1962, 69–70, as quoted in Fraterrigo, *"Playboy" and the Making of the Good Life*, 34.

101. See, e.g., Morton Hunt, "Crisis in Psychoanalysis," *Playboy*, October 1969, 106ff.

102. Pauline Kael, "The Current Cinema: Sam Peckinpah's Obsession," *New Yorker*, 29 January 1972, 80–85; see 85 for the source of *Playboy*'s paraphrase.

103. William Murray, "Playboy Interview: Sam Peckinpah," *Playboy*, August 1972, 66. His responses throughout the interview were equally visceral and antagonistic.

104. Konrad Lorenz similarly argued that by "releasing" aggression through playing sports, an individual's aggressive drive would be temporarily depleted, and he would become less likely to commit a violent crime. This conception of violence fitted well with his "hydraulic" model of behavior, in which the urge to act would build within an individual, providing a powerful (if unconscious) motivation to exhibit certain kinds of behavior. Burkhardt, *Patterns of Behavior*, 311–15.

105. Murray, "Sam Peckinpah," 68.

106. Robert Ardrey, in Dear Playboy, *Playboy*, November 1972, 18.

107. The article Eastwood had in mind appeared the previous year: William Drummond, "San Quentin: An Inside View of Its Turmoil," *Los Angeles Times*, 18 March 1971, 1.

108. Knight, "Clint Eastwood," 72.

109. Murray, "Sam Peckinpah," 70.

110. Ibid., 192.

111. Pauline Kael, "On the Future of the Movies," *New Yorker*, 5 August 1974, 43–59 (quotation on 43).

112. Ibid., 43.

113. Ibid., 44.

114. Erika Lorraine Milam and Robert Nye, eds., *Scientific Masculinities*, *Osiris*, vol. 30 (Chicago: University of Chicago Press, 2015).

115. Edward Osborne Wilson, *Sociobiology: A New Synthesis* (Cambridge, MA: Belknap Press of Harvard University Press, 1975).

116. Ullica Segerstråle, *Defenders of the Truth: The Battle for Science in the Sociobiology Debate and Beyond* (New York: Oxford University Press, 2000), 177–98.

117. Ibid., 24–29. These debates took professional form in sociobiologists' dismissal of V. C. Wynne-Edward's theory that animals could self-regulate the size of their populations; Mark Borrello, *Evolutionary Restraints: The Contentious History of Group Selection* (Chicago: University of Chicago Press, 2010).

118. E.g., Anthony Leeds, Barbara Beckwith, Chuck Madansky, David Culver, Elizabeth Allen, Herb Schreier, Hiroshi Inouye, Jon Beckwith, Larry Miller, Margaret Duncan, Miriam Rosenthal, Reed Pyeritz, Richard C. Lewontin, Ruth Hubbard, Steven Chorover, and Stephen Jay Gould, "Against 'Sociobiology,'" *New York Review of Books* 22, no. 18 (13 November 1975).

119. Scot Morris, "Darwin and the Double Standard," *Playboy*, August 1978, 109–11, 160, 208–12.

120. Ibid., 109. Feminist anthropologists quickly pointed to the gendered stereotypes embedded in sociobiological arguments. E.g., Elaine Morgan, *The Descent of Woman* (New York: Stein and Day, 1972); Sally Slocum, "Woman the Gatherer: Male Bias in Anthropology," in *Towards an Anthropology of Women*, ed. Rayna R. Reiter (New York: Monthly Review Press, 1975), 36–50, originally published under the name Sally Linton, in *Women in Cross-Cultural Perspectives*, ed. Sue Ellen Jacobs (Urbana: University of Illinois Press, 1971); Nancy Tanner and Adrienne Zihlman, "Women in Evolution, Part I: Innovation and Selection in Human Origins," *Signs* 1, no. 3 (1976): 558–608; and Adrienne Zihlman, "Women in Evolution, Part II: Subsistence and Social Organization among Early Hominids," *Signs* 4, no. 1 (1978): 4–20.

121. S. Morris, "Darwin and the Double Standard," 160.

122. Ibid., 160, 290.

123. Ibid., 160.

124. *Penthouse* published its first American issue in 1969 and quickly became *Playboy*'s main competitor; see also Lambert, "My Son, the Pornographer."

125. S. Morris, "Darwin and the Double Standard," 110–11.

126. Hefner tried to market *Playboy* as an organization in line with the

feminist revolution and joined forces with the National Organization of Women in 1978 to raise money for the continued legal debates over the Equal Rights Amendment. *Playboy*'s position as a supporter of feminism, however, was fundamentally challenged in the pornography battles of the 1980s. Fraterrigo, *"Playboy" and the Making of the Good Life*, 167–204.

127. The use of new technology to disseminate sexually explicit material constitutes another significant scholarly conversation along these lines. See, e.g., Jonathan Coopersmith, "Pornography, Videotape, and the Internet," *IEEE Technology and Society Magazine* 19, no. 1 (Spring 2000): 27–34; and Sherry Turkle, "TinySex and Gender Trouble," in *Life on the Screen: Identity in the Age of the Internet* (New York: Simon and Schuster, 1995), 210–32.

128. Terrall, "Salon, Academy, and Boudoir"; Londa Schiebinger, "Gender and Natural History," in *Cultures of Natural History*, ed. Nicholas Jardine, James Secord, and Emma Spary (Cambridge: Cambridge University Press, 1996), 163–77; and Geoffrey C. Bunn, "The Lie Detector, 'Wonder Woman' and Liberty: The Life and Work of William Moulton Marston," *History of the Human Sciences* 10 (1997): 91–119.

129. Desmond Morris, *Intimate Behaviour* (New York: Random House, 1971); and interview with Desmond Morris, *Gallery*, April 1978, 31–36.

130. D. Morris, "Naked Ape." No letters from scientists were published in response.

131. Darnton, *Forbidden Best-Sellers*; and Darnton, *Literary Underground*.

132. On the history of *Omni*, see Patrick McCray, "Omnificent," in *Visioneers: How a Group of Elite Scientists Pursued Space Colonies, Nanotechnologies, and a Limitless Future* (Princeton, NJ: Princeton University Press, 2013), 113–45.

133. David Kaiser, *How the Hippies Saved Physics: Science, Communication, and the Quantum Revival* (New York: W. W. Norton, 2011), xii.

Part Four: Legacies

10 Alloyed: Countercultural Bricoleurs and the Design Science Revival

Andrew Kirk

The bricoleur is a "jack of all trades" who, with cunning and resource, ransacks the "ready at hand" to create something new . . . based on the continuous reworking of the received elements of the world.[1]

In March 1969, industrial designer J. Baldwin pulled off a dusty New Mexico road just outside the abandoned town of La Luz. He and six of his design school students from Southern Illinois University, all acolytes of iconoclastic designer Buckminster "Bucky" Fuller, unpacked themselves from Baldwin's small Citroen sedan and wandered up to the historic La Luz Pottery Factory and a newly constructed village of geodesic "Zomes" (fig. 10.1). The exhausted crew had arrived just in time for an unusually interesting countercultural happening dubbed "Alloy" for its blending of insights and sensibilities toward a stronger, new composite whole. Alloy was organized by a group of influential "design outlaws" assembled by passive solar technology pioneer Steve Baer and promoted by the editor of the recently launched *Whole Earth Catalog*, Stewart Brand.[2]

Avant-garde photographer and filmmaker Robert Frank was on hand to capture the one hundred fifty or so attendees discussing, building, smoking dope, and

FIGURE 10.1 Image of Alloy in *The Last Whole Earth Catalog* (1971). The bricoleur tribe arrive at the old La Luz Pottery Factory at the beginning of Alloy, in March 1969. Whole Earth Catalog Records, published catalog edition collections, Department of Special Collections and University Libraries, Stanford University Libraries.

communing over several long days and sleepless nights.[3] Frank's photographs and film provide some of the only visual evidence of one of the lesser-known important legacies of the counterculture. Alloy was the most active effort to that point to launch a "design science revolution" that Bucky Fuller had been promoting since the 1930s. The organizers and many of the attendees, mostly unknown outside a small circle, were on the cusp of interesting and unlikely careers as proponents of what came to be known as the appropriate-technology and ecological design movements. Their humble early efforts at Alloy and in the communes and urban neighborhoods where they lived and worked ultimately helped shape American environmentalism and set the stage for the still-unfolding, twenty-first century wave of concern for environmental "sustainability." Alloy was also a creative cohort's important effort to use their academic scientific training and technical expertise outside the confines of the university or corporation. This ecologically inclined wing of the counterculture wanted to learn to do science with fewer restrictions by "ransacking the ready at hand" and "reworking the received elements of the world."[4]

Buckminster Fuller provided early inspiration for this significant effort to revive a human-centered and pragmatic environmentalism that united human ingenuity, thoughtfully designed stuff, and care for nature.[5] Fuller was a perfect "crazy scientist" uncle for the hippies who gathered in New Mexico that spring. His ability to survive as an iconoclastic researcher outside the normal parameters of scientific research and academic standards served as a model for their quest to recapture a sense of excitement for small-scale research and invention that many felt was lost during the Cold War, when megasystems and megamachines squelched older American traditions of garage R&D. By 1969 Fuller was forty years into his efforts to train students, and anyone willing to listen, to become "omnidisciplinary" soldiers in his somewhat ill-defined design science revolution.[6] While critics found Bucky's notions obtuse, he produced a remarkable run of useful designs and technologies and successfully marketed his products to world's fairs, corporations, and the US military. His idealistic dream of making the world "work in the shortest possible time, through spontaneous cooperations, without ecological offense, or the disadvantage of anyone," fit the countercultural mood of the 1960s perfectly.[7]

Most significantly, Fuller's holistic philosophy of design science offered a link between the thing-world of technological design science and emerging trends in the ecological sciences and environmentalism.[8] Fuller inhabited a rich material world of tools and thoughtfully made stuff all intended to make the world a more ecologically functional place. From the grand to the small, all of Fuller's products and designs were alive with his idealistic excitement for the liberating potential of science practiced right. At a time when the hard sciences were tracking students into narrow specialization with little room for big questions and even less for philosophical debate, Fuller's model appealed to a generation of countercultural scientists and designers who wanted to take on the big questions of their disciplines without playing by rules they no longer believed in.[9]

The emphasis that Fuller and his Alloy disciples placed on materiality, along with their shared enthusiasm for using design to achieve ecological balance, represented a very different kind of environmentalism from the ascendant wilderness movement of the late 1950s and early 1960s. Likewise, their love of material things, invention, and scientific expertise put to practical use placed this group of counterculturalists at odds with the political take-it-to-the-streets New Left. This better-known faction of the counterculture was generally disdainful of those willing to remain enmeshed in the making and selling of things. The hippies who gathered at Alloy craved their shot at the old American dream of building a city on a hill. They wanted to remake, literally remake with their brains and hands, the material world into a place that balanced nature and culture.

Countercultural cartoonist R. Crumb captured the Alloy vision perfectly in his most famous cartoon, "A Short History of America." This fifteen-panel set of rich line drawings starts with a picture of a pristine forest untouched by humans and then moves through the stages of technological development culminating with the Industrial Revolution. All the action takes place in the same scene, which changes from wilderness to a cluttered urban landscape dominated by clunky infrastructure, with the simple caption "What next?!" The final three, larger panels show three possible answers; the "worst case scenario: Ecological Disaster"; the "Fun" science fiction future, "Techno-Fix on the March!"; and finally, the "Ecotopian Solution." This last panel, obviously Crumb's choice, was the Alloy vision realized with a lush landscape of trees and animals living in harmony with bike-riding hippies, buildings powered by the sun, and Fuller's geodesic domes tucked into meadows. Crumb's "history" helped graphically explain Fuller's design science philosophy to a wide audience. Collaborations between scientists, designers, artists, and underground publishers fostered alliances across the generations, ultimately expanding the reach of the Alloy outliers.[10]

Fuller wasn't the only unlikely older guru adopted by the Alloy crew. Counterculturalists looking to make a new world for themselves in the 1960s were actively rediscovering the work of older contemporaries and early American pioneers and thinkers like the inventive Ben Franklin and the conservation-minded Theodore Roosevelt. Because many of their activities were centered in the American West, Native Americans, whose genius for living ecologically while making the most of what was at hand, and inventive pioneers, who built windmills for water and strung barbed wire where there was no wood, also served as inspirations. In his excellent study of ecological design, historian Peder Anker identifies the ecologically informed design science of the European Bauhaus modernists as another foundation for the later blending of art and science espoused by Fuller and his countercultural design science devotees and the ecological design movement they helped revive.[11] Thus, gatherings like Alloy, so seemingly of the frivolous hippie moment, were actually tapping long-standing trends in the history of science and technology, and the proposals that emerged from this particular happening represented a thoughtful revival of these trends rather than a clean break with the past.

When Baldwin (fig. 10.2) and crew arrived at Alloy, they were greeted by a crowd of scruffy hippie types whose outward appearance belied some impressive résumés as scientists, engineers, and designers.[12] Like Baldwin and his students, they had traveled far from across the United States and Canada to meet others like themselves who shared the alternative intellectual sensibilities of the counterculture but without the

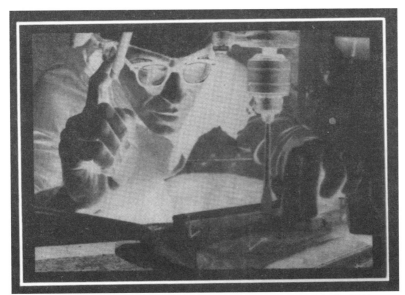

FIGURE 10.2 J. Baldwin depicted in an evocative negative for his now-classic article "One Highly-Evolved Tool Box," first appearing in *CoEvolution Quarterly*, Spring 1975. Oversized folder, box 32, Whole Earth Catalog Records (M1045), Department of Special Collections and University Libraries, Stanford University Libraries.

technophobia popularly attributed to the hippies. "The Alloy conference was amazing," Baldwin recalled. "It was the first get-together and recognizing that we were a 'we' of counterculture builders."[13]

Hippie Bricoleurs

In the 1970s anthropologist Claude Lévi-Strauss's notion of "savage bricoleurs" and rational engineers was turned upside down by countercultural scientists, engineers, and designers like the new tribe that gathered at Alloy. The green-minded innovators who attended Alloy linked "bricolage" (the creation "of a structure or structure of ideas achieved by using whatever comes to hand") with ad hoc participation in research as an antidote to formal modes of science and learning that they found too rigid and dogmatic. The ecological designers of the 1960s and 1970s counterculture were messy bricoleurs working outside the mainstream of science and engineering and expressing their accomplishments through the "medium of things and images."[14] Like earlier environmentalists such as Thoreau and Muir they were rendering their version of nature through prose, tools, and technologies alongside efforts to simply preserve more traditionally natural places like wilderness and parks.

The reconciliation of the human capacity for toolmaking and design with a love of nature was the Alloy mission. Historian Dan Flores has traced this conundrum all the way back to the Neolithic technological revolutionaries of six to ten thousand years ago, wondering where "along the lengthy but connected trail that follows tool-making out of the pre-human primate's rocks and sticks to our bulldozers . . . should we or could we have stopped?"[15] When the tents were stuffed in their sacks and the Volkswagen camper vans putted away down the dirt roads leading from La Luz at the end of the Alloy conclave, participants like J. Baldwin, Steve Baer, and Stewart Brand left with a new sense of community and, most importantly, a renewed faith in the ability of human intelligence to solve the environmental crisis through ideas and things already "at hand." Brand captured Alloy's spirit of reconciliation in an influential article, "Who Were They? (Who Are We?)," in his *Last Whole Earth Catalog* two years later. We were "Doers, primarily," he wrote, scientists and designers "with a functional grimy grasp on the world. World-thinkers, dropouts from specialization. Hope freaks."[16]

These certainly were not the hippies that Theodore Roszak so famously defined by their shared disgust with science and technology. Likewise, they were not the peace-loving flower children so adored by the media and loathed by conservatives. They were inventive bricoleurs: doers who shared a desire to unite creative design science with ecology to bring about a hipper, greener, more sustainable future. This sensibility was more than rhetoric; it drove meaningful research that resulted in the creation of viable products and knowledge that significantly contributed to the "everyday environmentalism" of the twenty-first century. The legacies of countercultural design science and the ecopragmatism of Alloy are visible in many places but most significantly in the areas of industrial design, shelter, and applied ecology as exhibited in the careers of individuals like J. Baldwin and Steve Baer.

Tool Freaks

Born in 1934, James T. Baldwin was the son of a successful AT&T engineer and grew up in the affluent New Jersey suburbs.[17] In many ways, Baldwin's life and work epitomize the rise of an ecological design movement and alternative ecological sensibility within the design community. Ecological design is, quite simply, the application of ecological insights to the research and design process to enable "sustainability." Most commonly associated with architecture, it applies equally to all aspects of material production and has been until recently the realm primarily of the bricoleur. Through projects across America and Canada, Baldwin established himself as a leading voice of ecopragmatism and ecological design before the field had a name. A true jack-of-all-trades and born bricoleur, he became a pivotal figure in the countercultural design community. Baldwin was a writer, a teacher to generations of design students, and an inventor of products like the "Pillow Dome." Baldwin's career was rich with valuable hands-on experience that he shared with hundreds of students at several universities between the 1950s and the new millennium. He was an early contributor to the *Whole Earth Catalog* and served for twenty years as the editor of the *Catalog*'s "Nomatics" and "Shelter" sections, which focused on science, engineering, tools, and design.

In the *Whole Earth Catalog*, "nomatics" meant living lightly on the land with a high degree of mobility. In his capacity as Nomatics editor Baldwin shared his design science insights with readers via discussions of the latest camping and commune gear. One of Baldwin's first jobs was designing tents for the innovative Bill Moss. Moss's company was among the first to offer serious outdoor gear to the post–World War II generation. His design studio emphasized small, light, and elegantly simple at a time when heavy military surplus was the norm. Moss shared Baldwin's growing interest in simple and environmentally friendly designs for a new generation of outdoor enthusiasts eager for tools enabling escape. Baldwin's tendencies as an open-minded bricoleur were honed at Moss, where designers used discarded or repurposed industrial materials to craft ingenious products like the Pop-Tent.

Best remembered by *Whole Earth Catalog* readers, however, were Baldwin's elegant essays on the history, use, and meaning of tools. He was a self-described "tool freak" who educated readers about the pleasures of simple tools like vise-grip pliers and the last-a-lifetime posthole digger with its handlebars that "encourage authoritative twisting through hard spots and past villainous stones."[18] In simple tools, newly designed or those that had stood the test of time, Baldwin found hope for "a benign technology useful for shaping a new environment" (fig. 10.3).[19] His love of tools went back to his early childhood, when he realized that most of the people around him "just didn't know how to look at tools. They didn't know how to think about tools. When I was a kid, in my rich neighborhood, when somebody burned out a light bulb, they called an electrician to come fix the light. I couldn't believe it."[20] Throughout his life Baldwin sought out others who knew tools and wanted to use them as entry points for debates about smart design. He met many of these folks in the military.

FIGURE 10.3 Baldwin's tools in negative from *CoEvolution Quarterly* production page proofs. Both tools and proofs are evidence of the do-it-yourself spirit of the moment. Oversized folder, box 32, Whole Earth Catalog Records (M1045), Department of Special Collections and University Libraries, Stanford University Libraries.

Like several of his Alloy colleagues, Baldwin was a veteran. Before he joined the ranks of the counterculture, he had served three years in the army's 10th Mountain Division stationed in Alaska at Eielson Air Force Base.[21] In Alaska Baldwin worked in a maintenance yard, taught soldiers to drive tanks and snow vehicles, and gained hands-on experience with industrial designs of all sorts. During this time, he watched repeatedly as poor designs caused injury or death to the unlucky. The lesson that "bad design can kill" was reinforced during his years in Alaska, where one of his jobs was to clean up human remains after the frequent plane crashes at the air base where he was stationed.

After leaving the army, Baldwin took a job with a major design firm in Ann Arbor, Michigan. He spent a brief, but significant, period learning about the world of industrial design as practiced in corporate America. One of his assignments, to design a fast-food chicken restaurant, led to an epiphany. Baldwin later said that he was not philosophically opposed to fast-food chicken, but he did not like the designs and wanted no part in foisting bad design on a landscape already littered with the postwar clutter of roadside attractions. Baldwin was teaching design part-time at local colleges and thought that "as a 'Coop' designer I wouldn't be able to retain the respect of my students."[22]

During the 1960s, Baldwin joined a handful of nomadic engineers, scientists, and designers who bounced between prestigious graduate programs, communes, and odd jobs. Baldwin spent time in seminar rooms studying but lacked the patience to stay away from the hands-on for long. It was during this period that he first encountered Fuller and again rethought his path in higher education. He later recalled Fuller's critique of housing design in particular as indicative of the insights he wanted to help spread. Fuller "showed the way a furnace is chosen; you design the building and then say, 'how big a furnace does it need?' instead of from the beginning, saying, 'Let's make the building in a way that doesn't need a furnace, or needs the smallest possible furnace.' In other words, built-in wastrel philosophy was being done as a matter of course, and was indeed being taught at University of Michigan."[23]

In order to combat the deficiencies he saw in formal education, Baldwin designed a "tool truck" to serve as a mobile classroom/workshop. Custom-made for ecological design work, the walls of Baldwin's truck folded out so that the workspace could expand for large projects like solar panels or construction of structural elements of a geodesic dome. The truck was "a three-dimensional sketch-pad—a place to make the first physical manifestation of an idea," he recalled.[24] Baldwin's well-used tool truck embodied the spirit of the countercultural bricoleur. Over the years the simple truck was a critical resource for many fledgling ecological design groups that lacked the resources for a basic stock of tools.[25] In both formal and alternative learning environments, Baldwin himself became an important resource and served as teacher and mentor for hundreds of students between 1967 and the early 2000s. Over the years, his ability to translate design science insights into plain language and with obvious enthusiasm contributed greatly to the growth of the field of ecological design. Baldwin joined others, such as writer John McPhee and ecologists Eugene Odum and Howard Odum, who were working to bridge gaps between science and culture for a new generation. From a shared sensibility among a dispersed group of individuals, design science joined other trends in environmental pragmatism transcending the confines of the industrial design community while contributing a material dimension to the American environmental movement.

Prototypers

Alloy organizer Steve Baer didn't fit the popularly accepted portrait of a hippie either. By the time of the New Mexico gathering, he had served in the military and had years of college under his belt from Amherst College and UCLA and a stint studying mathematics at the Eidgenössische technische Hochschule in Zurich, Switzerland. Baer was a skilled welder and designer with practical, on-the-job training in engineering and design problem solving.[26] When he organized Alloy, he had his own design company, Zomeworks, with partners Barry Hickman and Ed Heinz, and his design innovations and philosophy were circulating widely through the counterculture via a self-published booklet, *The Dome Cookbook*.[27]

Baer's later descriptions of Alloy as "a meld of information on Materials, Structure, Energy, Man, Magic, Evolution, and Consciousness" made for nicely countercultural quotes, as did portraits of the attendees as "outlaws, dope fiends, and fanatics."[28] But for Baer, Alloy was no Trips Festival, and like most of the people present, he was serious about using emerging insights from science and design to solve the pressing environmental issues of his day and helping pull together a community around these ideas. Not unlike David Kaiser's psychedelic physicists,[29] the designers at Alloy used the chemical tools and cultural sensibilities of the countercultural enlightenment to push serious and sophisticated debates about science toward new research results. In the process they opened a path for environmental advocacy very different from the mystical vision of nature that had propelled the environmental movement to unprecedented but polarizing success by the late 1960s. When Baer took the floor in the center of the Zomes in the New Mexico desert in the spring of 1969, he was a thoughtful ringmaster of proceedings that were groovy and hip, but in substance and sophistication this circus had more in common with a graduate seminar at Caltech or Berkeley than a hippie gathering in Haight-Ashbury.

Before Alloy, Baer and his collaborators Hickman and Heinz were best known for their well-documented design and construction of the alternative energy structures at the iconic Drop City commune near Trinidad in southeastern Colorado, where they and visiting design enthusiasts used that remote outpost to explore alternative architecture and appropriate-technology designs. Inspired by a Bucky Fuller lecture in Boulder, Colorado, they constructed a landscape of multicolored domes and innovative passive solar prototypes made from junked cars and other found materials on seven acres of windswept Colorado farmland.[30] Baer's Zomeworks, based in Albuquerque, New Mexico, converted his Drop City experiments into marketable, ecofriendly products (fig. 10.4).

In 1974 the Environmental Protection Agency, inspired by the work of Farm Security Administration photographers during the Great Depression, dispatched a corps of photographers around the nation under the banner of "DOCUMERICA"; their assignment was to record the era's environmental crisis and capture its most creative responses. Photographer Boyd Norton was sent to New Mexico to take a look at the communes and outposts of innovators gathering there. When Norton arrived in New Mexico, he sought out Baer and in a notable series of images captured the flavor of the Alloy-inspired blending of science, engineering, and countercultural sensibility. Norton's photographs reveal a productive and innovative aspect of the counterculture rarely depicted by the mainstream media. They also offer a sense of the thing-world of the ecopragmatists. One of Norton's series caught Steve

FIGURE 10.4 Zome house using solar heating built near Corrales, New Mexico. The modular, interconnected units are hexagon shaped with polyhedra roofs. Aluminum construction has an inner core of urethane foam for insulating efficiency. Glass walls (covered at night) pass sunlight to heat blackened fifty-five-gallon drums filled with water. Photograph by Boyd Norton for DOCUMERICA Project, 4/1974. Record Group 412: Records of the Environmental Protection Agency, 1944–2006, 555306, National Archives.

Baer in and around his workshop/laboratory and innovative off-the-grid home (fig. 10.5).[31]

Appearing thin and intense in Norton's photographs, Baer embodied the tough, practical, resourceful hippie bricoleur who knew how to use his scientific education to craft thoughtful structures and things. New Mexico in the 1960s was rich with scientists who worked in the atomic labs and testing regions scattered throughout the state. Taos and Santa Fe were nodes of countercultural activity, and the combination was perfect for innovators like Baer who straddled the world of formal and lay science. The hippie innovators of New Mexico fused their formal academic training in the sciences with the bricoleur willingness to embrace the reworking of existing materials and tools in light of new scientific theories and technological discoveries. As Norton wandered around Baer's expansive compound, he saw examples of inventions like Baer's patented "Double Bubble Wheel Engine," a heat engine that rotated "on a shaft set at an angle to the lines of force of a gravitational or centrifugal field," generating energy as it turned.[32] Like prototypes from all labs, the bubble engine was a rough but ingenious apparatus. It was only one of many examples of energy collection and generation devices that lurk in the backgrounds of Norton's historic images.

FIGURE 10.5 Steve Baer, founder and operator of Zomeworks of Albuquerque, New Mexico, described by Boyd Norton as "one of the pioneers in solar energy applications." Photograph by Boyd Norton for DOCUMERICA Project, 4/1974. Record Group 412: Records of the Environmental Protection Agency, 1944–2006, 555300, National Archives.

The landscape of Baer's Zomeworks recorded by Norton's photographs looks a lot like a high desert version of R. Crumb's Ecotopian future, with ecologically minded technologies woven into a beautiful natural landscape. In the pinyon-covered hills of New Mexico, Baer had brought to life the Alloy ideal of pragmatic environmental science undertaken by talented iconoclasts producing "brutally simple" designs, with marketing to wider audiences of consumers always in mind. Writing of Baer's notable efforts to make Alloy ideas reality, Stewart Brand asserted that Baer's research was "not academic, but . . . dedicated to developing and selling marketable systems." Baer explained that his Zomeworks business model was to foster a countercultural version of the "crusty old gents who design . . . in machine shops or invent all alone and pissed off." Baer's embrace of the marketplace was common among his cohort, and this attitude set them apart from popular understandings of counterculturalists as uniformly shouting down the materialism and mass culture of the 1950s and 1960s. Fiercely independent, Baer likewise was not a victim of savvy Madison Avenue types marketing hippie style as a way to cash in on psychedelic enthusiasms. Baer and his design science colleagues were somewhere in between. They were committed to putting their ideas and designs into practice and willing to embrace the contradictions and complexities of their day in order to make that happen.[33]

Baer's true passion was the sun. Throughout his formal education he had been fascinated with solar convection heating and conducted experiments in streams to determine how solar heat changed temperatures, depending on angle. Lewis Mumford's writings on the individual's relationship to technology and Farrington Daniels's *Direct Use of the Sun's Energy* were inspirations.[34] Baer's early alternative-energy designs, like a "solar chimney" at the Drop City commune made from recycled car windshields, were remarkable examples of functioning elements of an ecological architecture based on sophisticated science but rendered with simple found materials.[35]

During the 1970s Baer produced and marketed a series of viable solar products. He highlighted his ecological designs through striking photo-essays of his own in a self-published book, *Sunspots*. Photographs include images of products like his "Beadwall System" (US Patent 3,903,665), invented with student collaborator Dave Harrison.[36] This scalable window technology consisted of double-panes of clear glass fitted with a vacuum system that could fill the gaps between panels completely with insulating beads at the push of button. Like Frank's Alloy photographs, Baer's images showcase the sensibility of the bricoleur by illustrating found objects ingeniously repurposed or waiting for rebirth at the hands of a design scientist. *Sunspots* was lushly illustrated with crisp black-and-white photographs of junkyards, machine parts, and tools. Line drawings by the Drop City–based Criss-Cross Art Workshop clearly render the connections between Baer's science, the material world around him, and his ecopragmatic philosophy. Baer was a master ransacker of the ready at hand, and his text and images challenged his countercultural design peers by showing myriad ways that the necessary elements for an Ecotopian revolution were available in abundance to those who looked.

Material Culture

In their thinking about working with the material world "at hand" and taking their scientific training out to the field, the counterculture bricoleurs like Baldwin and Baer grasped the complexity of the nature/culture relationship in a way that the wilderness-based environmental movement often obscured. Bricoleurs by necessity looked to the land and resources around them for solutions to problems and in turn constructed buildings and objects that were of and from their place. Logically there

was a place for the human animal in this scheme. Thus, counterculture bricoleurs developed vernacular traditions of making that were in
sync with local environments. In their willingness to see the natural in
the cultural and the cultural in the natural, they were making do with
what was at hand—the messy mix of made and born that surrounded
them. The counterculturalists who adopted this age-old method challenged the tenets of progress and infinitude on which twentieth-century
modern industrial design was founded—not through a rejection of the
material world but through a shared desire to be makers of better stuff
in touch with place, ecosystems, and individual needs.[37] In short, for the
countercultural ecopragmatists like Baer and Baldwin, invention of ecologically sensitive or beneficial technologies, buildings, and objects in
service of sustainable living became a part of a broader desire to protect
and sustain nature everywhere, in cities as well as the countryside. These
countercultural makers were pulling the pragmatist's understanding of
the relationship of material entities to political engagement out of the
dustbin of history even as other hippies were raging against the materialism and consumer culture of the 1950s.

The counterculture engaged with the environmental movement in
many ways. Popular depictions of countercultural greens often simplified this engagement as a naïve return to nature exemplified by the
back-to-the-land movement; flower children, acorn gatherers, and nudists cavorting in a meadow were familiar tropes of these generalizations.
Unfortunately, historians have often perpetuated these stereotypes. [38]
The efforts of the Alloyed bricoleurs to explore the science and craft
of ecological technologies and green things complicated prevailing assumptions that objects and materiality are "antithetical to social and political engagement."[39] Sociologist Bruno Latour and colleagues formally
challenged "the idea that the rise of industrial objects—from technology
to commodities and scientific objects—should be held responsible for
the demise of sociability and politics." More recently Noortje Marres
concluded that the rise of the tech- and science-friendly sustainability
sensibility fostered by the counterculture bridged gaps in the history of
material production. In her formulation, sustainable-living technologies
and green-tech science link John Dewey's enthusiasm about the pragmatic power of thoughtfully designed objects to empower and educate
from the early part of the twentieth century with the later efforts of the
countercultural bricoleurs.[40]

The embrace of the material world has long been considered the
Achilles heel of the counterculture. In 1967 the Diggers, a San Francisco–
based underground theater group, were already declaring the "death"

of the hippie in response to rapid commodification of the counterculture.[41] Critics from all sides dismissed the counterculture in part because
of the market savvy of its adherents and their willingness to vigorously
jump into the world of commerce when it suited their needs and desires.
For critics, any whiff of commercialism was considered a sad sellout
for hippies who were assumed to be political radicals or aesthetics who
had no business making and selling things. These generalizations have
long obscured the complexity of the counterculture(s) and the ideas and
actions of the ecopragmatists in particular. The pragmatic wing of the
counterculture was rarely revolutionary but also never "sold out." They
simply assumed that market economy was here to stay and sought their
place within it even as they chaffed against its norms. This willingness
to make peace with capitalism while striving to literally make a better
future harkens back to the early part of the twentieth century and the
idealistic materiality of the American pragmatists.

Ecopragmatists

Countercultural bricoleurs like Baer and Baldwin recognized the material world of made things as part of nature and saw promise rather than
disaster in the artifacts of the Industrial Revolution. This simple insight
from the 1970s was a critical shift in environmentalism. The Alloyed
bricoleurs revived a potent strain of American environmental thought
and set the stage for an important move toward ecopragmatism in the
coming decades. The foundation of pragmatism rests on a belief in
tools—tools, as Louis Menand reminds us, even as simple as "forks and
knives" that can be used to accomplish a purpose.[42] The pragmatists
loved material things created by inventive people. John Dewey spoke
often of the power of good tools for effecting social change. American pragmatists, like Dewey and William James, described a practical,
problem-solving, tool-using, environment-transforming strain of American culture that designers like Baer and Baldwin clearly embraced.[43]
    American pragmatism in Dewey's day was characterized by an enor-
mous optimism in our ability as a people to effect change when we put
our minds to it. Dewey's ideas resonated with the Alloy crew because
they offered a hopeful vision of the future during a period of astounding
change. Optimistic pragmatism was a cornerstone of the conservation
movement tied directly to ebbs and flows of enthusiasm for science.
Dewey spoke directly to the relationship between nature, the material
world, and science and would have appreciated Crumb's graphic illustration of a future where the three worked in harmony.[44] Between the

1920s and 1960s the grassroots conservation movement worked closely with resource scientists toward a more balanced relationship between culture and nature. It was assumed by most that middle grounds and hybrid landscapes were the norm and that pure preservation of the "pristine" was special and rare. Ecological pragmatism was, in other words, the most common form of environmental advocacy for the first half of the twentieth century.

This changed in the late 1950s when wilderness preservation moved to the center of American environmental advocacy. By 1960, wilderness was both a captivating ideal and an increasingly practical bipartisan legislative tool effectively used by environmentalists advocating for expansion of federal control of certain public lands.[45] The remarkable success of these advocates at sounding the alarm for the wild ensured that the largest and most important generation of environmental advocates leaned toward safeguarding the "choice morsels" of nature and away from the more human-centered ecopragmatist views of the conservation movement. The idealist wilderness movement leading to the landmark Wilderness Act of 1964 meshed perfectly with the back-to-the-land and return-to-nature sensibilities of the emerging counterculture. And yet, at the height of the wilderness moment, the old optimism of the conservationists, their pragmatism and willingness to compromise between human made and nature born, struck a chord with a significant faction of the counterculture. The best evidence of this enthusiasm can be seen on the pages of the *Whole Earth Catalog* and in the writings of its editor Stewart Brand.

Cataloging Bricolage

Alloy and the material world of the countercultural bricoleurs was nowhere more evident than on the pages of the *Whole Earth Catalog*. The most significant edition, *The Last Whole Earth Catalog* (1971), featured a major retrospective on Alloy and some explanations of the importance of this event and its participants. Thinking that this would truly be the "last" of his astonishingly successful catalogs, Stewart Brand used a history of Alloy to explain why he thought his publishing efforts had struck such a chord. It was Alloy, he wrote, that exemplified the materiality of the catalog, the logic behind its seemingly eclectic mix of tools and ideas, the environmental sensibility of the catalog, and its goal of promoting ecological sustainability by uniting traditional environmental stewardship with technology and scientific know-how.[46]

Brand illustrated this significant article with photographs featuring the bricoleurs and their tools and resources. For the uninitiated, the mix makes little sense. Like Baer's *Sunspots* photographs, Brand chose stark images of objects and prototypes strewn across a natural landscape. The photo-essay is the opposite of much of the environmental photography of the 1970s. Where the Sierra Club and other such groups focused on the flotsam and jetsam of the industrial age to point to the ways human things had destroyed nature, Brand's photographs seem to suggest that one man's trash is another man's treasure. The essay opens with an image of Steve Baer holding forth in a barn surrounded by pieces of Zomes. A still life with tools follows, dominated by an open chest surrounded by scattered materials clearly used in the construction of the scene around La Luz. Other images show hippies hard at work building frames for structures or piecing together junk to make a passive energy device.

This collage of materials, tools, and labor conveyed a message that the countercultural readers and contributors to the *Whole Earth Catalog* understood: people working with their minds and hands to repurpose their inherited material world were the best hope for ecological harmony. The 450 illustrated pages that surrounded Brand's Alloy essay offered hundreds of examples of new and old science, green technologies, and ecological design and living all aimed at reconciling environmentalism and material production. More specifically the *Catalog* urged readers to think hard about the nature of things and enlisted individuals "disinterestedly excited about" objects to explain how even common goods could become tools toward an apolitical design science revolution, where thoughtful attention to the front end of production could preemptively solve potentially detrimental environmental consequences on the back side.[47] This constructive human-centered environmental philosophy tapped deep into conservative impulses like thrift, ingenuity, technical know-how, tinkering, and individual responsibility and agency.[48]

The various editions of the *Whole Earth Catalog* became the field guide for a countercultural maker-movement and, in print and graphics, codified the sensibilities of the Alloyed bricoleurs. Hippies were known in their day and to a large extent now as hedonistic dropouts and do-nothings, but for people who were supposed to be indolent they sure were busy. The pages of the *Whole Earth Catalog* were loaded with books for self-study, tools for work in the fields or the shop, gear for global exploration, and a universe of useful goods and do-it-yourself (DIY) information. Virtually everything in this "bible of the counterculture" was aimed at doers and doing. The counterculturalists drawn to the *Catalog*, numbering in the millions, were searching for "an anti-dote to the rationalist, industrialized work" that characterized America in the 50s and 60s.[49] While they rejected the status quo of corporate drudgery, they hungered for the "good work" of William Morris and his aesthetic movement. They sought to link older craft traditions and celebrations of "authenticity" and vernacular design with the new materials, techniques, and scientific insights of their own Cold War age.[50]

The desire for an authentic life and "good work" drove the back-to-the-land movement, and for many readers of the *Whole Earth Catalog* the search for the authentic meant a genuine, often ill-fated, return to primitivism. While some communards plowed naked with ancient tools, the Alloyed bricoleurs used their training in science, engineering, and technological design to build and redesign things and places aiming for a more practical blending of the new and the old in harmony with place and nature. Baldwin and Baer lived and labored in commune settings but took their cues from Hewlett and Packard and spent much of their time in urban shops and garages tinkering or out in the field.[51]

The goal of these counterculturalists was a hybridization of "advance technology and a return to nature," and the skills and knowledge required were on the leading edge of the possible. The Alloy types differed from their suit-wearing counterparts in corporate design only in their desire for ecology to be inherent in their work and in their belief that design needed to be freed from the Cold War corporate and government control. The "idea of restraint, of existing within limitations, was informed by a sustainable ethos," and the ad hoc ingenuity of the Alloy tribe of designers and ecopragmatists came to life most visibly through the international self-build and shelter movement.[52] Industrial designers like J. Baldwin forged alliances with "outlaw" architecture students and vernacular architecture advocates like Alloy veteran Lloyd Kahn, who worked with Baldwin, Baer, and Brand before launching a wildly successful publishing venture of his own. Kahn was a part of an emerging global movement represented not only on the pages of the *Whole Earth Catalog* but also in a range of iconoclastic design publications like Kahn's own *Shelter* in the United States and *Casabella* in Italy that promoted ecopragmatism and bricolage through architecture and design.[53]

Shelter

One of the most significant ways in which the countercultures expressed interest and enthusiasm for science and technology was through sustainable architecture. It is not an exaggeration to say that the entire field of ecological design emerged from countercultural reevaluations of making as a means of addressing environmental concerns. The DIY ethos and desire to reclaim and refashion the ecopragmatist mantle of an earlier generation were manifested most significantly in the area of shelter and self-build housing.

One of the first projects to benefit from J. Baldwin's tool truck was a green architecture project at the Farallones Institute. Farallones was established by pioneering green architect and self-described "outlaw builder" Sim Van der Ryn in Occidental, California.[54] Van Der Ryn was a professor of architecture at the University of California–Berkeley, author of a best-selling book on waste composting (*The Toilet Papers*), and a pioneering ecological designer.[55] Through his teaching career, he influenced several generations of architects, whom he taught to place ecological concerns at the center of their building designs. During Jerry Brown's first administration as governor of California (1975–83), Van der Ryn was appointed state architect and designed passive solar government buildings and helped push tax initiatives that funded appropriate-technology development in California.[56] In 1969 he founded the Farallones Institute to create a teaching laboratory for ecological design. Baldwin and his tools helped build the sustainable buildings and structures that made up the modestly sized but innovative institute.

Farallones was more than just an experiment in green architecture; it was a full-fledged attempt to build a complex of buildings and structures that were as self-sustaining as possible with the technology of the day. At Farallones, the buildings were designed to produce their own energy, recycle waste and reuse it as fuel, and provide for some of the food needs of the institute through organic gardens using the latest techniques in sustainable agriculture.[57]

All the protagonists of Alloy moved in part or whole toward ecological design of structures. The focus on shelter and domestic technologies among countercultural designers tapped a deep tradition of placing the reinvention of the home at the heart of the cultural revolution.[58] No other aspect of the counterculture, in fact, captured the spirit of the age better than the simple desire to provide one's own shelter. From innovations like Fuller's geodesic dome to reevaluations of ancient techniques like adobe building, the counterculture was obsessed with reinventing the ideal of the American home and, in the process, creating autodidactic models for architecture and design aimed at enabling practical ecological living. At communes and neighborhoods across the world, counterculture designers developed new technologies for ecological living that melded insights from the latest alternative-technology research with an older craft tradition.[59] Much of this ecological design activity happened in the American West, and it was not just hippies fueling interest in environmental design. The massive migration to the Sun Belt in the decades following World War II provided strong motivation for creative solutions to housing suited to fragile, arid environments.

Western-focused shelter discussions were a key feature of the Alloy gathering and the early years of the ecological design movement. These efforts built on the strong foundation provided by New Deal regionalists like Arthur Carhart, who celebrated regionally sensitive design in magazines like *Sunset,* and later decentralist-distributist movement proponents like New Mexico solar architect Peter Van Dresser.[60] Architectural historian Chris Wilson points to Van Dresser as a protocounterculture ecopragmatist. Van Dresser started building solar houses in the early 1950s and penned an important statement on alternative design with his *Development on a Human Scale* (1972). Van Dresser's research presaged the countercultural desire to link scientific research and new design materials with regional traditions in an effort to ensure that "decentralized economic development and local culture be mutually reinforcing."[61]

Celebrated countercultural design laboratories, like Drop City, Libre, Red Rockers, Paolo Soleri's Arizona Arcosanti experiments in hive-like living, the New Alchemy Institute, the Pacific High School experiment, the Farallones Institute, and Steve Baer's Zomeworks, to name just a few, received wide coverage in the popular media. Popular attention tended to focus on the psychedelic weirdness of these efforts, but students of design history viewed these experiments as efforts to revive traditions rather than spark revolution. The hippie weirdness was only a sideshow. These various efforts led to popular publications on the expanding universe of self-build that transcended region.[62] *Mother Earth News,* founded by John and Jane Shuttleworth in 1971, dedicated a substantial portion of its pages to evolving theories about shelter and became the preferred how-to guide for persistent communards through the 1990s.[63]

Of all the self-build publications, none was more influential than Lloyd Kahn's best-selling *Shelter,* a classic of DIY and hand-built homes from around the world.[64] Like Brand's *Whole Earth Catalog,* Shelter used lush illustrations to present the material world of the bricoleur to a wide audience of readers. *Shelter* was a subtle assault on the replication and conformity of postwar domestic architecture and a celebration of precedents that "evolved across centuries or millennia."[65] Rather than present design science or engineering research in cold prose, *Shelter* used photographs and art to show how technologies evolved. This compelling presentation of complex science and engineering was one of Alloy's lasting legacies. Kahn's successful efforts to disseminate design science insights to a wide audience through trade publications preceded best sellers by Gary Zukav and by Fritjof Capra that similarly popularized academic research through innovative trade books that presented hard science through the lens of a countercultural sensibility.[66]

Kahn founded Shelter Publications in 1970 and produced a series of successful design classics over a thirty-year period.[67] Prior to *Shelter*, his two *Dome Books*, along with Steve Baer's *Dome Cookbook*, were the standard instruction manuals for Buckminster Fuller's design science revolution, ensuring that the geodesic dome or some variant became the standard architectural form for communes, mavericks, and iconoclasts all over the American West. More mainstream architects and designers, inspired by renegade amateurs, began incorporating counterculture design concepts into complex and well-funded projects, resulting in wider acceptance of the field of ecological design.

These counterculture shelter trends are significant because the linking of alternative technology with older bioregional architectural traditions is perhaps the best example of the design science revival of the 1970s counterculture. The simple insights that emerged from efforts to rethink domestic architecture came to characterize the sustainability movement of the turn of the millennium, when ecopragmatism reached full bloom and went mainstream. It was the hippies who linked environmental advocacy with local and daily application of scientific and design research on a human scale. In so doing, they greatly expanded the constituency of the environmental movement at a time of critical retrenchment. The effort to bring sophisticated design science wedded with the latest insights from ecological research into the home, as architectural historians Christine Macy and Sarah Bonnemaison remind us, provides a counternarrative to the "nineteenth century proposition that nature must be isolated and purified of human activities in order to be protected."[68] This sensibility also countered the polarized environmental politics attached to wilderness, endangered species, and resource preservation and offered a middle ground for people from different political persuasions who nonetheless shared a pragmatic interest in innovative design for a sustainable future.

Design Ecologies

In 2009 Stewart Brand revisited the Alloyed environmentalism of his *Whole Earth Catalog* in a controversial new book, *Whole Earth Dis-*

cipline: An Ecopragmatist Manifesto.[69] The book caused a great deal
of debate primarily because of Brand's untimely advocacy for nuclear
power on the eve of Fukushima. The bulk of the book, however, focused
on somewhat less controversial legacies of the design science revolution
with an eye toward future promise. In particular, Brand surveyed the
science of restored wildlands and engineered landscapes, which he felt
were the logical next frontier for innovative bricoleurs. Restoration of
wildlands and the acceptance of the synergy between nature and culture
that this activity requires were squarely in the wheelhouse of the countercultural ecopragmatist sensibility that fostered reconciliations of science and environmental activism. Ecological design sought to reconcile
architecture with environmental limitations. Environmental restoration
took the notion a step further, suggesting that architecture was only
a "mediating juncture between a natural order and a synthetic, man-
made order."[70] The next step for designers was a "weedy architecture"
focused on restoration of the lands and places that might not include
buildings at all. Proponents argued that wasted lands on the fringes of
the wild and abandoned urban lots could be redesigned and restored
to productive ecosystems and hybrid landscapes. The insights of the
designer blended with the theories of the ecologist could result in places
refashioned as cultural/natural zones of utility, ecological diversity, and
beauty.[71]

Environmental restoration was the yin to wilderness preservation's
yang. Aldo Leopold's *Sand County Almanac* (1949), which immortal-
izes his famous Sand County restoration project, is the standard text
on the science and philosophy of human engineering of ecosystems, an
"ancient art, practiced and malpracticed by every human society since
the mastery of fire."[72] Leopold is best remembered as a pioneering advocate of the wilderness ideal exemplified by the National Forests, yet
it was his restoration work on a tortured piece of land heavily modified
by humans that provided the material for his most inspired writing. Environmentalist Barry Lopez later argued that environmental restoration
of damaged lands and crippled places was the most American form of
environmental activism. "It is no accident," he wrote, "that restoration
work, with its themes of scientific research, worthy physical labor, and
spiritual renewal, suits a Western temperament so well."[73]

Restoration emphasized human ingenuity and directly linked technological prowess and scientific knowledge with environmental health.
The ecopragmatists consistently favored environmentalism that focused
on places where people lived and worked or had lived and worked.
They spent little time advocating the preservation of "pristine wilder-

ness untrammeled by man," not because they disagreed with the wilderness movement per se, but because they tended to have a more historical conception of nature that left little room for the notion of the "pristine" and "untrammeled." The "partnership with nature," as ecologist Howard Odum explained, the "coupling" of nature and culture, is the goal of ecological engineering and only by focusing on the connections can we understand the systems of "fantastic complexity" that comprise the Earth.[74] Writing in the early 1970s and using his research from the 1950s and 1960s, Howard Odum, along with his brother and fellow ecologist Eugene, provided the ecological science foundation for the ecopragmatists, just as Fuller had provided the design science foundation. Likewise, Ian McHarg's *Design with Nature* (1969) and D'Arcy Thompson's *On Growth and Form* (1917), among other publications, linked design science with landscape architecture and broader discussions of ecology and design that influenced countercultural thinking. The *Whole Earth Catalog* prominently featured these works and encouraged readers to digest them alongside Fuller's oeuvre. Brand also pointed fans of Alloy to William Thomas Jr.'s *Man's Role in Changing the Earth* (1955) and Christopher Alexander's *Notes on the Synthesis of Form* (1964). All these books argued for the connections between culture and nature and assumed that smart and motivated people could use scientific insights and technical know-how to change things for the better.

Howard Odum's hope for "partnership of man and nature" required a fusion of ecology and technology beyond architecture and objects.[75] As Henry Trim's essay in this collection (chapter 5) demonstrates, the New Alchemy Institute was one of the most celebrated ecological design and environmental restoration efforts of the 1970s. By the mid-1970s, the grassroots efforts of scattered communards and individuals were eclipsed by far grander and complex efforts in "Ecotechnics," like Biosphere II in the Arizona desert and NASA space colony research, that had support beyond anything the countercultural bricoleurs could have dreamed.[76] Space colonies and billionaire-funded science experiments like Biosphere II split the counterculture bricoleurs and tested the limits of their shared enthusiasm for ecology united with technology. Moreover, these monumental endeavors seemed, to some, like a return to the megamachine, a move that appeared to negate the efforts of the bricoleurs to create a small-scale, everyday, and accessible ecopragmatist revival. Having launched his catalog based on the assumption that the view of Earth from space would spark an environmental revolution, Stewart Brand relished all aspects of science, big and small, that inspired holistic thinking about "Spaceship Earth."[77] While space colonies and the closed-system ecology they relied on sparked intense debates that permanently divided the coalition who had gathered at Alloy, other forms of space-based environmental research were embraced by most of this cohort as critical for ecological understanding and were accepted, if grudgingly, as high-tech examples of working with what's "at hand" on a planetary and galactic scale.

The partnership of made and born, science and craft, nature and culture, all working toward sustainability, was captured in the masterwork of the Alloyed bricoleurs: *The California Water Atlas* (1978). The *Atlas* was one of the first concrete environmental tools to emerge from space research, a beautiful large-format book produced by a remarkable group of design science illuminati. The effort brought together scientists, ecologists, environmentalists, astronauts, and Alloy veterans. They used high-resolution satellite imagery to graphically illustrate watershed relationships and bioregional connections in California; these images were framed by accessible essays translating the latest insights from environmental science research for a lay audience.[78] The bricoleur manuals *Shelter*, *Whole Earth Catalog*, and *Sunspots* were clearly the model for this important publication. The team who researched and crafted this wonderful artifact used their scientific training and technical expertise to create an illuminated manuscript of science that perfectly captured the blending of old and new, craft and high tech, that the counterculturalists appreciated and promoted.

Speaking with characteristic boldness about the accomplishments of the Alloy veterans, Stewart Brand, who served as the advisory group chairman for the *Atlas*, said simply, "we owe it all to the Hippies."[79] Maybe we do not really owe it *all* to the hippies, but as the lasting legacies of the Alloy conclave illustrate, history does owe these hippies something more thoughtful than an offhand dismissal as dropouts or hapless flower children. In a sometimes flawed and always ad hoc way they helped revive critical knowledge and practices, refashioning them in light of scientific and technological research and presenting their insights in creative new forms to important new audiences. Their accomplishments left a legacy that reverberates today and remains visible in the built, printed, and natural environment.

The practice of scientific research is often perceived as a cycle of eureka moments followed by revolutions. A closer look almost always reveals a slower model of collaboration and collective discovery sometimes lasting for decades, even generations, before the eureka arrives. As David Kaiser demonstrates with his history of hippies and quantum physics, some science requires the close attention of maverick thinkers and outliers working on the far fringes to revive important ideas long dismissed because they went against the perceived wisdom of their day.[80] It may seem counterintuitive to look to the countercultures of the long sixties for insights about environmental science and thoughtful ecopragmatism, but it would be difficult to understand our twenty-first-century environmental culture, green design enthusiasms, and pervasive "natural capitalism" without a closer look at the countercultural iconoclasts who "with cunning and resource" ransacked the ready at hand to "create something new . . . based on the continuous reworking of the received elements of the world."[81]

Acknowledgments

Sincere thanks to editors Patrick McCray and David Kaiser for the opportunity to revisit the history of countercultural design science that I first wrote about in my book *Counterculture Green: The "Whole Earth Catalog" and American Environmentalism* (Lawrence: University Press of Kansas, 2007). Since that book was first published, a remarkable run of wonderful new scholarship on design science and the counterculture has emerged, greatly to the benefit of those of us interested in the intersections of the counterculture and the science of "sustainability." Some of the material in this essay covers familiar ground from my earlier work, but all has been reenvisioned and expanded in light of this new scholarship, which has revealed new sources and brought together an even more eclectic interdisciplinary perspective than was available just five years ago.

Notes

1. Glenn Adamson and Jane Pavitt, eds., *Postmodernism: Style and Subversion, 1970–1990* (London: Victoria and Albert Publishing, 2011), 113.

2. J. Baldwin, interview by the author, 13 February 2006, Penngrove, CA, 39. Passive solar is simply the design concept of using the elements of a structure to collect and store energy from the sun during winter months for heat and to shade and deflect excessive solar heat during the summer months.

3. For details on some of these groups, see Simon Sadler, *Archigram: Architecture without Architecture* (Cambridge, MA: MIT Press, 2005), 150–51. Steve Baer's design company Zomeworks was invited later that spring to represent the United States at the prestigious Paris Biennale. See Simon Sadler, "Drop City Revisited," *Journal of Architectural Education* 59, no. 3 (Spring 2006): 5.

4. Sociologist Talcott Parsons coined the term "counterculture" to explain groups of like-minded iconoclasts who chafed against the organization men and megamachines of 1950s Cold War America. But it was historian Theodore Roszak who launched the term into the popular consciousness by giving the label to a colorful and controversial generation a decade later. In his of-the-moment history, *The Making of a Counter Culture: Reflections on the Technocratic Society and Its Youthful Opposition* (Garden City, NY: Anchor / Doubleday, 1969), Roszak identified what he saw as the unifying ideals of the messy and scattered countercultural "movement." In his telling, the young rebels were united most significantly by a rejection of technocracy and "Big Science" and by their shared appreciation of deeper American traditions of craft and self-sufficiency. Likewise, Charles Reich, in his controversial best seller *The Greening of America* (New York: Random House, 1970), also argued that rejection of technology was a fundamental component of the counterculture ideology and the reason these young rebels would usher in a new kind of environmentalism. These influential views obscured the work of the Alloyed wing of the counterculture.

5. R. Buckminster Fuller and Robert Marks, *The Dymaxion World of Buckminster Fuller* (Garden City, NY: Anchor, 1973), 12 (among the many books that chronicle Fuller's life and work, *Dymaxion World* is one of the most accessible and concise); Robert Marks, ed., *Buckminster Fuller: Ideas and Integrities* (Englewood Cliffs, NJ: Prentice-Hall, 1963); Robert Snyder, ed., *Buckminster Fuller: Autobiographical Monologue/Scenario* (New York: St. Martin's, 1980); E. J. Applewhite, *Cosmic Fishing* (New York: Macmillan, 1985); J. Baldwin, *Bucky Works: Buckminster Fuller's Ideas for Today* (New York: John Wiley and Sons, 1996); and Thomas T. K. Zung, ed., *Buckminster Fuller: Anthology for the New Millennium* (New York: St. Martin's, 2001).

6. Baldwin, *Bucky Works*, 62.

7. Ibid., v.

8. For an excellent study of Fuller in the context of the larger holistic movement, see Linda Sargent Wood, *A More Perfect Union: Holistic Worldviews and the Transformation of American Culture after World War II* (New York: Oxford University Press, 2010). See also Caleb Crain, "Good at Being Gods," *London Review of Books* 30, no. 24 (December 2008): 1–8.

9. For legacies of this sensibility, see Michael Guggenheim, "The Long History of Prototypes," *ARC Studio* (2010), http://anthropos-lab.net/studio/episode/03/; and Claude Lévi-Strauss, *The Savage Mind* (Chicago: University of Chicago Press, 1966), 16–30. Do-it-yourself and "Maker" legacies are explored in Peter H. Diamandis and Steven Kotler, *Abundance: The Future Is Better than You Think* (New York: Free Press, 2012), 123–31.

10. R. Crumb, "A Short History of America," *CoEvolution Quarterly* 23 (Fall 1979): 21–24.

11. Peder Anker, *From Bauhaus to Eco-house: A History of Ecological Design* (Baton Rouge: Louisiana State University Press, 2010), 4. See Anker's chapter "The Unification of Art and Science." I agree with his argument, but the vernacular, received-wisdom mode of thinking about ecological design was there throughout not just after the space-based cabin ecological model faded.

What's interesting is the split between the cabin model and the bricoleur model and their later reunification (129).

12. Stewart Brand, "History," in *The Last Whole Earth Catalog*, ed. Stewart Brand (New York: Portola Institute, distributed by Random House, 1971), 439; and Stewart Brand interview in Chris Zelov and Phil Cousineau, eds., *Design Outlaws on the Ecological Frontier* (Easton, PA: Knossus, 1997), 76.

13. Baldwin, interview, 39.

14. Lévi-Strauss, *Savage Mind*, 21.

15. Dan Flores, "Nature's Children: Environmental History as Human History," in *Human/Nature: Biology, Culture, and Environmental History*, ed. John Herron and Andrew Kirk (Albuquerque: University of New Mexico Press, 1999), 17.

16. Brand, *Last Whole Earth Catalog*, 112. See also J. Baldwin and Stewart Brand, eds., *Soft-Tech* (New York: Penguin, 1978), 8. For fans of short titles and the most active example of the *Whole Earth Catalog* reimagined by designers, see Rem Koolhaas, Bruce Mau, and Hans Werlemann, *S, M, L, XL* (New York: Monacelli Press, 1998); Bruce Mau, *Lifestyle* (London: Phaidon, 2000); and Bruce Mau, *Massive Change* (London: Phaidon, 2004). See also Lisa Tilder, "Ecologies of Access," in *Design Ecologies: Essays on the Nature of Design*, ed. Lisa Tilder and Beth Blostein (New York: Princeton Architectural Press, 2010), 41–59.

17. J. Baldwin, "Designing Designers," *Whole Earth Review* 86 (Summer 1995): 14–16. Baldwin explains his design teaching strategies and philosophy on becoming a working designer in his *Bucky Works*, 222–25. See also Lloyd Kahn, "Pacific Domes," in *DomeBook 2* (Bolinas, CA: Shelter, 1972), 20–33. For fantastic pictures of the construction of the domes by students, see Lloyd Kahn, "Pacific High School Revisited," and J. Baldwin, "Sun and Wind in New Mexico," both in Lloyd Kahn, ed., *Shelter* (Bolinas, CA: Shelter, 1973), 119, 164. See also Eva Diaz, "Dome Culture in the Twenty-First Century," *Grey Room* 42 (Winter 2011): 80–105.

18. J. Baldwin and Stewart Brand, eds., *Whole Earth Ecolog: The Best of Environmental Tools and Ideas* (New York: Harmony Books, 1990), 40.

19. Thomas P. Hughes, *Human-Built World: How to Think about Technology and Culture* (Chicago: University of Chicago Press, 2004), 88.

20. Baldwin, interview, 65–66.

21. Ibid. Baldwin bristled at the term "counterculture," arguing instead that he and his cohort were the "counter-counterculture" or just the vanguard of the culture, and it just took everyone else a long time to catch up.

22. J. Baldwin, "Making Things Right," Whole Earth Review 86 (Fall 1995): 6.

23. Zelov and Cousineau, *Design Outlaws*, 44.

24. J. Baldwin, "One Highly Evolved Toolbox Evolves Some More," *Whole Earth Review*, July 1985, 152. A very nice recap of Baldwin's thinking on tools and their role in a fast-changing world is J. Baldwin, "The Ultimate Swiss Omni Knife," *Whole Earth Review*, 30th anniversary issue, Winter 1998, 20.

25. Stewart Brand, "Introductions," in Baldwin and Brand, *Soft-Tech*. Brand provides a nice, concise overview of Baldwin's remarkable appropriate-

technology experiences and explains in brief how soft tech became a central feature of the *CoEvolution Quarterly* (4).

26. "The Plowboy Interview: Steve and Holly Baer," *Mother Earth News* 22 (July/August 1973): 8; Steve Baer, *Sunspots: Collected Facts and Solar Fiction*, 2nd ed. (Albuquerque, NM: Zomeworks, 1977); and Steve Baer, "The Clothesline Paradox," in Baldwin and Brand, *Soft-Tech*, 50–51.

27. Steve Baer, *The Dome Cookbook* (Corrales, NM: Lama Foundation, 1969).

28. Brand, "History," 111–12.

29. See David Kaiser, *How the Hippies Saved Physics: Science, Counterculture, and the Quantum Revival* (New York: W. W. Norton, 2011).

30. Bill Voyd, "Funk Architecture," in *Shelter and Society*, ed. Paul Oliver (New York: Praeger, 1969); Peter Rabbit, *Drop City* (New York: Olympia, 1971); Timothy Miller, *The Hippies and American Values* (Nashville: University of Tennessee Press, 1991); Steve Baer, "Zomes," in Kahn, ed., *Shelter*, 134–35; Baer, *Sunspots*; and Sadler, "Drop City Revisited," 5–14. See most recently Mark Matthews, *Droppers: America's First Hippie Commune, Drop City* (Norman: University of Oklahoma Press, 2010); and Richard Fairfield, *The Modern Utopian: Alternative Communities of the '60s and '70s* (Port Townsend: Process Media, 2010). For the most recent examinations of Drop City and Zome architecture's legacies, see Erin Elder, "How to Build a Commune: Drop City's Influence on the Southwestern Commune Movement," and Amy Azzarito, "Libre, Colorado, and the Hand-Built Home," both in *West of Center: Art and the Counterculture Experiment in America, 1965–1977*, ed. Elissa Auther and Adam Lerner (Minneapolis: University of Minnesota Press, 2012).

31. The Boyd Norton Documerica Series on Zomeworks can be accessed at http://research.archives.gov/description/555297 (Boyd Norton Photographer; National Archives Identifier: 555297; Local Identifier: 412-DA-12845; Creator(s): Environmental Protection Agency). See also Bruce L. Bustard, *Searching for the Seventies: The DOCUMERICA Photography Project* (Washington, DC: Foundation for the National Archives, 2013).

32. Steven Baer, "Double Bubble Wheel Engine," US Patent 4,134,264, granted 17 January 1977; Steve Baer, "Gravity Engines and the Diving Engine," *CoEvolution Quarterly*, Summer 1974, 81–83; and Steve Baer, *Sunspots*, 77–84.

33. Steve Baer to Stewart Brand, 28 December 1971, Point Foundation Records (M1441, box 2:22), Department of Special Collections, Stanford University Libraries. For a discussion of technology, science, and government, see Baer, *Sunspots*. For contrasting views of counterculture market philosophy, compare Roszak's *Making of a Counter Culture* with Thomas Frank's *The Conquest of Cool: Business Culture, Counterculture, and the Rise of Hip Consumerism* (Chicago: University of Chicago Press, 1997). The design science tribe doesn't fit either of these better-known characterizations of the counterculture.

34. Baer, *Sunspots*, 8; and Farrington Daniels, *Direct Use of the Sun's Energy* (New Haven, CT: Yale University Press, 1964).

35. Sadler, "Drop City Revisited," 12.

36. David Harrison, "Beadwall System," US Patent 3,903,665, granted 9 September 1975.

37. The classic study of the vernacular tradition is Bernard Rudofsky, *Architecture without Architects: A Short Introduction to Non-pedigreed Architecture* (New York: Doubleday, 1969). See also Catharine Rossi, "Bricolage, Hybridity, Circularity: Crafting Production Strategies in Critical and Conceptual Design," *Design and Culture* 5, no. 1 (March 2013): 69–87. In the same issue, see Stephen Knott, "Design in the Age of Prosumption: The Craft of Design after the Object," 45–67.

38. The historic relationships between science, technology, and environment are a primary concern of the field of environmental history. This field shares many interests with the history of science and technology. In particular, both of these areas of study seek to reveal the interplay between technological systems and the natural environment and to understand the cultural forces that shape perceptions of these interactions. Environmental historians know the hybrid environment of the bricoleur and adhocists as "second nature" or, more recently, even "third nature." Put simply, these terms refer to the overlapping terrain of the cultural and the natural, and proponents of these ideas seek to move beyond long-standing divisions between the human and the natural, the made and the born, and revel in the connectedness and entanglement that define most places and ecosystems. See esp. William Cronon, ed., *Uncommon Ground: Toward Reinventing Nature* (New York: W. W. Norton, 1995); and Richard White, "From Wilderness to Hybrid Landscapes: The Cultural Turn in Environmental History," in *A Companion to American Environmental History*, ed. Douglas Cazaux Sackman (Oxford: Blackwell, 2010), 183–90.

39. Noortje Marres, *Material Participation: Technology, the Environment and Everyday Publics* (New York: Palgrave Macmillan, 2012), 10. The editors of the current volume both have excellent studies of the results of countercultural science enthusiasms. See Kaiser, *How the Hippies Saved Physics*; and Patrick McCray, *The Visioneers: How a Group of Elite Scientists Pursued Space Colonies, Nanotechnologies, and a Limitless Future* (Princeton, NJ: Princeton University Press, 2013). See also Diamandis and Kotler, *Abundance*.

40. Marres, *Material Participation*, 10. The association of these trends with the counterculture and Alloy folks is mine. For more on the reenvisioning of objects as potentially environmentally empowering, see William McDonough and Michael Braungart, *Cradle to Cradle: Remaking the Way We Make Things* (New York: North Point Press, 2002); Alex Steffen, ed., *World Changing: A User's Guide for the 21st Century* (New York: Abrams, 2006); Lance Hosey, *The Shape of Green: Aesthetics, Ecology, and Design* (Washington, DC: Island Press, 2012); and Tilder and Blostein, *Design Ecologies*. See also Bruno Latour, *The Politics of Nature: How to Bring the Sciences into Democracy* (Cambridge, MA: Harvard University Press, 2004).

41. Michael William Doyle, "Staging the Revolution: Guerrilla Theater as a Countercultural Practice, 195–68," in *Imagine Nation: The American Counterculture of the 1960s and '70s* (New York: Routledge, 2002), 71–98.

42. Louis Menand, *The Metaphysical Club* (New York: Farrar, Straus and Giroux, 2002), xi.

43. For David Wrobel's views, see his *The End of American Exceptionalism: Frontier Anxiety from the Old West to the New Deal* (Lawrence: University Press of Kansas, 1993.

44. John Herron, *Science and the Social Good: Nature, Culture, and Community, 1865–1965* (New York: Oxford University Press, 2010), 119.

45. James Morton Turner, *The Promise of Wilderness: American Environmental Politics since 1964* (Seattle: University of Washington Press, 2012).

46. For environmental history perspectives, see Richard White, *The Organic Machine: The Remaking of the Columbia River* (New York: Hill and Wang, 1995); Patricia Nelson Limerick, "Mission to the Environmentalists," in her *Something in the Soil: Legacies and Reckonings in the New West* (New York: W. W. Norton, 2000), 171–85; William Cronon, "The Trouble with Wilderness; or, Getting Back to the Wrong Nature," in Cronon, *Uncommon Ground*, 69–90; and Braden Allenby, *Reconstructing Earth: Technology and Environment in the Age of Humans* (Washington, DC: Island Press, 2005).

47. The quotation is from Charles Jencks and Nathan Silver, *Adhocism: The Case for Improvisation* (New York: Doubleday, 1972), 67.

48. For bricolage and adhocism, see Jonathan Hughes and Simon Sadler, eds., *Non-plan: Essays on Freedom, Participation, and Change in Modern Architecture and Urbanism* (Oxford: Architectural Press, 2000).

49. Jonathan Hughes, "After Non-plan: Retrenchment and Reassertion," in Hughes and Sadler, *Non-plan*, 180.

50. Jeffrey Petts, "Good Work and Aesthetic Education: William Morris, the Arts and Crafts Movement, and Beyond," *Journal of Aesthetic Education* 42, no. 1 (Spring 2008): 30–45.

51. Bruce Schulman, *The Seventies: The Great Shift in American Culture, Society, and Politics* (New York: Da Capo Press, 2001), 78–79.

52. Rossi, "Bricolage, Hybridity, Circularity," 72.

53. Casabella was the voice of the Global Tools collective of designers (ibid., 71). For Casabella, see http://www.milfiera.com/designers/alessandro_mendini/ (accessed 2 April 2013).

54. Sim Van der Ryn, "Abstracted in Sacramento: The Late Great State Architect Tells What He Learned about Power," *CoEvolution Quarterly* 22 (Summer 1979): 17. Van der Ryn taught a course on outlaw building at the University of California–Berkeley in the fall of 1971.

55. For more on his work and philosophy of design, see Sim Van der Ryn and Stuart Cowan, *Ecological Design* (Washington, DC: Island Press, 1996); and Sim Van der Ryn, *The Toilet Papers: Recycling Waste and Conserving Water* (Sausalito, CA: Ecological Design Press, 1999).

56. Van der Ryn, "Abstracted in Sacramento," 16–21; and Carroll Pursell, *The Machine in America: A Social History of Technology* (Baltimore: Johns Hopkins University Press, 1995), 305.

57. Baldwin, interview, 38, 45–46.

58. Helga Olkowski, Bill Olkowski, Tom Javits, and the Farallones Institute staff, *The Integral Urban House: Self-Reliant Living in the City* (San Francisco: Sierra Club, 1979); and grant to Bill and Helga Olkowski for urban-oriented ecological design/farming, *Point Grants*, vol. 2, in Huey Johnson

Papers, Resource Renewal Institute. There's a nice discussion of the Olkowskis and the integral urban house concept in Christine Macy and Sarah Bonnemaison, *Architecture and Nature: Creating the American Landscape* (New York: Routledge, 2003), 333–38.

59. Constance M. Lewallen and Steve Seid, *Ant Farm, 1968–1978* (Berkeley: University of California Press, 2004). See also Simon Sadler, *The Situationist City* (Cambridge, MA: MIT Press, 1998).

60. Arthur Carhart, *How to Plan the Home Landscape* (New York: Doubleday, 1935); Chris Wilson, *The Myth of Santa Fe: Creating a Modern Regional Tradition* (Albuquerque: University of New Mexico Press, 1997), 292–93; Peter Van Dresser, *Development on a Human Scale: Potentials for Ecologically Guided Growth in Northern New Mexico* (New York: Praeger, 1972); and Peter Van Dresser, *Homegrown Sundwellings* (Santa Fe, NM: Lightning Tree, 1977).

61. Wilson, *Myth of Santa Fe*, 293. For an earlier discussion of decentrists, see Jane Jacobs, *The Death and Life of Great American Cities* (New York: Vintage Books, 1992), 17–21.

62. For more on counterculture trends in self-build, see Jonathan Hughes, "After Non-plan," 178–80.

63. "The Plowboy Interview: John Shuttleworth," *Mother Earth News* 31 (January/February 1975): 6–14.

64. Kahn, *Domebook 2*. Kahn followed up with the amazing *Shelter*. For an updated version, see Lloyd Kahn, *Home Work: Handbuilt Shelter* (Bolinas, CA: Shelter, 2004).

65. James Marston Fitch, *Historic Preservation: Curatorial Management of the Built World* (Charlottesville: University Press of Virginia, 1990), 8. This book includes a very nice, concise discussion of prototypes versus replicas and the cultural implications of the democratization of American architecture.

66. Gary Zukav, *The Dancing Wu Li Masters: An Overview of the New Physics* (New York: HarperCollins, 1979); and Fritjof Capra, *The Tao of Physics: An Exploration of the Parallels between Modern Physics and Eastern Mysticism* (Boston: Shambhala, 1975).

67. See Lloyd Kahn, *Refried Domes* (Bolinas, CA: Shelter, 1989). This newsprint broadside has a nice, concise history of Kahn's publishing efforts.

68. For an excellent in-depth analysis of counterculture architecture and its relationship to the rise of the ecology movement, see Macy and Bonnemaison, *Architecture and Nature*. Their chapter "Closing the Circle: The Geodesic Domes and a New Ecological Consciousness, 1967" provides a detailed analysis of the geodesic dome's role in shaping and reflecting changing perceptions of architecture and nature at a critical turning point in American history. The quotation is from 339.

69. Stewart Brand, *Whole Earth Discipline: An Ecopragmatist Manifesto* (New York: Viking, 2009).

70. Peter Hasdell, "Pneuma: An Indeterminate Architecture, or Toward a Soft and Weedy Architecture," in Tilder and Blostein, *Design Ecologies*, 95.

71. B. R. Allenby and D. Rejeski, "The Industrial Ecology of Emerging Technologies," *Journal of Industrial Ecology* 12 (June 2008): 267–69.

72. Brand, *Whole Earth Discipline*, 235.

73. Barry Lopez, foreword to *Helping Nature Heal: An Introduction to Environmental Restoration*, ed. Richard Nilsen (Berkeley: Whole Earth Catalog / Ten Speed Press, 1991), v.

74. Howard Odum, *Environment, Power, and Society* (New York: Wiley-Interscience, 1971), 274–76.

75. Ibid., vii. For an elaboration of Odum's idea applied to environmental restoration, see Stephanie Mills, *In Service of the Wild: Restoring and Rehabiting Damaged Land* (Boston: Beacon, 1995).

76. John Allen, Tango Parrish, and Mark Nelson, "The Institute of Ecotechnics: An Institute Devoted to Developing the Discipline of Relating Technosphere to Biosphere," *Environmentalist* 4 (1984): 205–18.

77. For closed-systems ecology and space colonies, see Anker, *From Bauhaus to Eco-house*, 83–95. For the environmental impact of the view of Earth from space, see Robert Poole's wonderful *Earthrise: How Man First Saw the Earth* (New Haven, CT: Yale University Press, 2008). For Brand's view and contentious debate, see Stewart Brand, ed., *Space Colonies* (New York: Penguin, 1977).

78. William L. Kahrl, ed., *The California Water Atlas* (Sacramento: State of California, Governor's Office of Planning and Research, and California Department of Water Resources, 1978).

79. Stewart Brand, "We Owe It All to the Hippies," *Time Magazine*, 1 March 1995, 12.

80. Kaiser, *How the Hippies Saved Physics*.

81. Adamson and Pavitt, *Postmodernism*, 113.

11

How the Industrial Scientist Got His Groove: Entrepreneurial Journalism and the Fashioning of Technoscientific Innovators

Matthew Wisnioski

Like many other fine myths, the great Capt. Super Science has been zapped. But our hero is no victim of villainous radicals. He did himself in. . . . Is there hope for him? —John Steele and Ronald Neswald, 1972[1]

In the concourse of the US Airways New York–Boston shuttle, professional-managerial travelers read about the inner lives of celebrities from complimentary high-gloss magazines. *MoFo Tech*, the trends magazine from the law firm Morrison and Foerster, explains how Andreas Sundquist, the CEO of DNAnexus, has made it possible for geneticists to shift from studying 380 snippets of DNA to over 380 million by using the digital cloud.[2] The once specialist-oriented *IEEE Spectrum*, published by the Institute for Electrical and Electronics Engineers (IEEE), has slimmed down to become "The Magazine of Technology Insiders." Its senior editor introduces readers to Sehat Sutardja, the "beating heart" of the Silicon Valley firm Marvell, whose chips are used in high-end Android tablets and $100 One Laptop per Child computers alike. Blending theater metaphors, feng shui decorating principles, and economic market surveys, it recounts Sutardja's career from China through the American education system, internships at IBM, and graduate school at Berkeley.

FIGURE 11.1 Today's technoscientific innovators at Marvell Technology Group. Tekla S. Perry, "Marvell Inside," *IEEE Spectrum* 47, no. 11 (November 2010): 43.

Sutardja's story is peppered with the secrets to the family business he runs with his brother and wife, a software engineer and former elite basketball player (fig. 11.1). "Marvell Inside" shifts from technical explanation to notes on government advising and the company's $20 million donation to establish Berkeley's Center for Information Technology Research in the Interest of Society, founded to maintain US global competitiveness. But Sutardja eschews the jet-setter's lifestyle; he eats in the company cafeteria and carpools to work.[3]

At a glance these magazines appear as distractions from an Excel spreadsheet, e-mail backlog, or half-written grant proposal—destined to be left in a seat pocket. Such ephemera, however, perform significant conceptual labor for their readers as well as for the lucky few spotlighted in their pages, the knowledge workers those entrepreneurs employ, and the editorial staffs who curate narratives of life on technology's leading edge. Neither technical reports nor mass popularizations, these publications reinforce a set of beliefs about historical change and the people responsible for it. In the same issue of *IEEE Spectrum*, re-

searchers at the National Renewable Energy Laboratory explain how
"algae's entrepreneurs" are prospecting for "green gold" in the form
of biofuels. They outline the efficiencies of corn, soybeans, and olives;
discuss the Energy and Independence and Security Act; and explain the
basics of algae cultivation.[4] *MoFo Tech* is more direct. Under the head-
line "Driving Technology" Morrison and Foerster offers that its clients
are "innovators, scientists, and business leaders like you" and that "today more than ever, technology companies need the best legal advice to
drive their ideas to success in the global marketplace." Technoscience,
our well-heeled travelers read, is the principal agent of societal growth.
It operates in a fluid environment in which distinctions between industry, academia, government, and geography lose their meaning.

Viewed in the *longue durée*, the profiles of Sundquist and Sutardja
present a novel image of the technoscientist as "hip" innovator. This
persona differs in important ways from earlier idealizations of scientists
and engineers. Representations of lab-coated experts, corporate organization men, tweedy academics, and scientific statesmen that dominated
the Cold War era have been replaced by socially engaged entrepreneurs.
Success for these new self-made men (they remain largely men) is less
the result of a priestly duty to objectivity as it is the ability to cut paths
across disciplines and institutions in pursuit of their goals. Creative collaboration and improvisation are among the most coveted skills for
traversing occupations and organizations. This age's heroes are the scientific mavericks who navigate the uncertainty of a globalized technological world in complementary pursuit of wealth and service.

The origin of this composite, switched-on innovator is one that historians of science have come to associate with the coevolution of the
West Coast microelectronics industry and communal utopians in the
1960s and 1970s, on the one hand, and the triumph of commercially
oriented science policy in the 1980s, on the other. The former portrays
a set of hippie entrepreneurs working on the outskirts and sometimes
within the military-industrial apparatus, while the latter evokes technology transfer, venture capital, and neoliberal political philosophy.[5]

But the ideal of scientists and engineers as hybrid innovators is
linked more deeply to the Cold War than is generally recognized and reveals broader and relatively unexplored exchanges between alternative
political cultures and the scientific mainstream, which were to play an
important role in the science commercialization of the 1980s. Alongside
the countercultural communication network of Stewart Brand's *Whole
Earth Catalog*, a parallel media economy emerged within Establishment
enterprises of modern science.[6] Its publications became convergence

points for journalists, intellectuals, scientists, engineers, research administrators, and innovation theorists, in which new managerial and organizational theories were cultivated and disseminated. First appearing in
the early 1960s, its journalists helped scientists and engineers navigate
the politically charged world of the military-industrial-academic complex, its unraveling, and the evolving global economy of the late Cold
War. These tastemakers of technoscientific life prefigured, simultaneously invented, and borrowed the techniques and vision of their hippie
counterparts.

In other words, the predecessors of *MoFo Tech*, *IEEE Spectrum*, and
the like played a vital role in crafting and promoting epistemic and social virtues of a new "technoscientific self."[7] In much the same way that
Thomas Frank's study of advertising executives transformed our understanding of the inseparability of counterculture and commerce, a focus
on scientific journalism—the public image makers of science—forces
a reconsideration of what it meant to be both "groovy" and scientific,
who defined those meanings, and how the persona of the technoscientific innovator was mutually constitutive of new forms of expertise.[8]

Among the key transit agents in the fashioning of this technoscientific self was the short-lived magazine *Innovation*, in print from 1969
to 1972. Bathed in Day-Glo colors and featuring irreverent illustrations, *Innovation* honed now well-worn themes of *fluidity*, *uncertainty*,
creativity, *hybridity*, and *change* among industrial scientists, research
managers, venture capitalists, and academic administrators. *Innovation*
was promoted as a network rather than a magazine, and its publisher—
Technology Communication Inc.—and its staff, contributors, and readers dubbed themselves the "Innovation Group."[9] Through print, executive workshops, and electronic media, its members posited a new kind
of professional—the change manager—who was at turns a scientist,
an engineer, a humanitarian, and a shrewd businessman. This adaptive
agent would fight bureaucracy with collaborative creativity, launch entrepreneurial ventures, and save society from runaway technology, on
the one hand, and antimodernism, on the other.

Rather than from social protest or LSD, the Innovation Group and
its magazine shed light on how knowledge and identity in the 1960s
and 1970s were created out of a confluence of factors including the
evolving institutional boundaries between science and engineering; the
shifting meanings of "science," "technology," and "innovation";[10] and
the blurred lines between "counterculture" and "Establishment." *Innovation* was the eventual product of widespread efforts in the early 1960s
to give scientists a human face; changes in the publishing industry; the

organizational upheaval of the microelectronics industry during the 1950s and 1960s;[11] the migration of social scientists to corporate practice; deep anxieties about a societal backlash against technology in the late 1960s; and the remarkable blending of cultural and professional norms of the 1970s documented throughout this book. Its networked history demonstrates how the rise of entrepreneurial journalism simultaneously accentuated the virtues of the technoscientific life and contributed to its knowledge practices of innovation management during a period of intense cultural, political, and technological change.[12] Gurus who could explain the underlying mechanisms of this upheaval were in high demand, and *Innovation* answered the call, for believers, skeptics, and the ideologically hostile alike, with a magazine designed to be edgy in content and form. Within its pages managerial shoptalk merged with epochal visions about humanity's future in a powerful interpretation of contemporary knowledge work, one in which *uncertainty* became a central feature of a coherent program of innovation expertise and a novel ideal of flexible professional selves.

From Boating Industry *to an International Scientific Revolution*

When, in 1957, Margaret Mead and Rhoda Metraux asked 35,000 American high school students to describe scientists, they discovered a positive view of science combined with a deplorable lack of interest in scientific careers. *Science* was for mankind's benefit, but the *scientist* was a white-coated man in a laboratory, bald, tired, and unfit to marry. The problem, according to Mead and Metraux, was an inadequate portrayal of the "real, human rewards of science—on the way in which scientists today work in groups, share common problems." They suggested that educators and media produce "[p]ictures of scientific activities of groups, working together, drawing in people of different nations, of both sexes and all ages, people who take delight in their work."[13]

It was not just America's youth who were unsure about scientific careers. Scientists and engineers in the late 1950s likewise wondered what kind of life they had chosen. Technical fields experienced staggering growth in the wake of World War II. By 1960, engineering was the leading occupation for white-collar males in the United States.[14] Approximately 75 percent of these nearly one million knowledge workers were employed in just 1 percent of all firms, and corporations with a workforce of ten thousand or more employed 35 percent of America's engineers.[15] The rate of growth was even higher in the top ranks of American science, with physics outpacing all other fields. Prior to World

War II, one in three PhD-trained physicists worked in industry; by 1957, the ratio had risen to half. Scientists and engineers were more likely to be working together and to be doing so in a corporate environment.[16] At the same time, universities increasingly resembled corporations, with dedicated research centers and practitioners moving freely between academia, government, and industry.

The aspirations and anxieties of "Big Science" were fed by a booming market in academic and popular commentary. Social scientists from Robert K. Merton to William Kornhauser portrayed a conflict between the "scientific ethos" and corporate bureaucracy.[17] Embracing the critiques of C. Wright Mills and William Whyte, scientists also lamented that they had become organization men. On the other hand, in the aftermath of the atom bomb, scientists became nationally recognized personalities. This postwar fascination with science and technology coincided with explosive growth in the publishing industry, with specialty magazines more than tripling their circulation between 1943 and 1963.[18]

Prominent scientists and policy makers wished to channel the public's interest but lamented that "popularization" debased science. They thus sought to support the right kind of publications and to have a voice in constructing them.[19] The marquee effort was *Scientific American*, which cast its market as the "scientific layman: the growing community of US citizens who have a responsible interest in the advance and application of science, . . . the scientists themselves, the doctors and engineers, the executives and managers of industry and those engaged in the non-technical professions of teaching and the law."[20] The push for public understanding of science was an international movement, framed in nationalist terms. In 1964, for example, Britain's *New Scientist* went to press with inaugural articles by "leading scientists" presented "in language as free as possible from technicalities" as a resource for "men and women who are interested in scientific discovery and in its industrial, commercial and social consequences" with the goal of bolstering the nation's status as "a first-class economic Power."[21]

In 1960 William G. Maass, a publishing executive at the firm Conover-Mast, saw a latent opportunity in the global drumbeat for scientific communication. He created a "new kind of publication, . . . the first practical means for the scientist and engineer to keep up with the significant developments in areas outside his own specialty that may have important bearing on his work."[22] The premise of his magazine *International Science and Technology* (IST) was that the accelerating development of science-based technology had created an environment in which the time from "idea" to "utilization" had been reduced from decades to weeks. In an idea-based society in which "discovery can occur anywhere," the lines between disciplines and nations were eroding. "That we need a single word to mean both engineer and scientist, that the two form a single community," he wrote, "is obvious to anyone who sees engineering schools throwing handbooks out of their curricula and stressing science or who sees industrial employers advertising for mathematicians and theoretical physicists."[23] He conceived of his audience as the leading 10 percent of the world's scientists and engineers: "the technical men who are bosses—group leaders, division heads, company officers—and those key technical experts who make important decisions."[24] *IST* would draw on their experiences to interpret, communicate, and build this "single community" of technoscience.

Maass assembled a top-flight editorial team of hybrid journalist-engineers to achieve his vision. *IST*'s senior editor, Robert Colborn, had received a degree in civil engineering before embarking on a distinguished publishing career that included a stint as managing editor of *Business Week*. Executive editor Daniel Cooper had a PhD in nuclear physics from MIT and had worked at Bell Labs before becoming an editor of the journal *Nucleonics*. Among *IST*'s associate editors, David Allison joined after graduating from Rensselaer Polytechnic Institute with a degree in industrial management; Ford Park was an MIT mechanical engineering graduate, former professor at the University of Buffalo, and former editor at *Product Engineering*; Seymour Tilson was a lecturer in geosciences at New York University; and Ted Melnechuk was a chemist and poet who had studied classics with his friend Allen Ginsberg.

IST's graphic identity sought to convey a nonviolent revolution that was modern, creative, aesthetic, democratic, and international. Maass hired Will Burtin, famous for his postwar redesign of *Fortune* and his multimedia *Cell* and *Brain* installations for Upjohn, to establish a unique visual identity.[25] The cover of each issue featured an abstract painting representing a scientific theme that was commissioned by *IST*'s art director, who trawled Manhattan's galleries in search of talent.[26] Contributors were introduced with candid portraits and informal biographies. Articles included sketches in the margins to convey creativity in action. A "To Dig Deeper" section at the back of the magazine provided an annotated bibliography so as not to clutter articles with footnotes. Instead of a letters to the editor section, *IST* had a "Communications Center," where readers could pose interdisciplinary problems and receive answers from other readers. Maass even had unrealized ambitions of augmenting *IST* with electronic data centers and closed-circuit television networks.[27]

Conover-Mast was an unlikely home for this journalistic experiment. In business since 1927, it was one of the larger industrial trade publishers in the United States. The company was founded when its partners Harvey Conover and Bud Mast left editing jobs at McGraw-Hill's *Factory* to start the competing *Mill and Factory*. Other titles in the company's catalog included *Boating Industry* and *Volume Feeding Management*. In *Sputnik*'s wake, however, Conover-Mast evolved with the industries it covered. In 1958, for example, *Aviation Maintenance and Operations* was rebranded *Space/Aeronautics*.

IST also had an unusual business plan: Conover-Mast gave it away for free to the majority of its more than one hundred thousand readers. To cultivate an audience, the magazine bled cash, reporting a net loss of $1,500,000 its first year. It established credibility, however, with an editorial board consisting of nuclear physicist Edward Condon; chemist Louis P. Hammett; European Space Agency scientific adviser Sir Harrie Massey; executive director of the National Committee for the International Geophysical Year Hugh Odishaw; and the first president of the newly created IEEE, Ernst Weber.[28] Advertising from corporations, industrial research laboratories, and government agencies followed. By the end of its second year, *IST* was in the black.

From a combination of consultant contract surveys and readership polls, Maass determined that 96 percent of *IST* readers had college degrees and 36 percent had advanced degrees. A third were technical executives and over half were at the project management level or above. The audience was only 10 percent international, though over 25 percent of articles covered international topics pertaining to Britain, France, Russia, India, Sweden, Canada, Poland, or Pakistan. Maass noted with pride that thousands of potential readers had been declined for failing to meet *IST*'s standards and that there was a waiting list of thousands more who were qualified.

IST projected itself as less encyclopedic and more relevant than *Scientific American* and as more revealing about the practices and personal qualities of scientists than journals such as *Science*. By point of comparison, the first issue of Gerard Piel's retooled *Scientific American* featured articles on the Renaissance anatomist Vesalius and quantum physics research from the 1920s; whereas the first issue of *IST* spotlighted information retrieval systems, Soviet R&D policy, and the promise of fuel cells. Articles were intended to be "state-of-the-art" reports authored by leading experts. They were to be understandable to a wide audience of technical and scientific readers and to communicate the "special state of grace" that was the "chaotic" process of scientific research, the "awful queasiness, this rumbling around inside, this subconscious knowledge that *something* is going to happen."[29] Articles fell into the categories of technical overviews by scientists and engineers, extended interviews intended to reveal the human qualities and problem-solving styles of scientists, and features on the "innovative process."[30]

IST poured its energies into answering Mead and Metraux's call for positive images of the scientific enterprise. The editorial team had a penchant for capturing the ethos of postwar science by linking its characteristics to individual practitioners, real and imagined.[31] In interviews with prominent scientists, readers learned, for instance, that Nobel laureate Albert Szent-Györgyi once intended to become a doctor in the Dutch East Indies and started his biochemical work using apples and bananas. Individual quirks were deployed toward generalizable arguments about the scientific life. According to *IST*'s executive editor, "that horrible word—IDENTITY"—was critical in order to combat popular culture's "vulgarization" and "denigration" of science.[32] In 1967 a selection of the biographies were reprinted in *The Way of the Scientist*, which sought to capture a "sharp turn" in the "style of our society," of which one part was a dramatic transformation in the "life of the scientist or engineer."[33] I. Bernard Cohen wrote glowingly of the book in the *New York Times*: "We are able to see the institution of science itself in our times as a composite of flesh and blood people and not of stereotypes with which our senses have so often been flooded."[34]

It was in their coverage of the organization and entrepreneurial practices of science and technology, however, where *IST*'s staff found the agents of societal change (table 11.1). *IST* gave voice to a cast of research managers and organizational theorists. The 1964 article "From Research to Technology" highlights the character of this journalistic form. In it, Jack Morton, the Bell Labs research director behind the commercialization of the transistor, explored Bell's history and corporate structure to argue that the technology manager was himself performing scientific research and that the systems approach was "nothing more than a direct steal from the scientific method."[35] Two years later, he followed up with a report, entitled "The Microelectronics Dilemma," that stressed the growing national need for comprehensive studies of innovation (fig. 11.2).[36] This second article was prefaced by *IST* editor David Allison, who declared that the Bell experience amounted to "a story of change—how to survive it and how to grow with it. . . . [It was a] story of how a society can be great."[37]

As *IST*'s focus evolved from the federal "science brain trust" to high-tech industries, it increasingly presented a unified set of beliefs about

Table 11.1 *IST* **Articles on Research Management and Entrepreneurship**

The Industrial Scientist	Fear of Innovation	From Research to Technology
Creativity	To Promote Invention	From Materials to Systems
Blocks to Creativity	Loyalties	Science on Park Avenue
Freedom in Research	What's the Boss For?	Designing a Technical Company
Diversity in Research	Supervision	How the US Buys Research
Solving Problems	Science and the Organization	
Growth of Ideas	Basic Research in Industry	

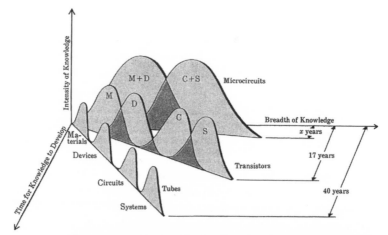

These curves suggest how knowledge intensifies and broadens as electronics progresses to microcircuits. Two things speed up the innovation process (17 yrs. for transistors vs. 40 yrs. for tubes): First, knowledge intensifies and broadens among specialists in materials, devices, circuits, and systems. Second, these specialists begin to share common knowledge as symbolized by the overlapping of the specialist areas on the curves.

FIGURE 11.2 Jack Morton and the science of innovation. Jack A. Morton, "The Microelectronics Dilemma," *International Science and Technology* 55 (July 1966): 42.

historical change. It found the source of progress not on the lab bench but among the lab's coordinators. In 1967 Allison wrote a composite profile of the "industrial scientist" that provided the clearest expression of *IST*'s synthesis. He first put the reader inside the mind of a young corporate scientist:

> He sits in his hard, gray chair and stares at the steel partition. His room is B 105, which he shares with another young physicist. Both joined the laboratory during the previous summer. As he swivels, aimlessly, alone in his two-man cubicle he tries to assess these past few months: What has he accomplished? How is he doing?

Allison included the details of corporate architecture to evoke a distinction between "real," academic science and its commercialized other, at the forefront of his protagonist's thoughts:

> [I]n the awful stillness, he is overcome with depression. He is no good. He will never make it here. "Dear God, what-ever brought me to this place?" The question actually comforts him. It helps him see that the place is wrong and he is right. He should never have come. . . . He is not an *industrial* scientist. He is a *scientist* goddammit, and proud of it!

Allison then inverted the stereotype:

> "If I am really too good for this place, why am I afraid of it? If I'm so good, why do I think I will never make it here?" . . . These people are good, very good. He must concede that his periodic depressions grow partly from the fear that the best of them are better scientists than he will ever be.[38]

Allison next described the hundreds of sociologists, economists, psychologists, and psychiatrists concerned with the characteristics of industrial scientists. This was a segue into the identification of the true heroes of industrial science: "the gamblers, the top managers. Nice guys and tough guys. Smart guys and dull guys. Well-heeled and fantastically well-heeled guys."[39] In this testosterone-laden ideal, the agents of innovation were risk-taking corporate managers. Summarizing Morton's insights about the development of the transistor and the revolutionary pace of the microelectronics industry, he concluded that without the Mortons of the world, the transistor would be just one more undeveloped discovery. The case of the transistor furthermore revealed general principles of innovation. The transistor "was *not* motivated by or directed toward the objective of a closely defined marketable product." Nor was it pursued via line-and-staff management. Rather, "a number of people, with different skills, contributed to the research, but the exact nature of the interactions among them could not have been predicted or planned in advance."[40] It thus was vital to cultivate experts who understood innovation's vagaries and could balance creative exploration with practical results. Such individuals forged collaborations between domain experts and utilized procedures for bringing imagined futures into reality.

The industrial environment was uniquely equipped for collaboration under uncertainty, but fostering it required national innovation policy. Engineering and science education should be directed toward industrial practice. Universities and government agencies also should emulate industrial organization to be more flexible and responsive. Hinting at the federal government's outsized role in high-tech R&D and growing discord in American culture, Allison concluded that "when one looks with *optimism* into the years ahead," it was obvious that science and technology would be critical to resolving challenges such as population growth and urban poverty. However, if industrial laboratories became overly cautious and ceded the "major technological risks" to government contracts, "the finest hour of industrial science will . . . have come yesterday."[41]

A Heraclitean Primer for Revolting Times

From the start, *IST* conveyed self-awareness of having tapped into world-changing energies. Maass maintained a dialogue with readers that expressed his astonishment over *IST*'s success. *IST* was "riding a wave, a wave of technical and social change," in which the entire globe was "under several kinds of pressure to become even more international than we are."[42] In 1964 *IST* used the marketing slogan "Change or Die!" in a film for corporate advertisers that showcased how technology was remaking society. Maass professed to his scientific readers that "the account of your role might seem more glamorous, more dramatic than you visualize it yourself" because the marketing and business side did not yet understand the revolution in which they were living.[43] This new epoch was a capitalist one. It was "competitive" and "unforgiving of mistakes"; however, for the "enterprise which is in tune" with the patterns of change, it was possible to "accelerate very very fast" with great reward.[44] Indeed, in July 1965 Maass left *IST* to become an executive at Cahners Publishing, a Boston-based competitor.

    IST was only one of many contributors to this vision of creative destruction. It was an emergent development across a range of disciplines and institutions with a basis in the structural changes of Cold War R&D in general and the microelectronics industry in particular. In the latter half of the 1960s, such "change" talk became deeply ideological.

    As visions of technology emerged as a fault line in the nation's escalating culture wars, the 1960s saw the rise of a new genre of writing on technology and society that channeled the concerns of a host of social movements. Engineers, industrial scientists, research managers, and other technoscientific practitioners in the 1960s reconciled their own visions of society and self with the shifting attitudes about technology portrayed in the writings of Jacques Ellul, Lewis Mumford, and others. A minority raised moral challenges in professional venues, often finding inspiration in critical theorists. Their dissent fueled introspection among the nation's top managers, academic administrators, and government officials about how to confront criticisms of technology.

Establishment visionaries portrayed technologists as those best equipped to manage change by maximizing technology's opportunities and minimizing its negative effects.[45] Academic social scientists, funded by corporate philanthropy, generated the most robust studies into the moral and political dilemmas of sociotechnical change, such as the IBM-funded Harvard University Program on Technology and Society.[46] At the same time, engineers and research managers were developing a similar analysis that put corporate technologists at the vanguard of managing the future. In 1966, for example, the National Academy of Engineering (NAE), National Science Foundation (NSF), and the US Department of Commerce sponsored a symposium at which J. Herbert Hollomon, a prominent engineer turned bureaucratic reformer, described "the entrepreneur" as society's leading change agent—"a person who is willing to take personal risk with respect to the changes he wishes to bring about and to stand behind the revolutionary character of these changes."[47]

Although there were tensions between academic, government, and corporate approaches to theorizing technological change, the approaches were mutually reinforcing. Interest in assessment, forecasting, and other strategies for the "management of change" populated engineering journals, business periodicals, and new futurist periodicals. Theories of technological change in this vein were hotly contested by the critics whose alternative visions they were designed to minimize.[48]

Ideological debates about technology and the human condition fell outside the purview of *IST*'s original mission, but its editorial staff watched the growth of hostility toward technology with unease and documented how it altered scientists' and engineers' conceptions of self. In October 1967, for example, *IST* published a "Looking Ahead" segment with Barry Commoner entitled "The Eroding Integrity of Science."[49] Colborn later editorialized that race riots stemmed from the fact that there would always be people "who can't fit themselves into technology's world," but he contended that fatalism was misplaced because "the newest technology may actually be re-introducing a human scale into the mechanisms of society."[50] What was needed to overcome technocratic and Luddite views was a new kind of professional who could master the forces of change. In their desire to aid this new man, the staff of *IST* took a groovy turn.

Inventing Innovators

One of the most widely read articles in the critical theory of technology owed its existence in part to a shakeup in the trade magazine industry. Paul Goodman's November 1969 *New York Review of Books* essay "Can Technology Be Humane?" surveyed the state of international unrest and its technological roots and argued for the wholesale revision of the structure of science and engineering. The scientific enterprise's problems extended beyond military co-optation and the neglect of domestic needs to the very essence of what it meant to be human. Technology, however, if decentralized, collaborative, reflective, and sensitive to the environment could play a critical role in a flourishing democratic society.[51] But before it became a touchstone of humanist academics, Goodman's article originally was published as "The Case against Technology," in *Innovation*, a novel magazine for the "technical man" (fig. 11.3).[52]

In 1968 Conover-Mast was acquired by Cahners, resulting in the second-largest industrial publisher in the United States and reuniting Maass with the *IST* team.[53] Maass and Colborn, however, jumped ship in the wake of the merger, taking most of *IST*'s staff with them. In consultation with Morton, they formed the venture Technology Communication, which they imagined as a publishing house, executive educator, and new media experiment. Maass would serve as publisher and Colborn as chairman of the board and senior editor. Their staff of a dozen associates included most of the *IST* group and other seasoned veterans of the technical press. Technology Communication drew on its *IST* connections to assemble a who's who of corporate researchers, government advisers, and technology theorists to serve on its advisory board.[54]

Technology Communication described *Innovation* "not so much as a magazine as a vehicle for initiating a process of interaction." Its team members saw themselves as "environmental creators" who would document the skills and norms needed to survive accelerating technological change.[55] *Innovation* would be a space for sharing managerial problems, debating government policy, and putting scientific entrepreneurs in contact with venture capitalists. *Innovation* was their principal medium, but the real product would be the Innovation Group network. A $75 a year fee gave access to the group through newsletters, a bibliographic service, seminars, and experimental conference calls.[56]

A recurring theme of *Innovation* was the search for identity in a

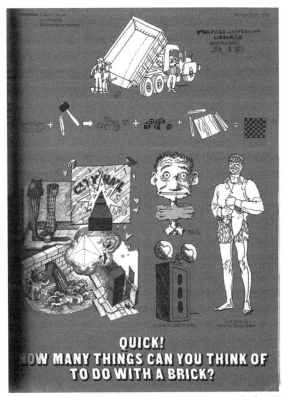

FIGURE 11.3 Cover of *Innovation* 8 (September 1969). Illustration by cartoonist Stan Mack; magazine design by Chermayeff and Geismar.

changing world. The magazine was marketed to a hybrid audience using the same technique Maass had employed with *IST*, but with ratcheted-up exclusivity that described the Innovation Group as a self-selected "set of men very important to this country." Participation appealed to attitude and action rather than credential or institution. "You may be part of it," advertisements beckoned, "or your boss may . . . or the man in the next office . . . or a key man who reports to you." What members held in common was the management of "technological change," whether in research, design, corporate reorganization, or federal policy. "Such men," Technology Communication boasted in rhetoric that echoed the cowboy mentality documented in Erika Milam's contribution to this volume (chapter 9), were "a lonely crowd" because "each one of them, all by himself, has had to learn an art that has no teachers or precedents or tradition."[57]

By 1970 the Innovation Group had 5,000 aspiring change managers, with Maass aiming for an upper bound of 25,000. Subscribers consisted of executives, staff engineers, senior scientists, and professors located in institutions ranging from Lockheed to the National Academy of Sciences. Colborn described *Innovation*'s readership as a "zestful crew who mostly felt good about the state of their lives," which he argued was contrary to behavioral research and popular opinion, which held that "engineers and applied scientists are the most discontented professionals in America."[58]

Whereas *IST* had pivoted off of *Scientific American* and other established publications, *Innovation* was cast as a radical break. The modernist firm Chermayeff and Geismar designed the magazine's layout and visual style. The result was an advertising-free, streamlined, pop-art experience. Inside its glossy cover, articles were set in Helvetica and numbered "1ne, 2wo, 3hree, 4our . . ." The table of contents provided only the authors' last names and abbreviated titles to cultivate an "in-group" sensibility. Text and illustrations bled together in articles, which were followed by short interviews with authors.

Innovation gave voice to a new world of technoscientific labor. Its charter issue was dedicated to "fluidity." The lead article, by University of Cincinnati president Warren Bennis, outlined the tenets of the "Temporary Society." Another article provided some of the best on-the-scene reportage of MIT's 4 March 1969 protest that exists in the historical record.[59] Book reviews examined John Kenneth Galbraith's *New Industrial State*, Jacques Ellul's *Critique of the New Commonplaces*, and Peter Drucker's *Age of Discontinuity*. The message was that "as this country heads into the 1970s, it is in a state of such rapid and intricate change that to understand your society you must think of it as a liquid in turbulent flow. You can't rely on solid organizational structures; they'll change. You can't locate any useful boundary between your area of responsibility and the turbulence outside; the outside comes flooding in."[60] Subsequent issues followed with articles on themes of "uncertainty" and "revolution." There was a strong emphasis throughout on global competitiveness (table 11.2).[61]

Innovation articles blurred together historical change, organizational management, and personal identity. In "And Now, the Creative Corporation," for example, Frieda Libaw, president of Cognitive Systems Inc., a small learning sciences start-up, described the evolution of corporate organization from the British East India Company to the present, with each era personified by the tools and knowledge of its time: the New World explorer with cartographic instruments; the railroad magnate and his industrial blueprints; the organization man with his systems

Table 11.2 Sample *Innovation* Articles

Robert Ayers, "We May Be Losing Our Technological Lead"
Warren Bennis, "How to Survive a Revolution"
Warren Brodey, "Building a Creative Environment"
R. Hobart Ellis Jr., "Seven Laboratories in Search of a Mission"
Lawrence Halprin, "Riding a Revolution: A Radical Experiment in Reorganization"
Emmanuel Mesthene and J. Herbert Hollomon, "Our Social Responsibility"
Robert Powell, "The Life and Times of a Scientist Turned Entrepreneur"
Sidney Simon, "From Aerospace to the Environment?"
Michael F. Wolff, "The Birth of Holography: A New Process Creates an Industry"

FIGURE 11.4 Personifying corporate evolution: composite of four images by Jerome Snyder appearing in Frieda B. Libaw, "And Now, the Creative Corporation," *Innovation* 23 (March 1971): 2–13.

diagrams; and, finally, the change manager, whose tool was networked social inquiry (fig. 11.4). The creative corporation was dedicated to "the solution of social problems," grounded in "behavioral, social and information sciences," and structured by a value system "that includes antimaterialism and the recognition that individual and community welfare are inseparable." Linking her experiences to the small group model used by Japan's Sony Corporation, Libaw described Cognitive Systems Inc. as a "company of activist scholars and scientists" that demanded that all its employees be "generalists." According to Libaw, this new social entrepreneurship had emerged from the revolt against hierarchy. She concluded that "participative management," based on humanist philosophies and "social problem solving," were the best hope for reenrolling a generation of disenchanted youth by providing "constructive and satisfying work."[62]

Innovation augmented its physical magazine with groundbreaking network forums. From the start, it hosted local seminars across the country with contributing authors such as Douglas Engelbart.[63] Describing itself as "the first magazine with an interactive feedback system," it then created an "electronic meeting place" around articles. These were hour-long "interactive" and "informal" sessions of "freewheeling discussion" with authors and were designed to "repeal geography." Costing $30 a session, advertisements boasted that Milton Friedman had participated from his "summer cabin" and judged the experience more productive than a physical meeting. "Complete strangers end up talking as though they've known each other for years—and wanting to call on each other's expertise again."[64] Technology Communication also held exclusive executive education workshops at a Long Island estate. The first, which was led by 22 distinguished panelists and limited to 200 participants, included TED-style theater talks; randomly generated conversations; and closed-circuit television seminars designed to create intimacy between a keynote speaker and individual audience members in their hotel rooms.[65]

What moral virtues and normative assumptions did this combination of magazine and network forum cultivate? A greening of industrial science? An appropriation of radical sensibilities for Establishment ends? Simply a source for thinking through tactics for conducting business in an uncertain environment?

Ideologically, the Innovation Group had a complicated relationship with the counterculture, critical intellectuals, and movement politics. On the one hand, *Innovation* published interviews with environmental policy maker Stewart Udall, but on the other, it filed positive reports on how systems analysis was resulting in "a better managed military-industrial complex."[66] There were a handful of technocommunitarians such as psychologist and computer programmer John Steele and journalist Ron Neswald, who claimed that elements of technocratic and "Aquarian" lifestyles could be productively merged. But their message was modulated for a corporate audience. In "How Science and Technology Just Might Survive in a Post-technological Age," they insisted that managers read Alvin Toffler's *Future Shock* and Charles Reich's *The Greening of America* because "any manager worth his survival must at least be aware of what's being said—and bought."[67]

Taken together, *Innovation* promoted a middle way similar to that espoused by the top managers it profiled. Editorials remarked on the cultural transformation in the developed world from overblown praise of science and technology to their outright rejection. Running throughout was the belief that once technology's critics and proselytizers recognized this middle way of seeing, social problems could be resolved through creative sociotechnical solutions. "The war is in your mind," Allison wrote, and "until there is enough enlightenment around to enable us to see what enormous good technology can do—and what enormous bad—until then we shall suffer the same old dreary irrelevant argument. . . . As though technology were spinach, and you could like it or hate it."[68]

But *Innovation* went beyond the proliferation of buzzwords and the managerial appropriations of countercultural imagery. The staff of Technology Communication cultivated a novel space for the development and communication of new expertise about innovation, one that attracted a wide range of influential stakeholders. Morton, who in 1971 produced the monograph *Organizing for Innovation*, is again illustrative. Likening the manager to "Maxwell's Demon," he followed Bell Lab's evolution into the 1970s, described how the systems approach provided the means for overseeing innovation as it changed the ecology of the organization that developed it.

Another case of *Innovation*'s knowledge making can be found in the contributions of Donald Schön, who used *IST* and *Innovation* to workshop his ideas.[69] Best known for his 1983 book *The Reflective Practitioner*, Schön's life mirrored the fluid world he theorized. He was at turns a pianist, a philosopher, a soldier, a consultant, a bureaucrat, a nonprofit innovation policy advocate, and an endowed management professor at MIT's Sloan School.[70] In a pragmatist philosophy that echoed Goodman's, Schön argued that humane technology required the constant remaking of professional selves. The disruptions of accelerating social and technological change undercut the norms and structures of the skilled professions, filling knowledge workers with psychic unease. "'I am a chemist,' 'I am a college professor,' 'I am a doctor,' 'I am a cook,'" Schön wrote; "we make such judgments not as tentative findings subject to change but as assertions about enduring aspects of the self. To be unable to make them, or to be ambiguous about them, is a matter for some embarrassment."[71]

Change was uncertain and it was not neutral, but it was inevitable. Adapting to it required an open mind and new expertise. Much as Libaw distinguished the "technological" from the "creative" corporation, Schön argued that "learning" systems of communication and organization were replacing those based on centralized diffusion. This evolution could be seen in military-industrial systems and social welfare programs but was especially characteristic of movement politics, in which

> [o]rthodox Marxism, the theory of power elites, the radical so-
> ciology of an Alinsky, the radical critique of a Chomsky, the
> doctrines of participation and advocacy, the Black Manifesto,
> the elaborations of guerrilla ideology and tactics, the rationale
> for mysticism and for the use of drugs, the philosophical radi-
> calism of Marcuse, the essays in radical politics of the new left,
> all flow and work together and change rapidly over time.

Analyzed in its entirety, "The Movement" was an identifiable entity of loose and evolving connections "in which centers come and go and messages emerge, rise, and fall." Its essence was a way of being, "in which transformation around the new is a value in itself."[72]

While Schön rejected the politics of revolt, he championed radical organizational modes. In the pages of *Innovation*, he analogized social protest to changes in corporate structure.[73] From the late 1960s through the 1980s, he argued for an "Ethic of Change" as a third option beyond the empty solutions of a romantic return to the garden and a confrontational revolt. Drawing management strategy from the "very process of change" would produce guidelines for maintaining one's humanity. To thrive in an uncertain world and confront its challenges required constant experimentation, living in "the here and now," and "seeking out the new"—a set of moral virtues that would provide "nuclei for identity and self-worth."[74]

Collectively, when seen not just as a magazine but as the network forum Technology Communication intended it to be, *Innovation* thus presented a focused logic for cultivating and disseminating both actionable innovation expertise and powerful images of the technoscientist as innovator. Its nucleus of journalists first provided technoscientific practitioners with an encompassing explanatory frame, "a unique picture of what is happening to your work environment."[75] That picture was shaped by looking at specific case studies augmented by interviews with innovators, such as microelectronics entrepreneurs Robert Noyce and Gordon Moore, who were to be admired and emulated. These explorations of past and present provided context for investigations of the shape of things to come: primers on time-sharing, explorations of holography at the intersection of art and engineering, accounts of struggles to adapt by traditional automakers, and an emphasis on solutions to environmental pollution that engaged industry as much as government regulators. Technology Communication then promised to equip members with managerial techniques to stay abreast of an "ever-changing environment." These ideas were cultivated first through exploratory seminars for Innovation Group elites and *Innovation* articles and later packaged as primers on innovation expertise in a book series via a partnership with the American Management Association.[76] The "message of the '70s" in this vision of innovation was a recapitulation of the axiom Technology Communication's founders hit upon a decade earlier: "change or die!"

Conclusion

As today's entrepreneurial technoscientists often discover, most innovative ventures fail. Sometimes the difficulty is transitioning from lab to market, at other times a poor business plan or a soured partnership or simple bad luck. Technology Communication's demise was marked by personal tragedy and financial trouble. Colborn died of cancer, at the age of fifty-nine, shortly after the formation of the Innovation Group. Then, in 1971, Morton, known for his hard-charging lifestyle, drove to a New Jersey bar after a late-night flight and was found dead the next morning, the victim of a violent assault. At the same time, the company explored a variety of means to avoid impending bankruptcy. In June 1972 *Innovation* was bought out by the economics journal *Business and Society Review*, ending its thirty-one-issue run.

What arguments of general sociological interest can we take from the Innovation Group's mercurial existence? First, when viewed in the context of its evolution from *IST*, images of the technoscientific self come into focus not as an Aquarian break from the Cold War order but rather as the outcome of a decade-long collaboration between journalists, industrial scientists, and academic theorists. This evolution of the scientist from paragon of intellectual purity and white-coated aloofness into a hybrid entrepreneur concerned with integrating technical knowledge in social systems came about through multiple paths. Fred Turner, John Markoff, Richard Barbrook, and Andy Cameron, among others, have convincingly shown how "the free-wheeling, interdisciplinary, and highly entrepreneurial style of work" among defense industry and university researchers was wedded with new communalist pioneers' utopian vision of liberation through information technologies as an inevitable democratic force.[77] But Brand's network forums were one variation in a set of norms that extended beyond California's borders and that predated the *Catalog*. The countercultural subcultures of the scientific enterprise in the 1970s, in other words, were deeply anchored in broader postwar transformations.

Second, Technology Communication and its *IST* precursors pio-

neered a hybrid journalism that shaped a persona that continues to impact technoscience's practitioners. These media were designed to appeal
to a broad public, but their consumers and producers were scientists,
engineers, and the myriad administrative and financial professionals
who worked in technoscientific organizations. By magazine-publishing
standards, *IST*'s peak audience of 150,000 was modest, and *Innovation*'s club-like readership of less than 10,000 was minuscule. However,
the means by which Colborn, Maass, and the editorial team assembled
world-class contributors, targeted their readership, and cultivated exclusivity and the network mode speak to the group's significance.[78] Moreover, when *Innovation* folded, its staff carried their message throughout
the 1970s in new positions as corporate communications managers at
Xerox and freelancers for *Science Digest, Research Management, IEEE
Spectrum*, and other scientific journals. Maass, the surviving member of
the founding partners, created his own company, the Executive Video
Forum, before returning to mainstream publishing to head *Electronic
Design* during the microcomputer revolution.

   IST and *Innovation* did not singlehandedly invent images of the
technoscientific innovator, but they prototyped techniques that were
widely appropriated and extended as those images rose to prominence
in the 1970s. While it is plausible to make connections between *Innovation* and late-1970s outlets such as *Omni* and early-1980s "hip"
popular-science magazines such as *Discover* and *Science 80*, its larger
impact was on the engineering and science establishment. Reading
through "mainstream" publications like *Mechanical Engineering, Civil
Engineering, IEEE Spectrum*, and *Technology Review* reveals a similar concern about out-of-control technology and technoscientists' new
identities. Companies also embraced the message and style of *Innovation*. In 1970, for instance, DuPont began its own general-interest
magazine for an external technical audience, likewise titled *Innovation*,
which mimicked Technology Communication's style.[79]

   Lastly, popularization, the fashioning of scientific personas, and
formal and informal innovation policy were deeply entangled in the
network cultivated by the Innovation Group. Throughout the 1970s,
books and magazines dramatized scientific entrepreneurs as the nation's
most important asset. Managerial guides to creativity, entrepreneurship, and organizational change based on Morton's and Schön's models
likewise flourished. For example, in Robert O. Burns's 1975 book, *Innovation: The Management Connection* (which frequently cited writers
in the Innovation Group network), the author defined his audience as
the "more than 75 percent of all active scientists and engineers, who

will devote about three-quarters of their productive lives to performing management functions."[80] These texts created a model of change management that constitutes a major share of today's market—for example, in books like *The Innovator's Toolkit* and *The Innovator's Cookbook*.[81] Likewise, we can see the legacy of the Innovation Group's multimedia experiments in TED talks and executive immersion seminars at Stanford's d.school and elsewhere.

It is this collection of guides to personal and organizational success amid unceasing change that returns us to today's shuttle concourse. As it was for their predecessors, contemporary profiles of technoscientific innovators in media like *MoFo Tech* and *IEEE Spectrum* are at once outlets for internal shoptalk among research managers, rhetorical ammunition to convince skeptical audiences, and resources to convince skeptical selves. The biggest difference from their predecessors, however, is that although today's readers are faced with many of the same sociotechnical challenges of half a century ago, edgy justifications for the belief that algae's entrepreneurs will solve the energy crisis or that smaller chips will put learning systems in the hands of the world's children are no longer necessary.

Acknowledgments

This chapter revises and expands upon Matthew Wisnioski, "The Birth of *Innovation*," *IEEE Spectrum* 52, no. 2 (February 2015): 40–45, 60–61; as well as on Matthew Wisnioski, *Engineers for Change: Competing Visions of Technology in 1960s America* (Cambridge, MA: MIT Press, 2012), 148–58; and Matthew Wisnioski, "'Change or Die!': The History of the Innovator's Aphorism," *Atlantic*, 12 December 2012, http://www.theatlantic.com/technology/archive/2012/12/change-or-die-the-history-of-the-innovators-aphorism/266191/.

Notes

1. John Steele and Ronald Neswald, "How Science and Technology Just Might Survive in a Post-technological Age," *Innovation* 27 (January 1972): 48–57.

2. Meryl Davids Landau, "First Mover: Decoding the Genome, Bit by Bit," *MoFo Tech*, Fall/Winter 2010, 11–13.

3. Tekla S. Perry, "Marvell Inside," *IEEE Spectrum* 47, no. 11 (November 2010): 40–43, 56–62.

4. Philip T. Pienkos, Eric Jarvis, and Al Darzins, "Green Gold," *IEEE Spectrum* 47, no. 11 (November 2010): 34–39.

5. Philip Mirowski, *Science-Mart: Privatizing American Science* (Cambridge, MA: Harvard University Press, 2011).

6. Fred Turner, *From Counterculture to Cyberculture: Stewart Brand, the Whole Earth Network, and the Rise of Digital Utopianism* (Chicago: University of Chicago Press, 2006).

7. Historians of science have utilized the concepts of the "scientific persona" and "scientific self" to explore the codevelopment of scientists' public representations with scientists' own aspirations of what constitutes a successful career. See Lorraine Daston and H. Otto Sibum, "Introduction: Scientific Personae and Their History," *Science in Context* 16, nos. 1–2 (2003): 2; Francesca Bordogna, "Scientific Personae in American Psychology: Three Case Studies," *Studies in the History and Philosophy of Biology and the Biomedical Sciences* 26 (2005): 95–134; Lorraine Daston and Peter Galison, *Objectivity* (Brooklyn, NY: Zone Books, 2007); and Paul White, "Darwin's Emotions: The Scientific Self and the Sentiment of Objectivity," *Isis* 100, no. 4 (December 2009): 811–26. The study of these "recognized social species" and their internalization by practitioners draws attention to the interaction between individuals and society, highlighting the role of cultural practices in the shaping of expertise and demonstrating that the supposedly autonomous "self" is created through collaborative interaction. See James A. Secord, "Knowledge in Transit," *Isis* 95, no. 4 (December 2004): 654–72.

8. Thomas Frank, *The Conquest of Cool: Business Culture, Counterculture, and the Rise of Hip Consumerism* (Chicago: University of Chicago Press, 1997). I draw on the extensive historiography of "public" and "popular" science. See, e.g., Andreas W. Daum, "Varieties of Popular Science and the Transformations of Public Knowledge: Some Historical Reflections," *Isis* 100, no. 2 (2009): 319–32; James Secord, *Victorian Sensation: The Extraordinary Publication, Reception, and Secret Authorship of "Vestiges of the Natural History of Creation"* (Chicago: University of Chicago Press, 2000); Peter J. Bowler, "Presidential Address, Experts and Publishers: Writing Popular Science in Early Twentieth-Century Britain, Writing Popular History of Science Now," *British Journal for the History of Science* 39, no. 2 (2006): 159–87; and Marcel C. LaFollette, *Making Science Our Own: Public Images of Science, 1910–1955* (Chicago: University of Chicago Press, 1990). But I explore an understudied dimension of such work, looking at media and communications aimed at practitioners rather than laymen. One of the few works to take up this thread of analysis is Maarten Van Dijck, "From Science to Popularization, and Back—the Science and Journalism of the Belgian Economist Gustave de Molinari," *Science in Context* 21, no. 3 (2008): 377–402.

9. Throughout this chapter, I will refer to Technology Communication when describing the activity of *Innovation*'s staff and to the Innovation Group when projects included the network that extended beyond the company's payroll.

10. Leo Marx, "Technology: The Emergence of a Hazardous Concept," *Social Research* 64, no. 3 (1997): 965–88; Eric Schatzberg, "*Technik* Comes to America: Changing Meanings of *Technology* before 1930," *Technology and Culture* 47, no. 3 (July 2006): 486–512; Rosalind H. Williams, *Retooling:*

A Historian Confronts Technological Change (Cambridge, MA: MIT Press, 2002), 14–19; and Benoît Godin, "Innovation: the History of a Category" (Working Paper no. 1, Project on the Intellectual History of Innovation, INRS, Montreal, 2008).

11. See, e.g., Ross Bassett, *To the Digital Age: Research Labs, Start-Up Companies, and the Rise of MOS Technology* (Baltimore: Johns Hopkins University Press, 2002); and Christophe Lécuyer, *Making Silicon Valley: Innovation and the Growth of High Tech, 1930–1970* (Cambridge, MA: MIT Press, 2006).

12. I build on Steven Shapin's dictum that the best way to understand industrial scientists is to take seriously what they have to say about themselves. But I offer a corrective to Shapin's interpretation that industrial scientists were pragmatists without the normative baggage of their academic critics. Steven Shapin, *The Scientific Life: A Moral History of a Late Modern Vocation* (Chicago: University of Chicago Press, 2008), 130–31.

13. Margaret Mead and Rhoda Metraux, "Image of the Scientist among High-School Students," *Science* 126, no. 3270 (30 August 1957): 384–90.

14. National Science Foundation and Bureau of Labor Statistics, *Employment of Scientists and Engineers in the United States, 1950–1966* (Washington, DC: National Science Foundation, 1968), 68–30.

15. Robert Perrucci and Joel Emery Gerstl, eds., *The Engineers and the Social System* (New York: Wiley, 1969), 3.

16. David Kaiser, "The Postwar Suburbanization of American Physics," *American Quarterly* 56, no. 4 (December 2004): 851–88.

17. Steven Shapin, "Who Is the Industrial Scientist? Commentary from Academic Sociology and from the Shop-Floor in the United States, ca. 1900–1970," in *The Science-Industry Nexus: History, Policy, Implications*, ed. Karl Grandin, Nina Wormbs, Anders Lundgren, and Sven Widmalm (Sagamore Beach, MA: Science History Publications, 2004).

18. Bruce V. Lewenstein, "Magazine Publishing and Popular Science after World War II: How Magazine Publishers Tried to Capitalize on the Public's Interest in Science and Technology," *American Journalism* 6, no. 4 (1989): 218–34. This is by far the best article on the mechanics of popular-science publishing.

19. Bruce V. Lewenstein, "The Meaning of 'Public Understanding of Science' in the United States after World War II," *Public Understanding of Science* 1, no. 1 (1992): 45–68.

20. "An Announcement to Our Readers," *Scientific American* 177 (December 1947): 244.

21. "This Is Our policy," *New Scientist*, 22 November 1964, 3.

22. William G. Maass, "New Information Services from a Not-So-Old Publishing House," *Journal of Chemical Documentation* 2, no. 1 (1962): 46–48.

23. William G. Maass, "From the Publisher: A Word of Introduction," *International Science and Technology* 1 (January 1962): front insert.

24. William G. Maass, "From the Publisher: After a Year," *International Science and Technology* 12 (December 1962): front insert. This cultivation of exclusivity through "controlled circulation" was an extension of a practice that Conover-Mast first innovated in the late 1920s.

25. R. Rodger Remington and Robert S. P. Fripp, *Design and Science: The Life and Work of Will Burtin* (Burlington, VT: Lund Humphries, 2007).

26. IST sold limited runs of its cover prints to readers; and in 1963 the originals were displayed in IBM's corporate art gallery on Fifty-Seventh Street, New York, NY.

27. Maass, "New Information Services from a Not-So-Old Publishing House," 47.

28. They were succeeded by Jerome Wiesner, then dean of science at MIT; Pierre Aigrain, a physicist in France's defense ministry; Charles C. Price, the head of the Department of Chemistry at the University of Pennsylvania and president of the American Chemical Society; and the Stanford rocket scientist Howard Seifert.

29. Daniel I. Cooper, "Pop Science," *American Documentation*, April 1966, 53–56.

30. Sumner Myers, "Attitude and Innovation," *International Science and Technology* 46 (October 1965): 91–96.

31. Prior to taking the editorial reins, for example, Colborn authored a nuclear disaster novel that used the marriage between a physicist-administrator and his politicking sex-kitten wife as a metaphor for the secrecy, lies, and excitement of science policy in the postwar era. As a preface to his novel, Colborn wrote: "I have simply invented an imaginary history—one which roughly parallels that of the real world but leaves me at liberty to appoint my own characters as laboratory directors, senators, and members of government commissions, in complete disregard of the people who hold such posts in reality." Robert Colborn, *The Future Like a Bride* (Boston: Beacon Press, 1958).

32. Cooper, "Pop Science," 56.

33. *The Way of the Scientist* (New York: Simon and Schuster, 1967), 7–8. The tapes of the interviews were donated to the Center for History and Philosophy of Physics, at the American Institute of Physics.

34. I. Bernard Cohen, "Questions and Answers: The Way of the Scientist," *New York Times*, 15 January 1967, BR4.

35. Jack A. Morton, "From Research to Technology," *Innovation*, May 1964, 35.

36. Jack A. Morton, "The Microelectronics Dilemma," *International Science and Technology* 55 (July 1966): 35–44. It's possible that this work was Morton's effort to understand how the transistor market had gotten away from Bell Labs.

37. David Allison, "In Our Opinion," *International Science and Technology* 55 (July 1966): 23.

38. David Allison, "The Industrial Scientist," *International Science and Technology* 62 (February 1967): 20.

39. Ibid., 22.

40. Ibid.

41. Ibid., 31.

42. William G. Maass, "From the Publisher: Janus' Other Face," *International Science and Technology* 25 (January 1964): front insert.

43. William G. Maass, "From the Publisher: Change or Die!" *International Science and Technology* 35 (November 1964): 9.

44. William G. Maass, "From the Publisher: Three Exciting Years," *International Science and Technology* 24 (December 1963): 6.

45. Matthew Wisnioski, *Engineers for Change: Competing Visions of Technology in 1960s America* (Cambridge, MA: MIT Press, 2012), 41–65.

46. Harvard University Program on Technology and Society, *First Annual Report of the Executive Director* (Cambridge, MA: Harvard University Program on Technology and Society, 1965), 1.

47. J. Herbert Hollomon, "Creative Engineering and the Needs of Society," in *Education for Innovation*, ed. Daniel V. DeSimone (London: Pergamon Press, 1968), 23–30.

48. In 1969 the philosopher John McDermott, for example, issued a harsh appraisal of the Harvard University Program on Technology and Society in the *New York Review of Books* titled "Technology: The Opiate of the Intellectuals" that accused this mode of thought as contributing to atrocities in Vietnam. John McDermott, "Technology: The Opiate of the Intellectuals," in *Technology and the Future*, ed. Albert H. Teich (New York: St. Martin's Press, 1990), 101, 116.

49. Barry Commoner, "The Eroding Integrity of Science," *International Science and Technology* 70 (October 1967): 51–60.

50. Robert Colborn, "In Our Opinion," *International Science and Technology* 70 (October 1967): 35; and Robert Colborn, "In Our Opinion," *International Science and Technology* 79 (July 1968): 17.

51. Paul Goodman, "Can Technology Be Humane?," *New York Review of Books*, 20 November 1969.

52. Paul Goodman, "The Case against Technology," *Innovation* 2 (June 1969): 36–47. This section builds on Wisnioski, *Engineers for Change*, 148–58.

53. At the time it was the largest merger in business press history. "Conover-Mast Acquired by Cahners in Merger of the Giants," *Industrial Marketing*, February 1968, 83.

54. The board consisted of *Robert M. Adams*, venture capitalist of New Ventures Division at WR Grace and Co., former vice president at 3M; *Warren G. Bennis*, provost at the State University of New York at Buffalo, former MIT Sloan School professor; *Emilio Daddario*, architect of the Office of Technology Assessment, former Connecticut congressman; *Eugene G. Fubini*, private engineering consultant, former IBM vice president, former assistant secretary of defense to JFK, former Rad Lab electrical engineer; *C. Lester Hogan*, chairman and CEO of Fairchild, a physicist who had been a Harvard professor and worked at Bell Labs and Motorola; *J. Herbert Hollomon*, MIT consultant, president of the University of Oklahoma, former assistant secretary for science and technology at the Department of Commerce; *Koji Kobayashi*, founder and president of Japan's NEC Corporation; *Warren Kraemer*, the corporate vice president at McDonnell Douglas; *Donald G. Marquis*, professor of industrial management at MIT, former Social Science Research Council associate, psychologist, director of the Office of Psychological Personnel in World

War II; *Emmanuel G. Mesthene*, director of the Harvard Program on Technology and Society; *Jack Morton*, Bell Labs vice president; *Gert W. Rathenau*, solid-state physicist at Philips Eindhoven; *Robert H. Ryan*, Gulf Oil Realty executive, responsible for the planning of Reston, VA; and *E. C. Williams*, British operations research pioneer.

55. Michael F. Wolff, "Says the Editor," *Innovation* 27 (January 1972): 1.

56. Nonmembers could pay *Innovation*'s domestic rate of $45 for the magazine. For comparison, membership in the American Society of Mechanical Engineers was $20 and a subscription to *Fortune* magazine was $14 in 1969.

57. "The Innovation Group!" (advertisement), *Wall Street Journal*, 25 April 1969.

58. Robert Colborn, "11even—Response," *Innovation* 4 (1969): 78–79.

59. Charles Horman, "The War against Research," *Innovation* 1 (May 1969): 30–37.

60. Robert Colborn, "Says the Editor," *Innovation* 1 (1969): 1.

61. Michael F. Wolff, "Says the Editor," *Innovation* 19 (March 1971): 1; and Michael F. Wolff, "Says the Editor," *Innovation* 23 (August 1971): 1.

62. Frieda B. Libaw, "And Now, the Creative Corporation," *Innovation* 23 (March 1971): 2–13.

63. Staff editors argued that session participants should then act as organizers for similar conversations in their own organizations. "11even—Retrieval," *Innovation* 31 (May 1972): 63–64; and Michael F. Wolff, "9ine—Response," *Innovation* 22 (June 1971): 62.

64. "11even—Retrieval," *Innovation* 31 (May 1972): 64.

65. Pat McCurdy, "The Innovative Group Innovates with TV," *Chemical Engineering News* 48, no. 7 (February 1970): 16–17; A. J. Parisi, "New Kind of Conference Focuses on New Ideas in Technology," *Product Engineering*, 2 March 1970, 22–24; and "Top Idea Men Trade Ideas," *Business Week*, 31 January 1970, 32–33. At another workshop, "The New Forces: Management's Challenge and Response," with a fee of $500, one of the main draws was Herman Kahn. Nancy Foy, "The Outer View: We're OK, Quality-of-Life-Wise," *Computer Bulletin* 16, no. 2 (1972): 71, 74.

66. Stewart Udall, "Technological Arrogance," *Innovation* 22 (June 1971): 12–17.

67. "After all it's not as though we're sending you out for a copy of Abbie Hoffman's *Woodstock Nation* (Vantage, 1969, $5.95 cloth, $2.95 paper) . . ." Steele and Neswald, "How Science and Technology Just Might Survive in a Post-technological Age," 57.

68. David Allison, introduction to *Dealing with Technological Change: Selected Essays from "Innovation," the Magazine about the Art of Managing Advancing Technology* (Princeton, NJ: Auerbach, 1971), 1. See also David Allison, "Measuring the Good and the Bad of New Technology," *Innovation* 9 (November 1970): 44–55; and Michael F. Wolff, "Says the Editor," *Innovation* 17 (January 1971): 1.

69. In 1964 Donald A. Schön published his first article in IST, "Innovation by Invasion," *International Science and Technology* 27 (March 1964): 52–61; his second appeared in 1966, "The Fear of Innovation," *International Science*

and Technology 27 (March 1966): 70–78. These would form the chapters "Innovation, Uncertainty and Risk" and "Ambivalence toward Innovation" in Donald A. Schön, *Technology and Change: The New Heraclitus* (New York: Delacorte Press, 1967).

70. This biographical summary is taken from Mark K. Smith, "Donald Schön: Learning, Reflection, and Change," in *The Encyclopedia of Informal Education*, accessed 10 January 2011, www.infed.org/thinkers/et-schon.htm.

71. Donald A. Schön, *Beyond the Stable State* (New York: Random House, 1971), 9–18 (quotation on 9–10).

72. Donald A. Schön, "The Diffusion of Innovation," *Innovation* 5 (October 1969), reprinted in Innovation, *Managing Advancing Technology*, vol. 1, *Strategies and Tactics of Product Innovation* (New York: American Management Association, 1972), 3–20 (quotation on 18).

73. Ibid., 19–20.

74. Schön, *Technology and Change*, 204–16.

75. Technology Communication Inc., "Change or Die!," *Electronic Design* 18, no. 13 (1970): 64.

76. Innovation, *Dealing with Technological Change* (Princeton, NJ: Auerbach, 1971); Innovation, *Decision Making in a Changing World* (Princeton, NJ: Auerbach, 1971); and Innovation, *Managing Advancing Technology*, 2 vols. (New York: American Management Association, 1972).

77. Turner, *From Counterculture to Cyberculture*, 4, 53; John Markoff, *What the Dormouse Said: How the Sixties Counterculture Shaped the Personal Computer Industry* (New York: Viking, 2005); and Richard Barbrook and Andy Cameron, "The California Ideology," *Science as Culture* 6, no. 6 (1996): 44–72.

78. Look into the footnotes of many histories of the microelectronics and personal-computer industries, and you are apt to find arguments built on information originally reported in *IST* and *Innovation*. See, e.g., Bassett, *To the Digital Age*; and Howard Rheingold, *Tools for Thought: The People and Ideas behind the Next Computer Revolution* (New York: Simon and Schuster, 1985), 180.

79. "Intimations of Validity," *Innovation* [DuPont] 1, no. 2 (1970): i. See also Don Fabun, *Dimensions of Change* (Beverly Hills, CA: Glencoe Press, 1971).

80. Robert O. Burns, *Innovation: The Management Connection* (Lexington, MA: Lexington Books, 1975), ix.

81. David Silverstein, Philip Samuel, and Neil DeCarlo, *The Innovator's Toolkit: 50+ Techniques for Predictable and Sustainable Organic Growth* (Hoboken, NJ: John Wiley and Sons, 2011); and Steven Johnson, *The Innovator's Cookbook: Essentials for Inventing What's Next* (New York: Riverhead Books, 2011).

12 When Chèvre Was Weird: Hippie Taste, Technoscience, and the Revival of American Artisanal Food Making

Heather Paxson

Cypress Grove Chevre in Arcata, California, produces goat cheeses with such names as "Humboldt Fog," "Purple Haze," and "PsycheDillic." The groovy names convey the countercultural roots of what in retrospect has come to be called an American artisanal cheese "movement." Cypress Grove founder, Mary Keehn, settled in remote Humboldt County in the 1970s, inspired by the back-to-the-land ethos that surrounded her first as a student in Berkeley and then for a time in Sonoma, California. Mary acquired her first goats as a source of "good milk" for her first daughter, whom she was weaning. The goats multiplied, Mary became overwhelmed with goat's milk—it seemed wasteful to dump it—and so Mary started experimenting in her kitchen, making goat's milk cheese much as she did her own ketchup, because (as she recalled in 2008) she aimed to be as "self-sufficient as we could be."[1] After her husband left her with four young children to raise, she built and licensed a creamery and went into the cheese-making business. But if her evocative cheese labels hint at nostalgia for the heady days of getting back to the land (and note that Humboldt Fog refers not only to the mists that wash this coastal stretch of California but also to that illicit countercultural crop the county is best known for!), days suffused with dreams of leaving the technocratic world behind, there is more to the story. In food making as in other arenas, the countercultural romance with nature did not stray far from a scientifically minded pragmatism. Scientific understanding of animal husbandry and technical tools for cheese making proved to be essential to the commercial success of artisans who, for the most part, continued to resist becoming overly dependent on technoscience.

Having majored in biology at the University of California–Berkeley and being fascinated by genetics, Mary Keehn became a skillful goat breeder; for years she ruled the goat show circuit. Like many of her countercultural cohort, Mary also drew tips and inspiration from Rodale Press publications on organic gardening, as well as the *Whole Earth Catalog*. "We *are* as gods and might as well get used to it," Stewart Brand famously began the inaugural, 1968 issue of the *Whole Earth Catalog* (in subsequent issues the line was revised to read, ". . . might as well get *good* at it" [emphasis added]). The catalog's purpose statement promoted a strategic reworking of British anthropologist Edmund Leach's call for the rise of expertise.[2] Opening a series of lectures delivered over BBC Radio in 1967, Leach had declared, "Men have become like gods. Isn't it about time that we understood our divinity? Science offers us total mastery over our environment and over our destiny, yet instead of rejoicing we feel deeply afraid. Why should this be? How might these fears be resolved?"[3] While adopting Leach's muscular rhetoric, Brand turned Leach's argument on its head by championing, rather than disparaging (as had Leach), the amateur, proclaiming:

> So far, remotely done power and glory—as via government, big business, formal education, church—has succeeded to the point where gross [defects] obscure actual gains. In response to this dilemma and to these gains a realm of intimate, personal power is developing—power of the individual to conduct his own education, find his own inspiration, shape his own environment, and share his adventure with whoever is interested.[4]

Mary Keehn and other amateurs who pioneered the farmstead cheese movement harnessed such personal power in reshaping their own local environments and, in so doing, remade American culinary landscapes. Having sought self-realization through the counterculture in the 1970s, in the 1980s people like Keehn—falling back on (and perpetuating) their largely middle-class upbringings—came to seek self-fulfillment through self-employment. Brand's anti-Establishment, ecologically minded, can-do sentiment—in no way antithetical, it turned out, to entrepreneurial capitalism—is at the root of the turn-of-the-century story of artisanal cheese's cultural and commercial flourishing in the United States. "At a time when New Age hippies were deploring the intellectual world of arid abstractions," Brand has written in retrospect, "*Whole Earth* pushed science, intellectual endeavor, and new technology as well as old."[5]

Through the pages of the *Whole Earth Catalog* and subsequent projects, Brand drew hippie homesteaders who traveled north from the San Francisco Bay Area into the same conceptual world as engineering and computer geeks who clustered around Stanford University south of the bay.[6] Just as the convergence of outside-the-box counterculturalism and cutting-edge technoscience led to the development of personal computing and vast fortunes in Silicon Valley, "American artisanal cheese" emerged as a food category, and "making cheese" as something one could do as an income-generating vocation, through a series of what Fred Turner calls the "network forums" manifest in countercultural publications, workshops, meetings, and digital networks, "within which members of multiple communities"—beginning, in the case of cheese, with rural communes, research universities, and urban markets—"could meet and collaborate and imagine themselves as members of a single community."[7] Back-to-the-landers with a few goats or a dairy cow and with scaled-down tools developed by and for the industrial food system converged, eventually, with a handful of renegade dairy scientists to put American artisanal cheeses on the map, via the menus of nouvelle restaurants.

Hippie cheese makers did not set out to make gourmet food. Food—its sourcing, composition, preparation, and distribution—gravitated to the center of countercultural life as hippies debated everything from what counts as a "natural" food to the gendered politics of domestic labor.[8] Alongside yogurt, cheese—an ancient means of preserving milk through processing by fermentation—was embraced as a quintessentially "natural" food. And yet, as I will detail, the countercultural cheese makers' "tools for access" (to adopt the *Whole Earth Catalog*'s language) were thoroughly technoscientific, in ways both self-consciously explicit and so tacit as often to have gone unnoticed.

In recounting the countercultural history of the "renaissance" in American artisanal cheese making, this chapter pays particular attention to how artisan food makers negotiated the authoritative language and practical reassurance of technoscience with the free-spirited artistic license of the countercultural ethos that led many of them into cheese making in the first place. The story I tell, drawing from broader ethnographic research that includes oral-history interviews with cheese mak-

ers who started commercial creameries or began teaching home cheese-
making workshops between 1978 and 1984, is one of *convergences* and
conversions. Artisanal cheese's growing popularity today is indebted to
convergences between back-to-the-land amateurs looking for a means
of making money and the industrial food system they imagined leaving behind. Those convergences occurred through a variety of media,
most centrally through communications media and via the instruments
and materials employed to make cheese. Various communications media spawned a "network forum" that operated much like the *Whole
Earth Catalog*. In teaching themselves how to transubstantiate milk
into cheese, amateur cheese makers of the 1970s and 1980s thought
of themselves as developing a practical art—but in perfecting that art,
and in discussing it within a heterogeneous community of practitioners,
they found themselves relying increasingly on the knowledge and practical tools of science. In some instances, neo-artisans reappropriated the
tools and materials of technoscience—the media of industrial cheese
making—for artisanal manufacture. In other instances, cheese makers
became aware that the use of scientific instruments, such as pH meters,
entailed a kind of craft practice that is not so very different from the
"art" of making cheese "by hand."

After providing an overview of the American revival of artisanal
cheese making by rural seekers of alternative lifestyles in the 1970s and
early 1980s, I will detail instances of the mediated convergence between
a freewheeling counterculture enamored of the power of "nature" and
technoscientific means to control that same "nature." Discussion of communications media will be followed by analysis of the use of instrumental
and material media: acidometers, pH probes, bacterial cultures, renneting
agents. In conclusion, I note that such practical convergences often led to
network members' conversion experiences. I view the "arrival" of American artisanal cheese—undoubtedly, it is more middle-class mainstream
today than markedly countercultural—as the result of craft conversions:
members of the cheese world have come to agree that the artistic approach of self-taught amateurs and the scientistic view of dairy scientists
(if not safety regulators) are more compatible than mutually exclusive.

Background: Strange Cheese

Although it is newly meaningful, the handcrafting of cheese is itself
nothing new in this country. The Puritans brought dairy cows and
cheese making to America in the 1600s, and by the eighteenth century,
New England farm women were crafting cheese for both domestic use

and commercial trade. Its quality was variable. Women worked in their kitchens without benefit of thermometers, let alone acidometers. As part of the modernizing drive of the mid-nineteenth century, cheese making was moved off American farms and into farmer-owned cooperative factories.[9] The factory system displaced home dairying—and hence women cheese makers—in a single generation. Lauren Briggs Arnold wrote in her 1876 manual for farmhouse cheese makers, "An occasional expert may be found in family dairying, but it is not possible to find one in every family," and so by pooling milk and centralizing fabrication, the factory system extended the reach of proficient cheese makers, who initially used much the same technique and tools as had farm women.[10]

Around the same time, in the late 1800s, scientists at the Pasteur Institute in France discovered that the seemingly magical process of curdling is not "spontaneous" but rather the outcome of microbial metabolism. It took a few additional decades to isolate and cultivate strains of the lactic acid bacilli that ferment milk and start the cheese-making process; only then could cheese makers begin to exercise the technoscientific control of pasteurization in working with milk. Aiming for increased consistency, product standardization, and more efficient economies of scale, in the 1930s artisan factories begin reformulating their recipes to enable the use of pasteurized milk; the rest of the twentieth century saw cheese factories increasingly industrialized and automated.[11]

Against the backdrop of the American food system's industrial homogenization, the obsolete became the authentic as cheese making was returned to dairy farms as an offshoot of the back-to-the-land movement by young people who were new both to farming and to making cheese. They were baby boomers, members of the generation born during the post–World War II demographic rise in births (between 1946 and 1960), whose middle-class cohorts were culturally associated with a rejection of traditional authority as represented by the patriarchal families and paternalistic ideals of industrial capitalism dominant during the 1950s of their childhoods. Beginning in the mid-1960s, clusters of young, white, educated people fled such cities as Boston and Berkeley (where many had attended college) to alight (if not permanently settle) in rural parts of western Massachusetts and coastal California, and beyond. Whether they thought of themselves as homesteading, creating rural communes, or simply escaping the seeming predictability of urban professionalism, rural living promised a measure of sought-after self-sufficiency. Like farm women of centuries ago, many of today's artisan food pioneers made cheese in their kitchens for home consumption and barter long before considering commercial trade. Often, it began with goats.

Leaving Los Angeles in 1972, Barbara and Rex Backus began raising goats in California's Napa Valley. Recalling those days, Barbara laughed at the naïvety of "these urban, overeducated types coming to the country"—meaning she and Rex and their friends. Back then, she told me, "most of the people coming up here were really ignorant of what it meant to be a farmer. This, of course, isn't farmland," she noted wryly, indicating the arid, hilly landscape outside her kitchen window.[12] Fortunately, they found that the land does sustain goats. In time, Barbara and Rex became goat dairy farmers, moving well beyond playing at it. In southern Indiana, Judy Schad—whose Capriole Farm has gone on to produce award-winning aged goat cheeses—initially had her heart set on a milk-giving cow, but her neighbors persuaded her (in 1977) that she wanted goats instead. Judy's children loved the goats but hated consuming their milk—until Judy discovered how to transform it into cheese. In southern Wisconsin, Anne Topham and Judy Borree so enjoyed living with a goat named Angie and making simple cheese in their kitchen from her milk that the women decided "to find our own farm, make goat cheese from the milk of our own goats, and hand that cheese directly to people who would buy it" at the Dane County Farmers' Market in Madison, the oldest farmers' market in the United States. "When Judy and I started" back in 1984, Anne told me in a 2008 interview, "we just didn't think there was anything we couldn't do. We just thought, 'Oh, well, we'll build a house! Oh, well, we'll build a barn! Oh, well, we'll learn to milk goats! Oh, well, we'll make cheese! Oh, well, we'll sell it.'"[13] Anne recalls that Judy "built the milk house with a hammer in one hand and a book in the other." They remodeled the 20- by 20-foot garage that stood on their fifty-acre plot of land to serve as a "cheeserie," as Anne calls it—"it seemed too small and modest to be called a licensed dairy plant, although that is what it legally became."[14]

It has become a mantra, an article of faith, among the hippie homesteader cheese makers that they all started out entirely on their own and were self-taught, by necessity. As Anne Topham said to me, "When we started, we were really working in isolation. There was nobody else doing it. There wasn't anybody else to talk to about it."[15] Another said, "The biggest problem I had was being self-taught. I never saw anybody else make cheese." To be sure, artisan factories across the country have been in continuous production since the early 1900s; the farmsteaders' sense that "there wasn't anybody else" reflects a gendered and class-based gulf between college-educated young people seeking self-sufficiency and the skilled tradesmen who were then crafting Cheddar and Colby in eighty-year-old factories.[16] Hippie cheese makers' reported sense of isolation and making-it-up-as-we-went-along is all the more striking because I was also told of the camaraderie of the "goat dairy network," about cheese-making demonstrations performed at dairy goat association meetings, and about secondhand cheese-making equipment circulating among members. As historian Warren Belasco writes, "One enduring goal of the 1960s was to upset the rule of 'experts.' 'Deprofessionalization' would return power and dignity to the grass roots, giving ordinary people a sense of worth and importance. Throwing away the few reigning cookbooks and conventional wisdoms, freaks adopted an 'anything goes' approach to food."[17] Like philosopher of science Paul Feyerabend, they would be "against method."[18] And like other counter-cultural communities, the fledgling farmstead cheese makers "embraced the notion that small-scale technologies could transform the individual consciousness and, with it, the nature of community. They also celebrated the imagery of the American frontier . . . , moving out onto the open plains in order to find a better life."[19]

Reflecting back on that sense of starting from scratch and all alone, Anne Topham commented to me, "That was both the good news and the bad news." The bad news was that they "didn't have a model" and didn't really know what they were doing: Anne and Judy tried this cheese and that cheese, this method of goat husbandry and that method, and painstakingly learned from what Stewart Brand once called the "wishful mistakes" of experience.[20] It took years to establish a dairy, a reliable cheese-making procedure, and a market for their fresh goat cheese in Wisconsin. The "good news" about all of this, according to Anne, was that absent a model of practice, absent the farmhouse cheese-making traditions of Europe—traditions that can be more dictatorial than suggestive—"what we developed was . . . just us." Allison Hooper, cofounder in 1984 of the Vermont Butter and Cheese Company, writes in her foreword to Jeff Roberts's 2007 *Atlas of American Artisan Cheese*, "Without the burden of tradition we are free to be innovative, take risks," suggesting that lack of tradition in regional cheese types and fabrication methods was a virtue rather than a deficit for Americans because it opened up possibilities for experimentation and self-expression.[21]

Eschewing the professional expertise of academic dairy science and cheese factories alike, the new cheese makers embraced a sense of themselves as pioneers in the wilderness, compelled to figure things out on

their own. As Anne's words indicate, this narrative echoes the countercultural ideal of democratization and nods toward gender equity, but as
I argue elsewhere, it also reflects a class-reproducing, ideological belief
in innovation as a source of value and self-making integrity; the countercultural food pioneers were after their own version of the American
Dream.[22] Mary and Barbara and Anne were not just eating counter-
culturally; they came to earn independent livings from producing the
"countercuisine" that countercultural peers purchased at food co-ops
and restaurants.[23] Not unlike how "the revival of midwifery" represented "a feminist defense against modern medical patriarchy," the way
these women capitalized on their ostensibly domestic labor and pursued
the craft of food making "reasserted female competence and control" in
a modern economy dominated by men.[24]

But if the lack of a cheese-making tradition was "freeing," the lack of
a culinary tradition for eating goat cheese posed a marketing problem.
Even out in California, as Mary Keehn recollected, "Goat cheese was totally weird. There was no goat cheese" in the 1970s.[25] Beyond learning
to make it, these women had to learn how to sell strange cheese. Joining forces with friends from the local goat dairy association, Barbara
formed a co-op that made Jack cheese from the pooled milk of their
goats to sell in health food stores and other alternative retail venues. Fed
up with their inability to interest a food distributer in their goat Jack,
one of those friends, Laura Chenel, traveled to France and learned how
to make fresh, French-style goat cheese. Alice Waters's Chez Panisse restaurant in Berkeley has been credited with launching Laura Chenel's
Chèvre company—and, by extension, American goat cheese.[26] On the
coasts, it was French-trained urban chefs who first bought hippie-made
goat cheese; what producers may have lacked in technique they made
up for in freshness. Yet even enthusiastic chefs warned the cheese makers that "goat cheese" would not appeal to diners; the cheese had to
be listed on their menus more obliquely, and seductively, by its French
name, *chèvre*. Allison Hooper's Vermont Butter and Cheese Company
was founded one year after political scientist Frank Bryan published
his book of humor, *Real Vermonters Don't Milk Goats*, which poked
fun at "flatlanders," the local term for newcomers to the state (Allison
hailed from New Jersey); note that the company name does not draw
attention to the fact that Allison produces butter and cheese from the
milk of *goats*.

Tellingly, however, many of the early farmstead cheese makers insist
that they were never making "French cheese." What makes an American cheese distinctively American, here, is that it presents itself as new,

different, unique—despite remaining inescapably indebted to European histories of practice and taste-making. Idiosyncratic cheese names, such as Humboldt Fog and Purple Haze, serve less to place a cheese within a familiar taxonomy or cheese "family" (e.g., Colby, Gouda, Brie) than to convey a sense of the personality of the maker. In a nod to her art history degree and special affection for Japanese art, Barbara Backus's cheeses included Hyku and Sumi. Cheese making is an ancient means of preserving milk; goat-keeping hippies did not invent the process. What they did help to create was a new source of cultural and economic value, one forged in opposition to the ecological costs of industrial agriculture and the culinary homogenization of highly processed foods. Craft foods offered a powerful symbolic antidote to Kraft Foods.

Communications Media and the New England
Cheesemaking Supply Company

Becoming a "self-taught" cheese maker meant reading books and pamphlets describing the process and then, through trial and error, working to translate written instructions into habituated, embodied practice. This is by no means a straightforward proposition, particularly given the dearth of information and of small-scale cheese-making equipment available at the time. What few printed resources were available in the 1970s spoke to radically different visions of expertise. In 1973 Garden Way Communication published a thirty-five-page "little yellow pamphlet" entitled *Making Homemade Cheeses and Butter*, written by Phyllis Hobson, whose other Garden Way publications include *Tan Your Hide! Home Tanning Leathers and Furs* (1977) and *Soybean Book: Growing and Using Nature's Miracle Protein* (1978). Hobson may have had firsthand experience with homesteading, but it is uncertain whether she ever made much cheese.[27] The basic information presented in the pamphlet would not have taken a serious home cheese maker very far. On the other end of the spectrum, in 1970 Cornell dairy scientist Frank Kosikowski (who went on to found the American Cheese Society to support small, farmstead producers) published the first edition of his two-volume, clothbound tome, *Cheese and Fermented Milk Foods*. This text was written for the dairy industry, and while a few home cheese makers with commercial aspirations and some scientific literacy report having found in it useful explanations of milk chemistry and curd development, others encountered a world of frustration in trying to process their goat's milk following guidelines written with industry-standard cow's milk in mind.

The one book that the early goat-cheese makers consistently praise is Jean-Claude Le Jaouen's *La fabrication du fromage de chèvre fermier* (*The Fabrication of Farmstead Goat Cheese*; fig. 12.1). Although also written by a research dairy scientist, Le Jaouen's publication was the first book (known to American cheese makers at this time) to speak directly to the concerns, needs, and facilities of the farmstead goat-cheese maker. The book includes chapters on how to design and equip a cheese dairy, including well-ventilated rooms for drying and curing cheeses. The problem with the book, for many Americans, was that it was written in French. (Another early publication, "by the Canadian nuns," as it was referred to me—*Fromage de chèvre: Fabrication artisanale*, published by Benedictines in Quebec—was similarly incomprehensible to Anglophone goat keepers.) Barbara Backus first learned of Le Jaouen's book from a Frenchman who turned up in Sonoma County "bootlegging" kitchen-made cheese to restaurants. Barbara ordered a copy through a local bookstore and hired a local Frenchwoman to read

FIGURE 12.1 Ricki Carroll's well-worn copy of Claude Le Jaouen's *La fabrication du fromage de chèvre fermier* (*The Fabrication of Farmstead Goat Cheese*), published by Institut de Élevage in 1977.

the text aloud, in translation, while sitting in Barbara's kitchen; Barbara audio recorded and transcribed the translation. The lay translator encountered technical words she could not translate, and according to Barbara's husband, Rex, there was at least one key error in the original French. But it helped. Eventually, the book was translated into English and, as Barbara said, "we all breathed a sigh of relief."[28]

Le Jaouen's book was published in English in 1987 by the New England Cheesemaking Supply Company, an accidental entrepreneurial concern with countercultural roots that has done much to meet (and generate) the rising demand for small-scale cheese-making tools and know-how. Begun by Ricki and Bob Carroll in 1978, the New England Cheesemaking Supply Company shares a trajectory with many early farmstead businesses. In 1975 Ricki and Bob were married so that they could sign a joint mortgage on a "hippie house" in Ashfield, Massachusetts, into which they moved with nine fellow travelers. I visited Ricki in 2013; she alone remains in the house, a high-ceilinged farmhouse surrounded now by vast lawns and perennial gardens, with African drums lining the dining room and art glass fronting the kitchen cabinets. Back in the 1970s, the rural commune grew and produced "ninety-eight percent" of what they ate, including soybeans to make tofu. A neighbor suggested they buy a pair of goats for the milk. Soon they were experimenting with making cheese—"we were making all kinds of weird cheeses and they weren't coming out so great," Ricki acknowledged—as well as experimenting with making a living. Associating "home cheese making" and "goat's milk" with European tradition, Bob wrote letters to US embassies around the world, asking to be put in touch with people who made cheese at home. When the embassy in Britain replied, in 1977 Bob and Ricki traveled to England, where they worked alongside a dairy farm family for two weeks; they returned with a new confidence and business plan.[29]

The day they left for England Bob placed a nine-dollar ad in the *Dairy Goat Journal* listing a cheese-making supply catalog, available by mail order for twenty-five cents. Bob's initial thought, likely inspired by the economies of scale enjoyed by the food co-op Ricki ran, was that their household could acquire cheese-making supplies cheaply if they bought in bulk and sold most of the inventory. Ricki thought he was crazy. "Who would want to do what we were doing?" she thought. Nonetheless, they returned from England to a mailbox stuffed full of quarters. Ricki and Bob set to work locating supplies and putting together a single-page price list featuring a cheese-making kit that they assembled from commercial rennet tablets, freeze-dried bacteria cultures, and precut lengths of butter muslin for cheesecloth. The Carrolls acquired much of their early stock directly from Chr. Hansen's, a Danish biotechnology firm that has supplied dairy processors with cultures and coagulants since the 1870s and has since expanded into probiotic supplements for human food and for livestock feed, for "rumen management."

Echoing the can-do sentiment of the *Whole Earth Catalog* but providing the direct sales of the L. L. Bean catalog, the New England Cheesemaking Supply Company catalog provided subscribers with the tools "not only to get a job done," as Turner writes of Brand's *Whole Earth Catalog* and the "network forum" it spawned, "but also to enter into a process. The process would accomplish tasks but also would transform the individual into a capable, creative person."[30] If Bob set out to sell cheese molds and rennet as a means of household management, the Carrolls soon saw the transformational potential in teaching people not only to make cheese but to become "home cheese makers." In addition to the mail-order business, Ricki and Bob began offering cheese-making workshops out of their kitchen, using raw goat's milk from their own goats. For years they took their workshop on the road, giving demonstrations at regional meetings of the American Dairy Goat Association and ending up in places as incongruous as a Las Vegas casino, stringing bags of wet curd from the chandeliers to drain.

In 1982, wanting to displace Garden Way's dreaded "little yellow pamphlet," Ricki and Bob published their own book, *Cheesemaking Made Easy: 60 Delicious Varieties*, also with Garden Way (fig. 12.2). Subsequently retitled *Home Cheese Making*, the book is now in its third edition and has sold more than 300,000 copies. The book came to be thanks to Annie Proulx, an attendee of one of the Carrolls' workshops. Before going on to become a Pulitzer Prize–winning fiction writer, Proulx coauthored a book with her father, Lew Nichols: *Sweet and Hard Cider: Making It, Using It and Enjoying It* (1980). After taking Ricki and Bob's workshop, according to Ricki, Proulx told her editor at Garden Way, "You need a book on cheese making; call these people—they're the ones to write it." The publisher instructed the Carrolls to keep it simple, saying, "'We don't want you to deal with pH and acidity, or any of that.' So we translated everything into time," Ricki told me. "We made it as simplified as we possibly could."[31]

The Carrolls' book could not have presented a starker contrast to Kosikowski and Le Jaouen, who revel in scientific detail and draw potentially mystifying distinctions between, say, lactic coagulation and rennet coagulation. In addition, the more scientifically minded guides put not-

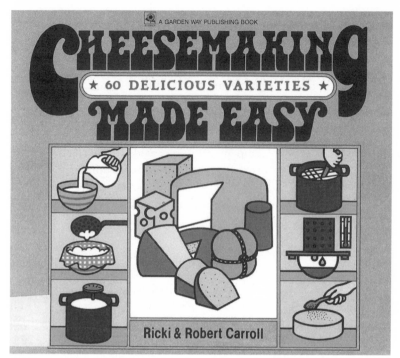

FIGURE 12.2 Ricki and Robert Carroll's home cheese-making book, published by Garden Way in 1982.

so-subtle pressure on the artisan cheese maker to invest in expensive, esoteric lab equipment (for example) for reading acidity ("The Dornic Acidometer is the Farm Producer's Number One Tool," reads a heading in Le Jaouen). Whereas Le Jaouen spends a full page (p. 81) discussing the optimum level of acidity at which milk might best be renneted (a discussion whose relevance rests on the cheese maker's ability to employ and read an acidometer; see below), Ricki Carroll simply asks the cheese maker to keep an eye on the clock and thermometer—scientific instruments that would not intimidate the serious home cook. Compare the instructions for coagulating and molding curd for the soft-ripened goat cheese Saint-Maure taken from (*a*) Le Jaouen's *The Fabrication of Farmstead Goat Cheese* and (*b*) Ricki Carroll's *Home Cheese Making*:

(*a*)
Coagulation of the Milk

- The milk temperature is 18–22°C.
- The temperature of the cheesemaking room is 20–22°C.

- The rennet dosage is 8–10 ml/100 liters of milk. (The rennet is 10,000 strength or 520 mg of active chymosin.)
- The average coagulation time is 24 hours.

Moulding

- The time to place the curd in the moulds is when the acidity of the floating whey is 45–55 Dornic degrees.[32]

(b)

3. Cover and allow the milk to set for 12–18 hours, or until it coagulates. The room temperature should not exceed 72°F.
4. Scoop the curds into individual goat-cheese molds (approximately 3¼ inches in height). When the molds are full, put them in a convenient place to drain. Drain for 2 days.[33]

Ricki's instruction, taking the reader's hand, is invitingly formatted to resemble a cookbook recipe. Indeed, in writing the book, Ricki made cheese in her kitchen and wrote down exactly what she did, when she did it, using the same tools she expected of her readership: a stainless steel pot, a cooking thermometer (sold through her catalog), and a clock. She turned practical knowledge into text and sold it as nonthreatening expertise.

In doing so, Ricki relied on "black-boxing" a scientific understanding of what was happening. The ingredient list for her recipes includes such items as "3 drops liquid rennet diluted in 1/3 cup cool, unchlorinated water" and an option between "1 packet direct-set mesophilic starter or 4 ounces prepared mesophilic starter." Liquid rennet can refer to animal rennet (derived from an enzyme extracted from the lining of the fourth stomach of a suckling calf), to a mold-derived version created in a biotech lab (often sold as "vegetable rennet"), or to a genetically modified version of chymosin (the active enzyme in animal rennet). All three types are described in *Home Cheese Making*, although the New England Cheesemaking Supply Company, espousing an anti-GMO position, sells only animal and mold-derived rennet. The business still buys in bulk from industrial concerns and divides the supplies into packages suitably sized for home use.

No less than the inventory of 250-gallon cheese vats that for a time cluttered Ricki's front rooms and spilled out on to the wraparound porch, domestic- and artisan-sized packages of industrial-grade rennet and bacteria starter cultures are tools—like many offered in the *Whole Earth Catalog*—that served what Turner has called a "linking function," by drawing "readers into performing actions within which they could link the insights of multiple networks." Turner gives the example of a *Whole Earth* reader who, by fixing a Volkswagen, could first "perform the role of the amateur engineer, managing a technological 'system' from above, and, once the car was running, perform the role of traveling hippie nomad."[34] Likewise, a *Home Cheese Making* reader could perform the role of Earth Mother, providing "natural" yet homemade sustenance to her flock, and then, by commercializing her domestic art, become a self-employed and empowered entrepreneur.

Instrumental Media: The Convergence of Industrial Tools and Craft Practices

Although pitched in opposition to the bland homogenization and industrial standardization of industrially processed food, the homesteading cheese makers' do-it-yourself "natural" food making relied, from the beginning, on technoscience developed for and by the industrial food industry. Before they discovered cheese making, goat-keeping homesteaders sold milk to industrial processors (such as Meyenburg in California) that dried goat's milk for use in industrial applications. Home cheese makers may have thought of themselves as buying bacteria cultures and rennet "from Ricki," but the Carrolls sourced their bulk goods directly from multinational biotechnology concerns. And home cheese makers who went on to commercialize their domestic art faced US and state regulations that demanded new degrees of technoscientific convergence.

Because the back-to-the-landers favored small ruminants (i.e., goats), many of them started out making chèvre, fresh goat's milk cheese. Since 1949, the US government has required fresh cheeses (indeed, any cheese ripened for fewer than sixty days) to be made from pasteurized milk. And so, because they had to, the hippies invested in pasteurizers and thus—though they may not immediately have thought about it—in technoscience. Moreover, to make cheese from pasteurized milk, the milk must be reseeded with bacterial cultures. It is possible to maintain a "mother culture" of lactic acid bacteria, much like a sourdough starter culture, but most cheese makers have preferred the ease and consistency afforded by laboratory-isolated, freeze-dried strains of proven bacterial cultures. There was always a microbiological machine in the hippies' farmstead garden.[35]

Le Jaouen taught goat-cheese makers the importance of careful microbial hygiene, writing, for example, "The farmstead cheesemaker could be compared to a captain leading an army of lactic acid producing bacteria against an army of objectionable bacteria while guarding against excessive ardor on the part of his troops."[36] But to be sold commercially, cheese had to be made from the milk of a Grade B certified dairy, in a certified creamery or cheese plant. One's kitchen would not suffice. This is not to say that early farmstead creameries could ever have been mistaken for industrial plants manufacturing commodity cheese, even at a smaller scale. Absent a ready-made equipment supplier for a not-yet-existent artisanal industry, and finding virtue in thrift, early producers got creative. For cheese molds, David and Cindy Major punched holes in the bottom of cheap plastic bowls they bought at a big-box store; the rind of Major Farm's Vermont Shepherd cheese continues to bear the rustic seam line of imperfectly fitted molds. Cheap and versatile PVC tubing was a popular material. David Major fashioned cheese weights from lengths of PVC tube filled with food-grade salt. Anne Topham and Judy Borree's tiny milking parlor featured head gates fashioned from lashed-together lengths of PVC.

Through the mid-1980s, local dairy and food safety inspectors were relatively lenient with jury-rigged and repurposed equipment.[37] Up until her retirement in 2013, Anne kept track of her pasteurization records through manual charting; today, state inspectors require tamper-proof, automatic, digital-recording devices for auditing purposes. In northern Vermont, Laini Fondiller, whose small goat farm and creamery ran on solar power backed by a small generator, fought for years with the state of Vermont to make goat cheese for commercial sale despite not being able to afford the equipment that regulators demanded. Eventually, she was given a dispensation to pasteurize milk in a large stockpot over a gas burner, provided she build a dedicated 10- by 10-foot room in which to do so. Laini described the room as "like a jail cell"; in 2002, with a 4 percent loan from the Vermont Community Loan Fund, she built a freestanding cheese house and outfitted it with a commercial-grade vat pasteurizer, still running it using off-grid, solar energy.[38]

Artisan cheese makers were selective in what elements of technoscience they allowed into their cheese-making practice; the use of commercial starter cultures did not inextricably lead to the purchase of a mechanical stirring device, or vice versa. If countercultural tools were intended not only to manufacture and fix things but also to generate new self-knowledge, artisan cheese makers have derived a range of personal values from their use of tools. The question is not merely what tools may appropriately be used by "artisans" but what sort of *artisan* is realized through the process of working more or less "by hand." In this regard, two of the most telling pieces of equipment, used by some artisans but not by others, have been the mechanical stirring device (affixed to a cheese vat to gently agitate curd) and the acid titrator or pH probe (used to gauge the acidity of curd as it develops in the vat).

As cheese-making enterprises grew larger, time- and laborsaving technologies, such as the mechanical stirrer, began to look more appealing. For decades, Taylor Farm's Jon Wright, who left his prep school background behind to farm in Vermont in 1975, hand-stirred curd to make Gouda in his 4,000-pound-capacity vat. In 2007, when I met him, he was looking to upgrade to a mechanized agitator because hand-stirring ties up an hour and a half of labor each day, and "While you're stirring, you're thinking about all the things that need to get done" but that you can't get to, tied to the vat. Still, reflecting on his long-standing practice he mused, "One thing I really like about [hand-stirring] is that I can really watch the curd, can adjust the rate of stirring. I like it"—for the craftsmanship it affords—"but," he concluded, "I could get over it!"[39] Wright articulated an artisan-scale version of "the craft-convenience conflict" that women of the counterculture confronted domestically. As Belasco points out, while cooking from "scratch" using unprocessed ingredients was de rigueur for the "natural foods" of the "countercuisine"—and a point of pride for many a domestic goddess—pressure cookers undoubtedly saved time in preparing dried beans to feed a crowd.[40] Some cheese makers came to look upon the mechanical stirring arm as a sort of commercial-scale pressure cooker: no eater is likely to detect from tasting the finished product whether or not the tool was employed, but its use significantly affects the producer's daily life and relationship to her work. Once the food maker becomes also a business manager, "time saved" comes to mean something new. Since Cypress Grove Chevre's inception in the early 1980s, Mary Keehn's business has grown to employ about forty people in an expanded factory producing for national distribution; along the way, Mary has introduced a number of automated technologies out of concern for her employees' physical well-being as well as a business owner's imperative to cut costs (e.g., workers' compensation). As Mary describes it, limited use of automation eases her employees' performance of manual labor (while arguably reskilling it toward practical knowledge of machinery rather than curd) while allowing Mary to perform the role of effective—efficient yet caring—business manager.[41]

The acidometer is a different sort of technology. It saves neither time nor manual strain but instead offers a scientific tool for reading the acidity of milk and curd. Knowledge of milk acidity can indicate to the cheese maker how much starter culture or rennet to use in initiating the process of fermentation and coagulation, as well as when the coagulated curd is ready to be molded (if making a soft-ripened cheese) or "cut" so as to separate the liquid whey from the solid curd (in making a hard-aged cheese). Milk is not a standard substance unless it is submitted to thorough processing: not only pasteurization but also homogenization (to break down the fat globules to a uniform size), skimming (to regulate butterfat content), and the mixing of early- and late-lactation milk through staggered, year-round breeding. This is what has fundamentally distinguished artisanal from industrial cheese making: artisanal cheese making begins with minimally modified milk. Artisan cheese makers, then, did not strictly follow a recipe; they have had to exercise reflexive adjustment to fluctuating materials and conditions.[42] Acidity levels, gauged by titrating sodium hydroxide into milk or curd or reading a pH probe thrust into it, are data points that cheese makers have used in adjusting their practice when working with changing materials under fluctuating environmental conditions. Understanding acidity—how to measure it and what it indicates for practice—opened the black box contained in Ricki Carroll's *Home Cheese Making* and, for some, helped to distinguish the professional artisan from the domestic amateur.

Although Le Jaouen devotes a full page to explaining how to use the acidometer to titrate sodium hydroxide with a facing page of step-by-step photo illustrations, I can attest that the procedure is not made entirely clear to someone with little lab science experience. Like Ricki Carroll, many of the early farmstead cheese makers eschewed the injunction to record and respond to acidity readings. Barbara Backus told me, "I started out using a—titrating the acid, like Mr. Le Jaouen says, 'maybe you want to do this,'" but after a while she gave it up, describing the extra step as "a waste of time because it was much easier to gauge what I was doing" through direct empirical assessment of her curd, using her bare hands. "Basically," she said, "I'm a cheese maker that uses her senses." She has gone by "what I see, what I handle, what I smell, what I taste," based on experience working with her own goat's milk in her cheese house. She understood that the acid titrator offered her another empirical tool for apprehending what her curd was doing, but she indicated, too, that the readings it gave her were meaningless without knowledge of how to interpret them; "I had absolutely *no* science background," she noted.[43]

Years ago, Laini Fondiller purchased a pH meter. "I took a reading. It didn't mean anything to me." A younger cheese-making friend explained to her that reading a pH meter takes practice and that if she were to begin using it routinely, she would figure out what it was telling her.[44] Digital pH probes, while more straightforward to use than acidometers, are notoriously fickle; they must be recalibrated routinely. Marcia Barinaga, coming to cheese making from a professional science background in 2008, explained to me that sometimes when taking a pH reading, "you look at it and say, 'I think there's something wrong with the pH meter. My gut tells me this cheese is right and I don't trust my pH meter.' You have to be *ready* to not trust your pH meter."[45] Marcia's more experienced friend Barbara (who makes cheese "using" her senses) remained skeptical. "Of course now everybody uses a pH meter," she told me, "which I laugh at because they break down so often." But Marcia, like Laini's friend, views the pH meter as she does her own senses—as a tool that one has to gain "a feel" for using, through experience. In 2012 Laini sent an e-mail message telling me she had purchased a new pH meter and had taken classes on understanding pH values. "I still observe the curd," she wrote, "but the pH meter now makes me feel more confident and I enjoy watching the consistency of the batches and taking the time to make more notes."[46]

In moving from home cheese making to commercial artisan production, former hippies incorporated the tools and knowledge of technoscience into their craft practice, both willfully and unwittingly. Mechanical equipment may be incorporated into artisanal practice when it is guided by experienced, reflexive assessment of materials at hand.[47] Peter Dormer writes, "Craftspeople can be defined generally as people engaged in a practical activity where they are seen to be in control of their work. They are in control by virtue of possessing personal know-how that allows them to be masters or mistresses of the available technology. . . . It is not craft as 'handcraft' that defines contemporary craftsmanship; it is craft as knowledge that empowers a maker to take charge of technology."[48] Artisanal skill includes the ability to use tools in a way that extends the artisan's body into the environment.[49] Such tools include cheese knives for cutting curd and various implements used to stir it, as well as devices that enhance one's understanding of the contingent materiality of milk and curd: thermometers, acidity titrators and pH meters, and computerized spreadsheets for data collection.[50]

Conclusion: Craft Conversions and Their Legacies

Facilitated by communications and instrumental media, convergences between industrial technoscience, academic dairy science, and rural entrepreneurship laid the foundation for what came to be hailed as the American artisanal "renaissance"—a renewal that occurred in home kitchens as well as commercial creameries. Through such convergences, moreover, artisans and scientists alike underwent conversion experiences. While from the beginning the farmstead cheese makers relied inescapably on the products of technoscientific knowledge, only gradually did many come to embrace scientific tools as ones that could *aid* their craft rather than displace or undermine it. Meanwhile, far from leaving behind their research labs to take up artistic endeavor, a handful of dairy scientists at Cornell, the University of Wisconsin, and the University of Vermont instead carried their labs into the field, recognizing that "cheese" is not so much a controlled experiment on milk but rather a *food* that is made by some people to be enjoyed by others and that has social significance beyond serving as a nutrition delivery system. While their colleagues in physics and electrical engineering contemplated turning their backs on the military-industrial complex and "reconverting" to civilian projects in the late 1960s and early 1970s, these agricultural and dairy scientists began to question the hyperhygienic Pasteurianism of the commercial-industrial food system.[51] It is not the case that scientific and artistic members of the cheese-making community were persuaded to leave one camp or worldview for the other. Rather, both sides came to share the persuasion that the "science" and "art" of making cheese are synergistic, not antithetical to one another.

The American Cheese Society (ACS) has embodied this synergy. In 1983 Cornell University's Frank Kosikowski (author of *Cheese and Fermented Milk Foods*) hosted a network forum–type conference in Ithaca, New York, bringing together 150 small-scale and home cheese makers, research scientists, retailers, and cheese enthusiasts. An established academic whose work contributed knowledge and new products to the industrial dairy industry, Kosikowski had also traveled widely and witnessed the effectiveness of "traditional" cheese-making methods from the remote mountains of the Mediterranean to the arid villages of Iran and Afghanistan. Convening the inaugural meeting of the ACS just three years before his retirement, Kosikowski seemed intent on diversifying the possibilities for "American cheese." Meanwhile, numerous attendees of that first meeting made the trek to Ithaca after receiving a brochure from Ricki Carroll, who had used the New England Cheesemaking Supply Company's mailing list to advertise "Dr. Frank's" conference. The ACS was founded as a grassroots organization to support small-scale, nonindustrial cheese making. For years, Ricki published the ACS newsletter out of her "hippie house."[52]

A cheese competition was introduced at the third ACS meeting, in which a technical judge from academia (versed in industrial standards) teamed up with an aesthetic judge often drawn from the retail sector (which relished artisanal exceptionalism). Thirty cheese makers entered 89 cheeses in that initial ACS competition.[53] In 2013, 257 companies from Alabama to Washington State (and Quebec, Canada) submitted 1,794 products to the judging and competition. Hugely successful in retrospect, the organization's development did not proceed smoothly. University of Vermont dairy scientist Paul Kindstedt was, as Frank Kosikowski's graduate student in 1983, a reluctant participant at the inaugural meeting; he remembers looking askance at the so-called "cheese makers" coming out of the woods with their handmade offerings.[54] If Kindstedt viewed the ACS as a waste of his time, the home and homesteader cheese makers at that first meeting had no idea what to make of the visiting scientist from China who gave a presentation on distilling whisky from cheese's by-product, whey. Their first thought was, "How cool is that?" The hippies anticipated not only making cheese but moonshine, too! It turned out, though, that deriving whisky from whey was a very high-tech, industrial process, not something for the home kitchen. Kosikowski's vision of a network forum notwithstanding, mutual incomprehension ruled at the first meeting of the ACS. A decade later, however, Kindstedt sensed a change. Not only had he come around to the view that artisanal cheese making *had* a method and, therefore, is a legitimate means of food making, but he also found that farmstead producers were no longer freaked out by talk of acidification and pH values.[55] They were beginning to express an interest, even a curiosity, in what "science" could add to their own sensory apprehension of what was going on when milk became cheese—sometimes wonderful, sometimes wild.[56]

Today, chèvre is no longer weird. Artisanal is the new natural: its worth seems beyond question, even as people struggle to define what it is. American culinary landscapes have undergone an aesthetic conversion; plastic-encased food never touched by human hands has lost much of the appeal it had in the 1950s and '60s. Mottled, lopsided, local cheese features at farmers' markets in cities and rural communities across the country. It is no longer making a countercultural statement to ask where our food comes from. Indeed, perhaps nowhere is the legacy of the counterculture more apparent than in the world of food.

Commemorating the American Cheese Society's thirty-year history in 2013, Seattle cheese maker Kurt Dammeier produced for the annual conference a "rock of ages" video, which checked in over the decades with four of the ACS's founding members.[57] In a striking segment, Dammeier remixed the iconic 1984 Apple Computer Super Bowl television ad. Instead of smashing through IBM to break open the door to a whole new world, the athlete's hammer obliterates an image of Kraft Singles to make way for the New American Cheese: handcrafted by self-made artisans, harnessing the power of their personal potential. Chèvre is here to stay.

Notes

1. Mary Keehn, interview by the author, 18 August 2008.
2. http://www.wholeearth.com/issue/1010/article/195/we.are.as.gods (accessed 22 August 2013).
3. Edmund Leach, *The Reith Lectures: A Runaway World; Lecture 1, Men and Nature*, BBC Radio, Radio 4, broadcast 12 November 1967.
4. http://www.wholeearth.com/issue/1010/article/195/we.are.as.gods.
5. Ibid.
6. For the definitive account, see Fred Turner, *From Counterculture to Cyberculture: Stewart Brand, the Whole Earth Network, and the Rise of Digital Utopianism* (Chicago: University of Chicago Press, 2006).
7. Ibid., 5.
8. Warren J. Belasco, *Appetite for Change: How the Counterculture Took on the Food Industry, 1966–1988* (New York: Pantheon, 1989).
9. See Sally McMurry, *Transforming Rural Life: Dairying Families and Agricultural Change, 1820–1885* (Baltimore: Johns Hopkins University Press, 1995).
10. Lauren Briggs Arnold, *American Dairying: A Manual for Butter and Cheese Makers* (Rochester, NY: Rural Home Publishing, 1876), 17.
11. For notable exceptions, see Heather Paxson, "Cheese Cultures: Transforming American Tastes and Traditions," *Gastronomica: The Journal of Food and Culture* 10, no. 4 (2010): 35–47.
12. Barbara Backus, interview by the author, 30 July 2008.
13. Anne Topham, interview by the author, 9 July 2008.
14. Anne Topham, "Taste, Technology, and Terroir: A Transatlantic Dialogue of Food and Culture," paper presented at the European Union Center at the University of Wisconsin–Madison, 8 September 2000, http://www.fantomefarm.com/text.htm (accessed 22 August 2013).
15. Topham, interview.
16. Paxson, "Cheese Cultures."
17. Belasco, *Appetite for Change*, 44.
18. Paul Feyerabend, *Against Method: Outline of an Anarchist Theory of Knowledge* (London: New Left Books, 1975).
19. Turner, *From Counterculture to Cyberculture*, 74.
20. Stewart Brand, *The Last Whole Earth Catalog* (Menlo Park, CA: Portola Institute, 1971), 181.

21. Allison Hooper, foreword to *The Atlas of American Artisan Cheese*, by Jeff Roberts (White River Junction, VT: Chelsea Green, 2007), xiii.

22. Heather Paxson, *The Life of Cheese: Crafting Food and Value in America* (Berkeley: University of California Press, 2013).

23. On "countercuisine," see Belasco, *Appetite for Change*.

24. Ibid., 54.

25. Keehn, interview.

26. David Kamp, *The United States of Arugula: The Sun-Dried, Cold-Pressed, Dark-Roasted, Extra Virgin Story of the American Food Revolution* (New York: Broadway Books, 2006).

27. Ricki Carroll, interview by the author, 17 July 2013.

28. Barbara and Rex Backus, interview by the author, 30 July 2008.

29. Carroll, interview.

30. Turner, *From Counterculture to Cyberculture*, 83–84.

31. Carroll, interview.

32. Jean-Claude Le Jaouen, *The Fabrication of Farmstead Goat Cheese*, trans. Elizabeth Leete (South Deerfield, MA: Cheesemakers' Journal, 1987), 67.

33. Ricki Carroll, *Home Cheese Making: Recipes for 75 Homemade Cheeses*, 3rd ed. (North Adams, MA: Storey Way, 2002), 183 (first published in 1982).

34. Turner, *From Counterculture to Cyberculture*, 93.

35. Cf. Leo Marx, *The Machine in the Garden: Technology and the Pastoral Ideal in America* (New York: Oxford University Press, 1964).

36. Le Jaouen, *Fabrication of Farmstead Goat Cheese*, 78.

37. In light of growing numbers of artisan producers, more detailed regulations, which can vary from state to state, are appearing on the books.

38. Laini Fondiller, interview by the author, 11 October 2007.

39. Jon Wright, interview by the author, 4 July 2007.

40. Belasco, *Appetite for Change*, 54.

41. Keehn, interview.

42. I have described this process as "synaesthetic reason"; see Heather Paxson, "The 'Art' and 'Science' of Handcrafting Cheese in the United States," *Endeavor* 35, nos. 2–3 (2011): 116–24; and Paxson, *Life of Cheese*.

43. Barbara's husband, Rex, who had a degree in physics, served as her science adviser. Barbara Backus, interview.

44. Fondiller, interview.

45. Marcia Barinaga, interview by the author, 8 August 2008.

46. Laini Fondiller, e-mail message to the author, 10 January 2012.

47. Paxson, "'Art' and 'Science' of Handcrafting Cheese"; and Paxson, *Life of Cheese*.

48. Peter Dormer, ed., *The Culture of Craft* (Manchester, UK: Manchester University Press, 1997), 140.

49. Tim Ingold, *The Perception of the Environment: Essays in Livelihood, Dwelling, and Skill* (London: Routledge, 2000).

50. Paxson, *Life of Cheese*, 138–42.

51. On physicists' and engineers' "reconversions," see Matthew Wisnioski, "Inside 'The System': Engineers, Scientists, and the Boundaries of Social Pro-

test in the Long 1960s," *History and Technology* 19 (2003): 313–33; Matthew Wisnioski, *Engineers for Change: Competing Visions of Technology in 1960s America* (Cambridge, MA: MIT Press, 2012); and Cyrus C. M. Mody, "Conversions: Sound and Sight, Military and Civilian," in *The Oxford Handbook of Sound Studies*, ed. Trevor Pinch and Karin Bijsterveld (New York: Oxford University Press, 2012), 224–48. On "Pasteurianism," see Paxson, *Life of Cheese.*

52. "ACS's Rock of Ages: Unplugged, Uncensored, and Uncut!," session at the 30th Annual Meeting of the American Cheese Society, 3 August 2013, Madison, WI.

53. http://www.cheesesociety.org/about-us/history/ (accessed 22 August 2013).

54. "ACS's Rock of Ages: Unplugged, Uncensored, and Uncut!"

55. Ibid.

56. Paxson, "'Art' and 'Science' of Handcrafting Cheese."

57. "ACS's Rock of Ages: Unplugged, Uncensored, and Uncut!" The four panelists themselves epitomize the convergences and conversions discussed in this chapter: former homesteader Alyce Birchenough; Ricki Carroll, from the New England Cheesemaking Supply Company; University of Vermont dairy scientist Paul Kindstedt, who in 2004 cofounded the Vermont Institute for Artisan Cheese; and Vermont cheese maker Peter Dixon, who was raised by homesteading, cheese-making parents and went on to get an MS in dairy science, under Paul Kindstedt at the University of Vermont.

Afterword: The Counterculture's Looking Glass

David Farber and Beth Bailey

To the authors of *Groovy Science*, most of whom operate within the disciplinary boundaries of the history of science and technology—a scholarly realm that exists in an almost-parallel universe to that of their colleagues in the more densely populated thickets of political and social history—the American counterculture appears as a place of high purpose populated by risk-taking people of vision and occasional lunacy. Where others see hapless hippies descending into the far-out silliness of drug-induced magical thinking, they find astronauts of the inner realms and engineers (self-taught or otherwise) of alternative technologies. Their freaks dig science, know how to use a slide rule, and embrace the possibilities of systemic experimentation. These alternative intellectual travelers are not panhandling on the corner, seeking escape from adult responsibility; they're thinking, working hard, and figuring out how to apply carefully garnered wisdom to the possibilities of interspecies communication, self-discovery, and space travel while they ponder problems of social organization and material sustainability.

In ignoring or sidelining these visionaries, many of whom were fueled by a combustible mix of LSD and social alienation, mainstream historians of recent American history have missed a big story about how a substantial subset of Americans came to rethink how they eat, how they communicate, how they stay healthy, how they design and build, and how they have fun. These "new alchemists," communal midwives, acid-spooked scientists, stoned tinkerers, and many of the other straight-up hippies and freaks who sought to invent and master new forms of technology and challenge existing scientific understandings are the characters in a history we are only just starting to understand.

By including quirky lab-coated scientists, serious tech geeks, hippie midwives, surfboard entrepreneurs, goat-cheese innovators, LSD-inspired cryogenic enthusiasts, and many others under the big tent of "groovy science," the writers in this collection fundamentally change our understanding of "the counterculture." They disrupt the traditional narrative arc that begins in the fifties with Beat writers Jack Kerouac and Alan Ginsberg, travels into the sixties on Ken Kesey's magical bus, settles in the Haight-Ashbury for the Summer of Love, stops off at Woodstock and a few New Mexico hippie communes, and then—as the decade crashes and burns—falls to the insanity of the Manson Family and the violence at Altamont. In powerful contrast, *Groovy Science* tells a story looking at (or toward) the seventies—the new, historiographical "long seventies" that has begun to eclipse the bright light of "the sixties." While none of these authors dismisses the traditional story of "Hippie" entirely, all enlarge its boundaries. In *Groovy Science*'s telling, the counterculture does not end in despair. Instead, its promise is fulfilled in ways both strange and wonderful . . . and sometimes startlingly prosaic and matter-of-fact.

The photographer Irwin Klein, who took pictures of the "new settlers" in northern New Mexico in the late 1960s, argued that he was documenting the counterculture's movement "from innocence to experience," and that is, in part, what these authors describe.[1] The counterculture, amalgam that it was, always contained a strand that, for lack of a better term, might be called "productive," and that strand found its center after the chaos of the sixties began to subside. The countercultural morass of social visionaries and runaway kids, motorcycle clubs, charismatic gurus, communal farmers, and sex, drugs, and rock-'n'-roll included men and women with rare combinations of talent and imagination, passion, and persistence. Some were products of the best formal education America had to offer, including scientists with PhDs in hand; others, the sons and daughters of the working class, made "handy" by necessity, schooled in the inventiveness of making-do. In a wild variety of ways, some of them began to build new worlds: from the intimate to the cosmic, and all in between.

These are the "groovy scientists" who fill this volume. In writing their stories, the authors of this volume remind us that "counterculture" has always been an odd-job word that defies easy definition, and that identifying who made or contributed or belonged to this loose amalgam of impulses, beliefs, practices, and dreams in the sixties era is mostly a matter of artful claims based on fuzzy evidence. But they do not remain in that definitional morass. Art Kleiner, onetime editor of the *Whole Earth Catalog*, captures their version of the counterculture and the kind of people they wish to include in its grasp in his crazy-quilt work *The Age of Heretics*. Kleiner writes: "[The counterculture] was a movement of apostasy—of rebellion against the bitter impersonality of industrial culture and the split that this culture had driven between the mind, body, and vernacular spirit."[2] Most of the groovy scientists described by Kaiser and McCray's crew operate in this apostate realm. They sought to make science and technology a human-scaled, hands-on enterprise that touched, if not the soul—few of these characters were theistically inclined—then a secular impulse that reached for the sublime possibilities of both transcendence and connection.

The tech rebels and scientific visionaries assembled here—whether or not they wore flowers in their hair—did reject many of the precepts of the Cold War military-industrial-academic Big Science complex. As historian Timothy Moy wrote in his evocative take on the "geek" counterculture of the 1970s, "It was antiestablishment" but it was "not antitechnology."[3] To identify with the counterculture and to be an antiestablishmentarian meant, at least in part, that one wished to overthrow or circumnavigate both the institutions that governed American life *and* society's epistemological underpinnings. For at least some people who identified with the counterculture, loosely or otherwise, being anti-Establishment was a state of mind that allowed them to challenge the boundaries of legitimate scientific and technological exploration in the Cold War while still identifying with the underlying project of scientific investigation and technical development.

Frustrated by the disciplined and soulless demands of universities, corporations, and government bureaucracies that were organized around notions of hierarchy, order, and relentless rationality, some cultural dissidents, especially in the mid-1960s, as the cultural critic Theodore Roszak claimed, did embrace an antiscience worldview.[4] In *their* search for new ways of life that would promote peace, love, and a sustainable planet, they turned to astrology, magical thinking, and the mystical realms of the imagined "East." But others, often in the late 1960s and early 1970s, while not disregarding the multiple allures of the nonrational and the mystical, and while doing their best to reject, ignore, or manipulate the masters of Big Science, tuned in to the humming buzz of scientific and technical knowledge and sought their answers therein. As reported here, some of these technohippies and scientific rebels found each other, and in groups large and small—in what another generation would call a network—developed their alternative technologies, scientific questions, and even institutional bases.

They understood that in these disciplined realms, magical in their own ways, they might give shape and substance to their sacraments, their pleasures, and their adventures in inner exploration and community building. This component of the counterculture did not reject the utility or promise of scientific investigation and technological practice; they only reconsidered the purposes to which they believed these human tools had been reduced. They did not reject biology or obstetrics; they used them to refashion maternal care and childbirth. Material science and engineering were not enemies but allies, as they refashioned home building, energy systems, and even surfboards. And most certainly countercultural practitioners who wished to make LSD or grow useful strains of cannabis respected chemistry and agronomy. Similarly, even as these sorts of groovy scientists and tech activists rejected the narrow channeling of science and technology into for-profit corporate labs or military-sponsored research, they struggled to build their own counter-institutions, forums, networks, and even businesses that could raid the world's scientific and technological treasure trove and reignite the Promethean fire of innovation.[5]

The counterculture was not—as too many histories would have it—merely a stoned romp by white middle-class youths through the last few years of the 1960s; it was not a passing extravagance that left shallow footprints in history and a faint shadow on our present. Instead, as a wayward band of scholars have begun to explain, at least some of those rebels who believed themselves to be challenging the core beliefs and everyday practices of mainstream American society left a deep mark on American culture. How and what many of us eat; how we find spiritual solace and physical renewal; how we collaborate and communicate; how we craft both institutions and commonplace goods; and how we struggle to find safe passage—to use a sixties phrase—on Spaceship Earth: so the counterculture changed its future, and our own.

Notes

1. Quoted in Benjamin Klein and Tim Hodgdon, "From Innocence to Experience: Irwin B. Klein and the 'New Settlers' of Northern New Mexico, 1967–1971," *New Mexico Historical Review* 87, no. 1 (Winter 2012): 86.

2. Art Kleiner, *The Age of Heretics* (New York: Doubleday, 1996), 267.

3. Timothy Moy, "Culture, Technology, and the Cult of Tech in the 1970s," in *America in the Seventies*, ed. Beth Bailey and David Farber (Lawrence: University Press of Kansas, 2004), 210.

4. Theodore Roszak, *The Making of a Counter Culture: Reflections on the Technocratic Society and Its Youthful Opposition* (Garden City, NY: Anchor / Doubleday, 1969).

5. For more on the productive counterculture, see David Farber, "Building the Counterculture, Creating Right Livelihoods," *The Sixties: A Journal of History, Politics, and Culture*, May 2013, 1–24.

Contributors

BETH BAILEY
   Foundation Professor
   Department of History
   University of Kansas

D. GRAHAM BURNETT
   Professor
   Program in History of Science
   Department of History
   Princeton University

DAVID FARBER
   Roy A. Roberts Distinguished Professor
   Department of History
   University of Kansas

MICHAEL D. GORDIN
   Rosengarten Professor of Modern and Contemporary
      History
   History Department
   Princeton University

DAVID KAISER
   Germeshausen Professor of the History of Science
   Program in Science, Technology, and Society
   Massachusetts Institute of Technology

ANDREW KIRK
Professor
Department of History
University of Nevada–Las Vegas

WENDY KLINE
Dema G. Seelye Chair in the History of Medicine
Department of History
Purdue University

W. PATRICK MCCRAY
Professor
Department of History
University of California–Santa Barbara

ERIKA LORRAINE MILAM
Associate Professor of History
Department of History
Princeton University

CYRUS C. M. MODY
Professor and Chair in History of Science, Technology, and Innovation
Maastricht University
The Netherlands

PETER NEUSHUL
Visiting Researcher
Department of History
University of California–Santa Barbara

HEATHER PAXSON
William R. Kenan Jr. Professor of Anthropology
Department of Anthropology
Massachusetts Institute of Technology

HENRY TRIM
Social Science and Humanities Research Council of Canada Postdoctoral
Fellow
Department of History
University of California–Santa Barbara

NADINE WEIDMAN
Lecturer
Department of History of Science
Harvard University

PETER WESTWICK
Assistant (Research) Professor
Department of History
University of Southern California

MATTHEW WISNIOSKI
Associate Professor
Department of Science and Technology in Society
Virginia Tech

Index